New perspectives in clinical microbiology 3

New perspectives in clinical microbiology 3

Series editor **W. Brumfitt**

VOLUME 1. **New Perspectives in Clinical Microbiology**, edited by W. Brumfitt. Assistant-editor: J. M. T. Hamilton-Miller. Kluwer Medical - London/Martinus Nijhoff - The Hague 1978. ISBN 90-247-2074-5.

VOLUME 2. **Aspects of Slow and Persistant Virus Infections**, edited by D. A. J. Tyrrell. Martinus Nijhoff - The Hague - Boston - London 1979. ISBN 90-247-2281-0.

VOLUME 3. **Combined Antimicrobial Therapy**, edited by W. Brumfitt, L. Curcio and L. Silvestri. Martinus Nijhoff - The Hague - Boston - London 1979. ISBN 90-247-2280-2.

Combined antimicrobial therapy

Proceedings of the
International Workshop on Combined Antimicrobial Therapy
held in Rome, April 21-22, 1978, in collaboration with
Università di Roma, Facoltà di Scienze, 2ª Cattedra di Microbiologia.

Edited by
W. Brumfitt, L. Curcio, L. Silvestri

1979
Martinus Nijhoff Publishers
The Hague - Boston - London

The distribution of this book is handled by the following team of publishers:

for Italy

Isedi - Mondadori
Segrate (Milano)
P. O. Box 1824
20100 Milano
Italy

for the United States and Canada

Kluwer Boston, Inc.
160 Old Derby Street
Hingham, MA 02043
USA

for all other countries

Kluwer Academic Publishers Group
Distribution Center
P.O. Box 322
3300 AH Dordrecht
The Netherlands

Published in Italy by
Isedi - Mondadori
Segrate (Milano)
P.O. Box 1824
20100 Milano

Published in all other countries by
Martinus Nijhoff Publishers
P.O. Box 566 The Hague
The Netherlands

ISBN 90-247-2280-2

PRINTED IN ITALY

Contents

Contributing authors

Dr. Umberto ALIERI
M. D., Assistente, Laboratorio di Microbiologia, Ospedale S. Carlo Borromeo, Milano.

Prof. E. S. ANDERSON
M. D., D. Sc., F. R. S., (formerly Director) Enteric Reference Laboratory, London.

Dr. Vittorio ARIOLI
Direttore, Chemotherapy Department, Lepetit, Milano.

Dr. Franco AZZOLLINI
M. D., Ospedale S. Carlo Borromeo, Milano.

Prof. Ettore BARTOLI
M.D., Incaricato di Chemioterapia, Università di Sassari.

Prof. M. BASAGNI
Incaricato di Istituzioni di matematica, Facoltà di Scienze, Università di Milano.

Dr. Paola BEGGI
Incaricato di Istituzioni di matematica, Università di Milano.

Dr. Gabriele BENTFELD
M.D., Institute for Medical Microbiology, Hamburg University.

Dr. Moreno BONDI
Assistente supplente, Cattedra di Igiene, Università di Modena.

Prof. William BRUMFITT
M.D., Ph. D., F.R.C.P., F.R.C. Path., Professor of Medical Microbiology, Royal Free Hospital, London.

Dr. D. W. BULLOCK
M.B., B.S., Department of Medical Microbiology, Southmead Hospital, Bristol.

Dr. Piero CAPPUCCINELLI
Direttore, Istituto di Microbiologia, Università di Sassari.

Dr. Giuseppe CARUSO
M.D., Assistente di Medicina interna, Università di Napoli.

Dr. Riccardo CIOFFI
M.D., Assistente di Medicina interna, Università di Napoli.

Dr. Keith COMBER
M.I. Biol., Beecham Pharmaceuticals Research Division, England.

Dr. Pasquale CRINITI
M.D., Assistente ordinario, Clinica urologica, Università di Chieti.

Dr. Pasquale CRISTIANO
M.D., Assistente di Medicina interna, Università di Napoli.

Dr. Costantino DELLA BRUNA
B.D., Farmitalia, Milano.

Prof. Franco DI SILVERIO
M.D., Direttore incaricato, Clinica Urologica, Università di Chieti.

Dr. Chiara D'ONOFRIO
M.D., Assistente di Medicina interna, Università di Napoli.

Dr. Fiona DRASAR
Senior Lecturer in Medical Microbiology, The London Hospital Medical College.

Dr. Jacob DRIESSEN
M.D., Ph. D., Regional Public Health Laboratory, Dordrecht.

Dr. Alfred EMMERSON
Consultant Microbiologist, Whittington Hospital, London.

Prof. Ugo FABIO
Titolare Cattedra di Microbiologia, Università di Modena, Direttore Servizio microbiologia, Policlinico di Modena.

Dr. Rossana FAEDDA
M.D., Assistente volontaria, Istituto di Patologia medica, Università di Sassari.

Dr. Maria Luisa FARINA
M.D., Unità di cura intensiva neurochirurgica, Ospedale Maggiore, Milano.

Dott. Anna FERRARA
Tecnico Laureato, Istituto di Tisiologia e Malattie dell'apparato respiratorio, Università di Pavia.

Dr. Laura FERRARI
M.D., Assistente universitaria, Cattedra di Microbiologia, Università di Modena.

Dr. Ferenc FOLLATH
M.D., Division of Clinical Pharmacology, Department of Medicine, Kantonsspital, Basel.

Dott. Vincenzo GENTILE
M.D., Assistente incaricato, Clinica urologica, Università di Roma.

Dr. Mario GHIONE
M.D., Montedison Divisione Farmaceutica, Milano.

Prof. Giuliana GIALDRONI GRASSI
Direttore, Cattedra di Chemioterapia, Università di Pavia.

Dr. Beth GOLDSTEIN
Ph. D., Chemotherapy Department, Lepetit, Milano.

Prof. Enzo GORI
Direttore, Istituto di Farmacologia, Facoltà di Scienze, Università di Milano.

Prof. Carlo GRASSI
M.D., Direttore, Istituto di Tisiologia e Malattie dell'apparato respiratorio, Università di Pavia.

Dr. David GREENWOOD
Ph. D., Senior Lecturer in Microbiology.

Dr. Anthoula GRIGORIADOU-EDIPIDES
M.D., Lecturer at the Department of Hygiene, University of Thessaloniki.

Dr. Reuben Naphtali GRÜNEBERG
M.D., M.R.C. Path., Consultant Medical Microbiologist, University College Hospital, London.

Dr. J. M. T. HAMILTON-MILLER
Ph. D., M.R.C. Path., Senior Lecturer, Department of Medical Microbiology, Royal Free Hospital, London.

Dr. Richard LACEY
M.A., M.D., Ph. D., M.R.C. Path., D.C.H., Department of Bacteriology, King's Lynn Hospital, Norfolk.

Dr. Alberto MALACRIDA
Pharm. D., Reparto di Patologia, Ospedale di Rho.

Dr. Marinella MANDELLI
Biol., Istituto Mario Negri, Milano.

Dr. Giuliano MANICARDI
Aiuto universitario, Cattedra di Microbiologia, Università di Modena.

Dr. Umberto MARINI
M.D., Ospedale S. Carlo Borromeo, Milano.

Dr. Antonietta MAROCCHI
M.D., Ricercatore capo CNR, Patologia speciale medica e Metodologia clinica II, Università di Roma.

Prof. D. A. MITCHISON
F.R.C.P., F.R.C. Path., Director, Department of Bacteriology, Hon. Director, M.R.C. Unit for Laboratory Studies of Tuberculosis.

Prof. Silvio MONFARDINI
M.D., Vice-direttore, Divisione di Clinica Oncologica F, Istituto Nazionale Tumori, Milano.

Dr. Rachele NEGLIA
Assegnista, Cattedra di Microbiologia, Università di Modena.

Dr. T. OLAY
Instituto de Farmacologia Española, Madrid.

Prof. John PAPAPANAGIOTOU
Professor of Microbiology, University of Thessaloniki.

Prof. Athanassios PARADELIS
Professor of Pharmacology, University of Thessaloniki.

Prof. Franco PARADISI
M.D., Professore di Malattie infettive, Università di Chieti.

Dr. Arnold PINES
M.D., F.R.C.P.E., Consultant Phisician, East Herts Hospital.

Prof. Giampaola QUAGLIO
Incaricato stabilizzato di Igiene, Università di Modena.

Dr. M. C. R. QUECEDO
Instituto de Farmacologia Española, Madrid.

Dr. D. S. REEVES
M.B., M.R.C. Path., Southmead Hospital, Bristol.

Dr. Ermanno ROBOTTI
Direttore, Clinical Research Dept., Lepetit, Milano.

Dr. A. RODRIGUEZ
M.D., Instituto de Farmacologia Española, Madrid.

Dr. Paolo SALA
M.D., Istituto di Tisiologia e Malattie dell'apparato respiratorio, Università di Pavia.

Dr. Georges SALPIGIDES
Assistant, Department of Pharmacology, University of Thessaloniki.

Dr. Aurora SANFILIPPO
B.D., Farmitalia Carlo Erba, Milano.

Dr. Maria Pia SANGIOVANNI
Laboratorio di ricerca Zambelletti, Milano.

Dr. Daniela SASSELLA
Clinical Research Dept., Lepetit, Milano.

Dr. Giorgio SAVAZZI
M.D., Clinica medica II, Università di Parma.

Prof. Hans-Heinrich SCHASSAN
Institute for Medical Microbiology and Immunology, Hamburg University.

Dr. Giovanna SCHIOPPACASSI
B.D., Farmitalia Carlo Erba, Milano.

Dr. Franco SCROLLINI
Direttore, Laboratorio di ricerca Zambelletti, Milano.

Dr. Pietro SETTE
M.D., Assistente Enviro Control Program; Ospedale "L. Spallanzani", Roma.

Prof. Luigi SILVESTRI
Cattedra di Microbiologia e Patologia generale, Facoltà di Scienze, Università di Roma.

Dr. Georges STATHOPOULOS
Lecturer, Department of Hygiene, University of Thessaloniki.

Dr. Anicca SUKARDI
Institute for Medical Microbiology and Immunology, Hamburg University.

Dr. Robert SUTHERLAND
B. Sc., Beecham Pharmaceuticals Research Division, England.

Dr. Raffaele TENAGLIA
M.D., Assistente incaricato, Clinica Urologica, Università di Roma.

Dr. Rudolf THEN
Dr. rer. nat., Hoffmann-La Roche, Basel.

Dr. Giovanni TOGNONI
M.D., Istituto Mario Negri, Milano.

Dr. Marco TONI
M.D., Istituto di Microbiologia, Università di Parma.

Dr. Roberto VAIANI
M.D., Reparto di Patologia, Ospedale di Rho.

Dr. M. V. VICENTE
Instituto de Farmacologia Española, Madrid.

Prof. Giuseppe VISCO
M.D., Primario, Ospedale "L. Spallanzani", Roma.

Prof. Arturo VISCONTI
Primario, Laboratorio Microbiologico, Ospedale S. Carlo Borromeo, Milano.

Dr. Samuel VOZEH
M.D., Division of Clinical Pharmacology, Kantonsspital, Basel.

Prof. John WILLIAMS
M.D., Professor of Medical Microbiology.

Dr. Richard WISE
M.B., Ch. B., M.R.C. Path., Consultant Medical Microbiologist, Dudley Road Hospital, Birmingham; Senior Clinical Lecturer, University of Birmingham.

Dr. Stefania ZANETTI
Assistente volontaria, Istituto di Microbiologia, Università di Sassari.

Foreword

L. Silvestri

Department of Microbiology and General Pathology
Faculty of Sciences - University of Rome - Italy

Ladies and Gentlemen,

I believe that the best way of fulfilling the assignment given to me for this workshop is to explain why the Organizing Committee has chosen the subject of combined therapy.

The main reason for this choice has been the knowledge that many problems in the field of combined chemotherapy — about which specialists have conflicting ideas — still remain to be solved. A typical example of this situation is tuberculosis, which is treated by compulsory combined therapy. A physician prescribing a monotherapy would be criticized by all his colleagues. Meanwhile, outside the field of tuberculosis, many specialists consider monotherapy to be the ideal treatment for infectious diseases.

On the basis of these premises, the Organizing Committee has ignored the traditional barriers among disciplines and invited to this workshop both specialists in anti-tubercular therapy and specialists in infectious diseases. Furthermore, experts in antitumoral therapy have been invited, even though tumours are not infectious diseases. This is because striking analogies do in fact exist, chiefly with regard to the problem of drug resistance, which the investigators hope to control by means of combined therapy.

If the investigators assembled here succeed in making an important contribution, even if not providing a definitive answer, to the question of whether true biological and medical reasons exist to explain the differences between these two therapeutic approaches, then our workshop will not have failed.

One of the principal aims of combined therapy is to control the emergence of resistant strains. Nevertheless, ideas on resistant strains differ. According to some investigators, the diffusion of resistant strains constitutes a considerable potential danger, chiefly when seen in perspective. Other researchers feel that this danger has been exaggerated. They feel that the therapeutic agents available to date and in the future constitute a considerable force and that it will always be possible, among so many drugs, to find an agent that will maintain its efficacy. The opinion of these investigators is that the reduction of the use of antibiotics which show a decrease in efficacy, will cause the disappearance of the resistant strains, which will no longer be exposed to selective pressure, whilst the microbial population will recover its original sensitivity.

On the other hand, some investigators feel that the current practice which recommends monotherapy might be the principal if not the only cause of the diffusion of resistant strains. As has been demonstrated, the frequency of resistance reflects the frequency with which an antibiotic is employed.

There is another aspect of resistance which is worthy of discussion. In long course diseases, e.g. tuberculosis or some chronic urinary tract infections, the failure of therapy is determined by the selection of resistant mutants. On the other hand the treatment of an acute infectious disease may fail, on rare occasions, because of the selection of resistant mutants during treatment. In this type of diseases failures are generally due to treatment being started too late, when the lesions induced by bacterial proliferation are no longer reversible, whilst the natural host defences are no longer sufficient. It is to be presumed that during monotherapy there is always an increase in resistant strains controlled by the natural host defences.

It is likely that in the case of cure of the patient, the diffusion of resistant strains depends upon the diffusion into the environment of resistant bacteria during monotherapeutic treatments, even when these are successful.

If all these hypotheses are true, when comparing monotherapy to combined therapy we should take into consideration not only the advantages to individual patients, but also the epidemiological consequences, and the pressure which the different therapies exert on the evolution of microbial populations towards resistance.

1

The epidemiological study of enterobacteria carrying resistance plasmids

E. S. Anderson

(formerly Director) Enteric Reference Laboratory, Public Health Laboratory Service, Colindale Avenue, London, NW9 5HT England

Summary

The indiscriminate use of antibacterial drugs throughout the world has resulted in the emergence of multiresistant strains of important enterobacterial pathogens such as Shiga's bacillus and *Salmonella typhi* which have caused extensive and serious epidemics of the respective diseases. Multiresistant strains of *S. typhimurium* have also caused widespread and severe epidemics, predominantly in children. These outbreaks, which have high morbidity and mortality rates, may involve a number of countries. The bacterial strains concerned seem to be cycling in man, and the epidemics are clonal in origin. The epidemiology is at present obscure, except when the strains gain entry to hospitals, where nosocomial spread is rapid.

The multiresistant strains responsible for these outbreaks may have acquired unidentified plasmids coding for communicability and virulence.

Introduction

Combined antimicrobial therapy, in the sense that it involves the use of more than one antibiotic or synthetic antimicrobial drug (all will be called antibiotics hereafter) is of two sorts: the employment of two different antibiotics marketed in a fixed ratio mixture; and the use of multiple antibiotics, simultaneously or sequentially. The first use is in general regarded unfavourably nowadays, with the exception of cotrimoxazole, a mixture of sulphamethoxazole and trimethoprim. The use of separate multiple antibiotics is only too frequently the result of lack of response to the first drug(s) administered, in illnesses that may not be diagnosed beyond the point of their being febrile. The sequence may even involve changes of drug, in the event of apparent lack of response to each antibiotic in turn, until the infection finally subsides. The tendency is then to attribute cure of the illness to the last antibiotic employed. In fact, such resolution is usually the result of the infection having run its course, the end of which coincides with the last antibiotic administered. Many of these illnesses are viral in origin, and are untouched by antibiotics. And a high proportion of the patients are children, who are running the gauntlet of viral infections that contribute to the immunity of the adult. Antibiotics have even been used for routine prophylaxis in all children admitted to some hospitals, a nefarious practice that should be totally abandoned.

Of course, the doctor-patient relationship has been radically changed by the fact that everyone now knows about the wonder-drugs. And if a doctor has not used one of them within 24 hours of seeing a febrile patient - frequently a child - his professional competence is doubted, whether or not the ailment has been diagnosed. There is therefore heavy social pressure on the general practitioner to use antibiotics from the start, and as I have already indicated, if there is no apparent response within a day or two, he is tempted to change to another antibiotic. The same lack of therapeutic prudence also occurs in many hospitals.

I was a general practitioner in the 1930s, and I treated many patients with febrile illnesses. Children with viral diseases of then unknown aetiology usually recovered within three to five days. Treatment was symptomatic: there was no more specific therapy for these illnesses then than there is now. Yet we did not become alarmed about the persistence of fever for a few days, because experience had taught us that in the great majority of cases it would subside within a week. There were, of course, the specifically identifiable febrile diseases, such as streptococcal infections. for which we had no remedy until the introduction of the sulphonamides, an event that happened while I was in general practice. But an experienced eye could diagnose a streptococcal throat at a glance, and throat swabs were routinely taken for confirmation. Unfortunately, the indiscriminate use of sulphonamides soon limited their effectiveness, though they remain valuable drugs to this day.

Influenza was another recognisable disease, but it had to run its course, and we did not panic if the pyrexia lasted as long as a week, because most cases resolved uneventfully. We still have no effective drug for it, but prophylactic vaccination shows promising results.

The point I am making is that it was recognised at that time that the great bulk of febrile infections abated without the use of drugs, except perhaps where they were indicated on symptomatic grounds. We knew that most febrile infections resolved spontaneously, and we were not alarmed when they took a few days to do so.

The picture is very different today, when there seems to be a general impression that patients with persistent fever are at serious risk if antibiotics are not used: if the doctors do not have that impression the patients'relatives have, so the doctor is under heavy pressure to use antibiotics, whether he is convinced they are necessary or not. This pressure has added an important dimension to current medical practice.

Risks of combined chemotherapy

There are several risks involved in combined chemotherapy.

1. The action of the drugs may be such that one is rendered ineffective by the use of the other. For example, the penicillins act on exponentially growing bacteria, while chloramphenicol is bacteriostatic. It is therefore illogical to use these drugs together.
2. Antibiotics used together in a fixed ratio may be no more effective than each antibiotic used separately, and in any event, even if their results are additive or synergistic, the ratio in which they are combined may not be optimal.
3. The combined antibiotics may present more risk of adverse reactions than the respective antibiotics used alone.
4. The use of any antibiotic may result in the appearance of bacteria with chromosomally determined or plasmid-mediated drug resistance. The risk of multiple drug resistance increases with the multiplicity of antibiotics. Combined chemotherapy thus adds to the risk of the emergence of multiply resistant bacteria.

To this catalogue can be added the fact that the use of antibiotics for the therapy of infections caused by resistant bacteria favours the respective pathogens while inhibiting sensitive competing bacteria, so that in such instances the antibiotics will not only be useless against the infections, but may actually promote them. Moreover, when infection is caused by bacteria with multiple resistance, the use of any antibiotic represented in the spectrum of resistance will select for the organism with all the resistances - a serious matter when that spectrum may extend to as many as nine resistances.

Having set the stage for a number of cautionary tales relating to bacterial drug resistance - I am confining myself to the enterobacteria - I think it worth while to describe the first clearly defined major instance in which the indiscriminate use of antibiotics led to a prolonged outbreak of multiply drug-resistant bacteria in animals and man, caused by phage-type 29 of *S. typhimurium* [1], [2], [3].

Type 29 of *S. typhimurium* was first described by Callow [4]. Its incidence was low until the winter of 1963-1964, when it suddenly increased. After a short pause in 1964, it resumed its upward trend,

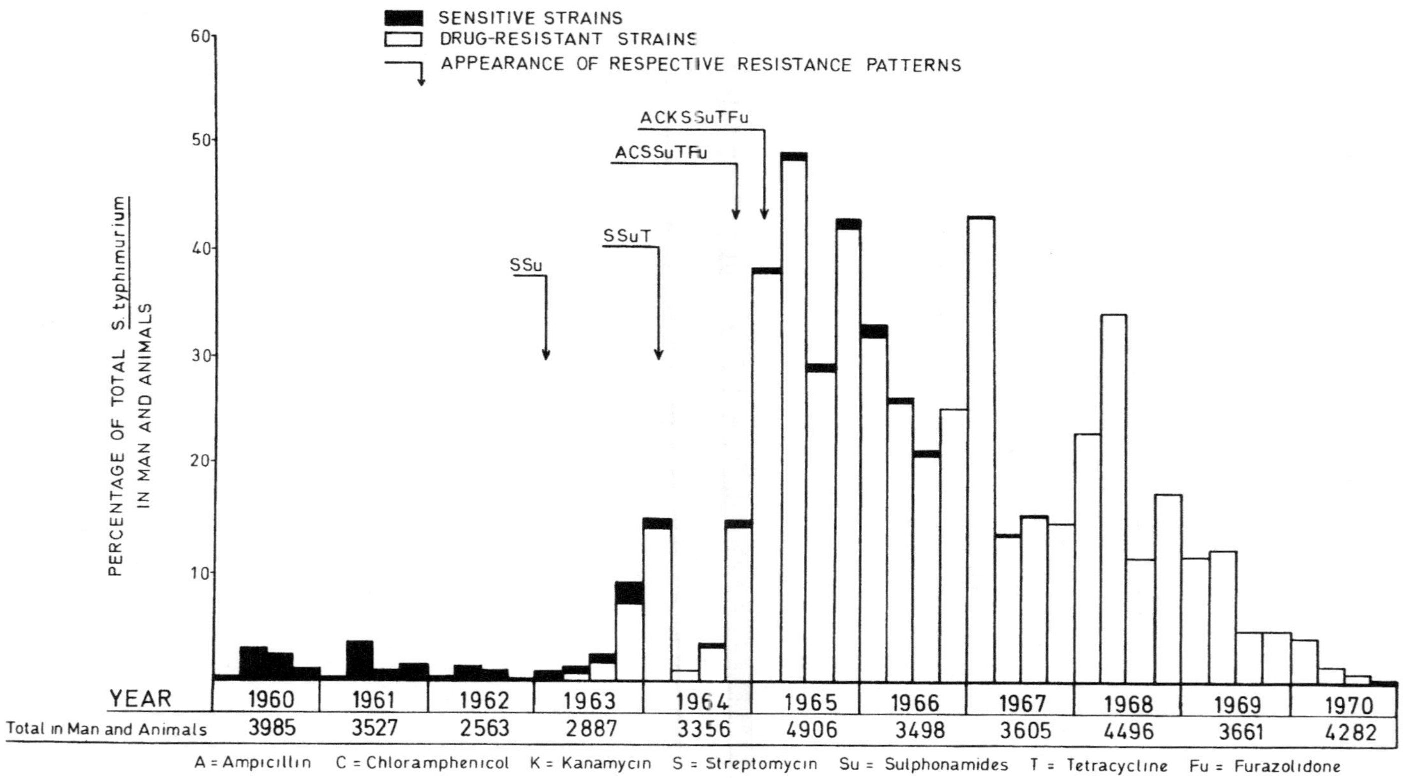

Fig. 1.1 - Incidence of type 29 as a percentage of total *S. typhimurium* 1960-1970.

which increased steeply the reafter, so that by the second quarter of 1965 it had reached almost 50 per cent of the total *S. typhimurium* isolated in animals and man. It maintained its high frequency until 1968, and thereafter declined, to disappear at the end of 1970. The course of this outbreak, which lasted for 7 years, is shown in Fig. 1.1.

It will be seen from this histogram, which covers the period 1960-1970, that type 29 was drug-sensitive until the start of the outbreak in 1963, and that the proportion of resistant strains of the type remained high throughout the epidemic period. Indeed, no drug-sensitive strains were encountered after the end of 1967. Moreover, the amplitude of the drug resistance increased rapidly during the initial phases of the epidemic, starting with streptomycin and sulphonamides (SSu) in 1963, adding tetracyclines (T) early in 1964, and attaining ampicillin (A), chloramphenicol (C), neomycin-kanamycin (K), streptomycin, sulphonamides, tetracyclines and furazolidone (Fu) by the beginning of 1965.

Epidemiological studies revealed that the great majority of animal infections were in calves, and followed the introduction of intensive farming in these animals. They were housed in large numbers under very poor conditions, so that when type 29 of *S. typhimurium* found its way into such a herd, it spread like a prairie fire. As isolation of infected animals was practised only minimally if at all, the entire herd was frequently engulfed by the disease. Attempts at treatment were made with antibiotics, which were rarely effective. And efforts to limit the spread of the disease were restricted to the treatment of whole herds with high prophylactic doses of antibiotics when infected animals were detected. The groups of antibiotics used are evident from the sequential acquisition of resistance shown in Fig. 1.1. Naturally, once the organism was resistant, use of the respective antibiotics provided a protective umbrella under which it could spread, while its drug-sensitive competitors were suppressed. Thus, the prophylactic use of the relevant antibiotics simply ensured that, once infection had appeared in a herd of calves, it would affect the entire herd.

I do not know how many animals died in this protracted outbreak, but the number was high and the financial loss was heavy. A substantial number of human infections occurred and there were some deaths, principally in young children.

Our enquiries, which were instituted as soon as it was apparent that an outbreak was in progress, clarified the epidemiological picture [3]. The calves concerned were being collected for resale by a single dealer, who was passing 500-700 animals through his premises weekly. They were then distributed in unsatisfactory transport over long distances to clients throughout Britain. By the time they reached their destinations many were moribund, and the remainder infected.

They were introduced into healthy herds, to which they rapidly passed their infection. The calves were being sent out with bags of furazolidone round their necks, the clients being advised by the vendor to administer the drug for some time after the animals arrival as a prophylactic against infectious diarrhoea, which was currently prevalent. Of course, the "infectious diarrhoea" was caused by type 29, which had become resistant to furazolidone early in the outbreak, so that the drug did nothing but harm. And many other antibiotics, notably ampicillin, were employed, sometimes simultaneously, for prophylaxis and treatment, with equally adverse results. There was in fact no way of dealing with this condition except by the prevention of calf infection, which could be done only by the employment of the animal health measures ordinarily employed to prevent and limit infectious diseases. But the farmers concerned, many of whom had little or no experience in this field, tried to substitute antibiotics for animal hygiene, and the result was disastrous.

The type 29 *S. typhimurium* outbreak subsided because the calf dealer who was distributing the infected animals was killed in a road crash early in 1967 and his firm ceased to operate. As he was the central source of supply, and as there was only restricted communication between the infected farms at the periphery, the outbreak burned itself out: but it can be seen from Fig. 1.1 that it took nearly four years to do so.

I have described this outbreak in some detail because it provided us with much important information, which has helped in the study of other epidemics of infection with resistant bacteria in animals and man. Primarily, it showed unambiguously that antibiotics were no substitute for animal hygiene. Secondly, it showed the futility of the use of antibiotics against resistant bacteria: a proposition I would have thought to be axiomatic. Thirdly, it demonstrated the need for rapid identification of the infecting organisms in communicable disease — another proposition I would have thought to be axiomatic, yet one which human medicine fails to observe on a scale at least as great as that of veterinary medicine.

The last lesson I propose to discuss, that resulted from studies of the type 29 outbreak, is that it was caused by a single strain, that is, a clone, of *S. typhimurium* over the whole period. This was established by the fact that the resistances to tetracyclines and furazolidone had special properties that served as genetic markers to identify the strain concerned: Fu was chromosomal in origin, and never transferred; and T, although probably plasmid-mediated, was mobilised by the transfer factor in the host strain at a very low frequency, probably less than 10^{-9}; however, once mobilised, it transferred with the frequency of the transfer factor itself, up to 10^{0}, in overnight crosses. This established that T was mobilised by a recombination event, superficially analogous to the formation of F-prime plasmids by the F factor in *Escherichia coli* K 12. All the tetracycline-resistant strains of type 29 in this outbreak showed this property. And all strains of the type isolated from the winter of 1964 onwards were resistant to furazolidone. From the history of the outbreak it seems probable that the conversion fo the strain to furazolidone resistance took place on the premises of the calf dealer who launched the outbreak.

Naturally, we have long known from the study of infectious diseases that single strains of communicable organisms cause discrete outbreaks. But the picture was complicated in the case of type 29 by the fact that a range of resistances appeared progressively in the strain, and because there was growing interest at that time in plasmid-mediated transferable drug resistance, there was at first an impression that the resistance was being constantly acquired in the field, and that many individual instances of such transfer were manifesting themselves. In fact, the number of transfers that contributed to the final resistance spectrum was very few: one for SSu and one for A; one for C (probably CSSu) and one for K. Each of these resistances or resistance linkage groups was present on a single plasmid, so the maximum number of transfers was four. As I have said earlier, the state of T is unknown, but its special mobilisation property indicated the clonal nature of its host strain. And Fu was chromosomal.

As their resistances, especially those carried by the A, SSu and T determinants and the chromosomal Fu resistance, were stable, the strain operated effectively as a multiresistant mutant: once the resistances were acquired, transfer became of only secondary importance.

The characterisation of plasmids and its epidemiological importance

As the result of our experiences with the type 29 outbreak of drug-resistant *S. typhimurium*, and with plasmid-bearing strains from other outbreaks of enterobacterial infection, we evolved a routine methodology for the characterisation of bacterial plasmids [5]. This methodology involves full genetic examination of the plasmids concerned, and has been carried to the molecular level. It has now been successfully applied to the study of many outbreaks of enterobacterial infection. Although "molecular epidemiology", as I have termed the subject, may need to be employed at its most sophisticated level on only a limited scale, the definitive information can be of crucial importance in the study of plasmid-bearing strains.

Table 1.1 Outbreaks of infection with plasmid-bearing enterobacteria.

Countries	*Years*	*Disease*	*Approximate number of cases*	*Organism* *	*Resistances* **	*Plasmid(s) carried*	*References*
Central America	1968-69	Shiga dysentery	? c. 125.000 deaths	*Shigella dysenteriae* Type 1	CSSuT	Group B	[10], [11], [12]
Mexico	1972-73	Typhoid	c. 100,000 c. 20,000 deaths	*S. typhi* degraded Vi-strain	CSSuT	Group H_1	[11], [12], [13]
India	1972-77	Typhoid	?	*S. typhi* Vi-types A, C5, D1-N, DVS, DVS14	CST, CSSuT, CSSuTFu, CSSuTNx	Group H_1	[11], [14], [15]
Vietnam	1973-75	Typhoid About 80% resistant by 1975	?	*S. typhi* Vi-types A, D6, E7, 53, 56, UVS1, UVS2, UVS3, UVS4	CSSu, CSSuT	Group H_1	[15], [16]
Thailand	1973-77	Typhoid About 30% resistant by 1975	?	*S. typhi* D1 variant, M1, 53, UVS3	CSSuT, ACSSuT, ACKSSuTTm	Group H_1	[15], [17]
Indonesia	1974	Typhoid	?	*S. typhi* Vi-type 53	CSSuT	Group H_1	[18]

Table 1.1 Outbreaks of infection with plasmid-bearing enterobacteria. (cont.)

Countries	*Years*	*Disease*	*Approximate number of cases*	*Organism **	*Resistances ***	*Plasmid(s) carried*	*References*
Latin America ***	1970 onwards	Salmonellosis	?	*S. typhimurium* phage type 193	Many different R-types. Commonest ACKSSuTNx	Group N *fi*+	[19], [20], [21]
Middle East ****	1969 onwards	Salmonellosis	Unknown but large	*S. typhimurium* phage type 208	Many different R-types, up to 8 resistances	F*me* A or AK and SSu determinants	[21], [22], [23]

* DVS = degraded Vi-strains showing characteristic cross reactions with Vi-typing phages. UVS = untypable Vi-strains, resistant to Vi II typing phages, but reacting characteristically with anadapted Vi-phages: categorised as UVS1, UVS2, UVS3 etc.

** A = ampicillin, C = chloramphenicol, Fu = furazolidone, K = neomycin-kanamycin, Nx = nalidixic acid, S = streptomycin, Su = sulphonamides, T = tetracyclines, Tm = trimethoprim.

*** Latin America: Brasil, Uruguay, Paraguay, Argentina, Perù, Mexico.

**** Middle East: India, Iran, Iraq, Israel, Jordan, Kuwait, Syria, Turkey (also carried to Britain in 1974).

Epidemiological studies of outbreaks caused by drug-resistant enterobacteria

In many resistant strains studied since the type 29 epidemic, most or all of the resistances, up to at least six, were present on a single plasmid, which was often, but not always, autotransferable. In addition, independent, mobilisable resistance determinants are often present that code for one to three resistances [1], [2], [6], [7], [8], [9]. I propose to outline the results of studies of human infection with enterobacteria carrying one or more of both types of plasmid. The outbreaks concerned are very large and are summarised in Table 1.1.

In 1968-69 an enormous outbreak of bacillary dysentery occurred in Central America, particularly in Guatemala [10]. The number of deaths that occurred is estimated at about 125,000. The disease was particularly severe in children and old people. The infecting organism was *Shigella dysenteriae* type 1 (Shiga's bacillus). It carried a drug resistance plasmid, of R-type CSSuT. This plasmid proved to belong to a new compatibility group, which was designated group B [11], [12].

In 1972-73 there was an immense outbreak of chloramphenicol-resistant typhoid fever in Mexico. It is now believed that there were over 100,000 cases with about 20,000 deaths. The source of this outbreak has not been determined with certainty, though a non-aerated fruit drink is under suspicion (Gangarosa, pers. comm.). It was some time before it was realised that the strain was resistant to chloramphenicol, so that early patients were treated with the drug, with unfortunate results. Ampicillin and co-trimoxazole were later used, but the mortality rate does not suggest that they were very effective.

The R-type of the Mexican strain of *S. typhi* was CSSuT, and the transferable plasmid that mediated the resistance belonged to a new compatibility group, which we designated H_1 [11], [12], [13], [15].

It is curious that, whenever typhoid fever resistant to chloramphenicol has been found in epidemic or hyperendemic conditions, the organism has proved to be carrying a group H_1 resistance plasmid coding for the R-type CSSuT. This has been observed particularly in South East Asia, specifically in India [11], [14], [15], Vietnam [15], [16], Thailand [15], [17] and Indonesia [18]. In Thailand a line of Vi-type D1 carries an H_1 plasmid coding for R-type ACSSuT; and a strain of type 53 carries not only the CSSuT group H_1 plasmid, but another autotransferring plasmid coding for ACKSSuTm (Tm = trimethoprim), which effectively endows the host strain with resistance to all drugs that might be used to treat typhoid fever.

Many Vi-types of the typhoid bacillus have now been found

carrying CSSuT group H_1 resistance plasmids in South East Asia. It is evident that in that region we are observing the interlocked epidemiology of the typhoid bacillus and of group H_1 plasmids, which apparently have a special affinity for it, for reasons as yet undetermined. Unfortunately, chloramphenicol resistance of this sort had become so common in *S. typhi* by 1975, that about 80% of strains were resistant, and chloramphenicol was no longer the drug of choice for typhoid fever.

Salmonellosis

Since the late 1960s a type of salmonellosis has appeared in a number of countries that is different in character from that normally encountered. The great majority of outbreaks have been caused by *S. typhimurium* and the organism is always resistant, usually to a large number of antibiotics. The disease is most common in children's hospitals and paediatric units; the morbidity rate is high, up to 50%; and the case mortality rate may reach 20-30%. Septicaemia and meningitis — the latter always fatal — are frequent in the most severe cases. Because of the organisms' multiple drug resistance, no antibiotic is effective against the disease. Cycling appears to be occurring in man, and examinations of local animals as possible sources of infection have been unproductive. The organism has spread over extensive land masses, as can be seen in Table 1.1 in the sections dealing with Latin America and the Middle East, but its epidemiology is obscure. And the outbreaks are due to single lines of *S. typhimurium.*

Drug-resistant *S. typhimurium* in Latin America

This disease is best exemplified in Latin America in the countries of the south eastern triangle: Argentina, Paraguay, Uruguay and Brazil. The *S. typhimurium* strain involved belongs to phage type 193, and 56% of 393 resistant cultures examined were resistant to six or seven antibiotics. All resistant cultures examined carried an fi^+ group N autotransferable plasmid [19], [20], [21]. This disease is limited in the north by the southern edge of the Mato Grosso, and in the west by the Andes. The countries beyond these natural barriers have similar types of infection but with other phage types of *S. typhimurium* carrying other plasmids, or with other drug-resistant salmonellae.

Drug-resistant *S. typhimurium* of Middle Eastern origin

An extensive outbreak of infection with *S. typhimurium* carrying resistance plasmids has occurred during recent years in the Middle East, as Table 1.1 shows. This outbreak is similar in character to that in South America, but the phage types are 208 and types closely related to it. The plasmids carried are of at least two sorts in all strains: a large, single-copy group F_I *me* plasmid, which was usually, but not always, non-autotransferring; and small, probably multicopy plasmids coding for A or AK and SSu resistances [11], [12], [21]. These small plasmids were always non-autotransferring, but could be mobilised by transfer factors we have routinely used for this purpose. The F_I *me* plasmids usually coded for at least five and frequently up to seven resistances. Our studies suggest that this outbreak is clonal in origin, that is, that only one strain of *S. typhimurium* is involved, and that variations in resistance pattern of the F_I *me* plasmids are the result of the gain or loss of translocons coding for the respective resistances.

A small outbreak of infection with this strain of *S. typhimurium* occurred in a burns unit in Birmingham. This was traced to a badly scalded patient who had been flown to England a san emergency from Teheran. He was a symptomless excreter of type 208.

An outbreak of infection similar in character but due to *S. wien* has occurred in Algeria, France and to a much lesser extent in Britain (McConnell, Smith, Leonardopoulos and Anderson, submitted for publication). The infections were similar in nature to those caused by the epidemic *S. typhimurium*, and it is probable that only one *S. wien* strain was responsible for all the infections in the countries concerned, that is, that this too was a single outbreak.

All these outbreaks of salmonellosis presented similar features: they were predominantly paediatric; they spread extensively; the causal organisms were highly communicable and were mostly resistant to at least six antibiotics; they caused severe disease; and, where animal investigations were carried out, there was no indication that the respective salmonella strains were of immediate animal origin, that is, cycling of the salmonellae was occurring in man.

Because of the special features of these widespread and protracted epidemics caused by multiresistant salmonella strains, the possibility must be considered that the organisms concerned carry, not only the resistance plasmids identified, but also plasmid-borne characters contributing to communicability and virulence, other than those responsible for synthesis of enterotoxins, haemolysins and adhesive surface antigens. The emergence of these strains is clearly a response to the widespread and indiscriminate use of antibiotics in their countries of origin.

References

[1] ANDERSON, E. S. and LEWIS, M. J. (1965), *Nature*, **206**, 597-583.
[2] ANDERSON, E. S. and LEWIS, M. J. (1965), *Nature*, **208**, 843-849.
[3] ANDERSON, E. S. (1968), *Brit. Med. J.*, **3**, 333-339.
[4] CALLOW, B. R. (1959), *J. Hyg.*, **57**, 346-359.
[5] ANDERSON, E. S. and THRELFALL, E. J. (1974), *J. Hyg.*, **72**, 471-487.
[6] ANDERSON, E. S. (1965), *Brit. Med. J.*, **2**, 1289-1291.
[7] ANDERSON, E. S. (1965), *Nature*, **208**, 1016-1017.
[8] ANDERSON, E. S. (1968), *Ann. Rev. Microbiol.*, **22**, 131-180.
[9] SMITH, H. R., HUMPHREYS, G. O. and ANDERSON, E. S. (1974), *Molec. Gen. Genet.*, **129**, 229-242.
[10] MATA, L. J., MENDIZABAL-MORRIS, C., GANGAROSA, E. J. and PERERA, D. R. (1970), *Tenth Int. Cong. Microbiol.*, Abstracts, p. 109.
[11] GRINDLEY, N. D. F., GRINDLEY, J. N. and ANDERSON, E. S. (1972), *Molec. Gen. Genet.*, **119**, 287-297.
[12] GRINDLEY, N. D. F., HUMPHREYS, G. O. and ANDERSON, E. S. (1973), *J. Bact.*, **115**, 387-398.
[13] ANDERSON, E. S. and SMITH, H. R. (1972), *Brit. Med. J.*, **3**, 329-331.
[14] PANIKER, C. K. J. and VIMALA, K. N. (1972), *Nature*, **239**, 109-110.
[15] ANDERSON, E. S. (1975), *J. Hyg.*, **74**, 289-299.
[16] BUTLER, T., LINH, N. N., ARNOLD, K. and POLLACK, M. (1973), *Lancet*, **2**, 983-985.
[17] LAMPE, R. M., MANSUWAN, P. and DUANGMANI, C. (1974), *Lancet*, **1**, 623-624.
[18] ANDERSON, E. S. (1976), *WHO Wkly Epidem. Rec.*, **41**, 317-324.
[19] ANDERSON, E. A., THRELFALL, E. J., CARR, J. M. and FROST, J. A. (1974), *Proc. Soc. Gen. Microbiol.*, **1**, 66.
[20] ANDERSON, E. S. (1974), *WHO Wkly. Epidem. Rec.*, **8**, 65-69.
[21] ANDERSON, E. S. (1977) in: *Topics in Infectious Diseases.* (Drews, J. and Högenauer, G. eds.), vol. 2, Springer-Verlag, Wien, New York, pp. 25-38.
[22] THRELFALL, E. J., CARR, J. M. and ANDERSON, E. S. (1976), *Proc. Soc. Gen. Microbiol.*, **3**, 88.
[23] ANDERSON, E. S., THRELFALL, E. J., CARR, J. M., MCCONNELL, M. M. and SMITH, H. R. (1977), *J. Hyg.*, **79**, 425-448.

2

Mechanism of antibiotic resistance in *Staphylococcus aureus*: with particular reference to the use of antibiotics in combinations

R. W. Lacey

Department of Bacteriology, King's Lynn General Hospital, London Road, King's Lynn, Norfolk PE30 5QD, England

Summary

Three general mechanisms of antibiotic resistance exist in *Staphylococcus aureus*: (a) plasmid mediated; (b) mutational. Resistant cocci arising by either of these mechanisms are selected by the use of antibacterial agents; (c) methicillin resistance. This trait may not be selected by the use of β-lactam antibiotics. Plasmid carriage is associated with reduced growth rate, and plasmid loss occurs on withdrawal of antibiotics. Resistance to several antibiotics occurs by mutation, notably to fusidic acid. Such mutants possess little, if any, disadvantage relative to the wild strains. Detailed suggestions are given for the use of antibacterials in combination that are relevent to each of these types of resistance.

The relationship of macrolide resistance in gram-positive organisms of veterinary origin to therapy of infections in man is discussed. Although additional information is required, such resistance appears at present to pose an extremely remote threat to man.

Introduction

Many of the approaches that we now adopt in prescribing antibiotics have been based on hasty interpretations of a relatively few clinical and experimental observations. This is particularly true for therapy against *Staphylococcus aureus*, that is still influenced by unsubstantiated fears prevalent during the 1950's. Since 1940, explanations for antibiotic resistance in this organism have swung wildly from one extreme to another. Initially mutation (that is focal alteration(s) in existing cellular DNA) accompanied by increased virulence due to passage between patients, was thought to underlie the problems of the resistant staphylococcus. This view was reasonable in the 1950's as mutation was considered the principle method that varied the heritable properties of this organism. Then, following the discovery of transferable resistance, first in the Enterobacteriaceae, then in other organisms including staphylococci, the presence of plasmids coding for resistance, together with a degree of inter-strain spread became accepted as the predominant, or even universal mechanism of resistance. Emotive terms such as "Infective Heredity" are still used.

During extensive laboratory experiments through the antibiotic era, the relevance of resistance to virulence has rarely been explored in a controlled way, although for resistance to be significant in the organism in question, that organism must also be pathogenic for the host. (Of course the threat of transfer of plasmids to new microbes must also be considered).

In this paper, I shall summarise our general understanding of antibiotic resistance in *Staphylococcus aureus*. The relevance of these resistance mechanisms to prescribing of antibiotics will be discussed, with an attempt to define the role of combination therapy. Of course considerations in prescription other than those relevant to resistance are important.

Plasmid-mediated antibiotic resistance in *Staphylococcus aureus*

Because of the technical difficulties in genetic studies, rather few workers have systematically studied this organism for plasmid mediated antibiotic resistance. These difficulties include (a) the problems of cell lysis where the only effective agent is lysostaphin; (b) uncertainty over the organization of its DNA-is there one, or are there several chromosomes?; (c) the absence of a true conjugal system, but see [1]; (d) transduction analysis is often difficult because of low frequencies attainable and the problems of interpretation. Thus many papers describing plasmid-inheritance in this organism present conclusions inappropriate for the data, notably those where plasmid inheritance is assumed established on the grounds of instability of the relevant trait. Despite these shortcomings, many staphylococcal plasmids have also been isolated physically. Although even here there is sometimes doubt as to whether the isolated DNA is the element within the cell responsible for the antibiotic resistance in question.

β-lactamase

In the great majority of strains examined that produce β-lactamase, this enzyme is plasmid mediated. Where chromosomally mediated lactamase production is seen, plasmid genes have probably become integrated into the chromosome [2].

Chromosomal determined lactamase might be expected to give the cell an advantage over plasmid control with the intensive prescription of β-lactam drugs, since only the necessary genes would be integrated into the chromosome with expenditure of the remainder of the superfluous plasmid DNA. Thus the relative rarity with which lactamase has been established as *bona fide* chromosomal is surprising. However, this may be an example of the relatively poor correlation - certainly as regards frequency — between genetic manipulations forced *in vitro* and those occurring spontaneously *in vivo* (it should be noted, however, that the failure to isolate a bacterium showing a particular phenomenom from clinical material does not necessarily indicate that the postulated event has not occurred, but could be due to the failure of survival of the progeny). In this instance therefore, we can conclude that the successful integration of the penicillinase plasmid genome into the chromosome occurs rarely in nature.

The penicillinase plasmids nearly always carry genes coding for resistance to some heavy metals (e.g. mercury, lead, cadmium, arsenate) and occasionally to other antibiotics, notably resistance to erythromycin or fusidic acid. In *Staphylococcus albus* there is evidence for linkage of other antibiotic resistances to penicillinase, but the evidence in *Staphylococcus aureus* is fragmentary. The reason for the penicillinase plasmids determining heavy metal resistance is inexplicable at present, but is certainly unlikely to be related to exposure of the human host to these metals (the suggestion that mercury resistance once gave the organism an advantage where mercurial diuretics were used was highly speculative and is now irrelevant because of the reduction in use of these agents). A number of genes contribute to the control of penicillinase; our understanding of the molecular mechanisms involved is very incomplete [3].

It is probably a reasonable approximation to say that globally, approximately 50-60% of staphylococci isolated from non-hospital patients and 70-80% from hospitalized patients produce penicillinase. Since plasmid-mediated resistance tends to be lost spontaneously [4] in the absence of selection pressure, there is therefore the inescapable inference that this high incidence of penicillinase production is maintained by the massive prescription of some β-lactam antibiotics e.g. penicillin, ampicillin, etc. Such prescription will include those antibiotics directed specifically at this pathogen, and also those directed at other pathogens (or no pathogens!), the commensal staphylococci being inadvertantly exposed.

Since about 1960, the "penicillinase-resistant" penicillins (i.e. methicillin, the isoxazolyl penicillins etc.) have been used against the penicillinase-producing staphylococcus. During this time, there has also been a decline in the preponderance of the Staphylococcus as a hospital pathogen and the manufacturers of these penicillins take the credit for this! However, the following points suggest that these penicillins may not *per se* be responsible for the relative decline in staphylococci sepsis that may well be only temporary.

1. Gram-negative organisms may have become an increasing problem in absolute terms; that is they may be now more virulent than they were (or the hosts more vulnerable).
2. The numbers of staphylococcal infections may well have been declining before the introduction of these penicillins.
3. Some of these penicillins are, in fact, sufficiently hydrolysed by penicillinase to cause therapeutic failure, e.g. flucloxacillin [5], [6], [7].

The cephalosporins have a role comparable to that of the penicillins. Whilst, some of the more recently introduced agents (e.g. cephalexin, cephradine) have good resistance to staphylococcal penicillinase, the early members of this group (e.g. cephaloridine) are significantly decomposed [6].

There are two main approaches for the improvement of β-lactam antibiotics: (a) modification of the structure to render the drug more resistant to penicillinase yet retaining full antibacterial capability. Unfortunately these seem almost mutually exclusive; (b) the addition of a specific β-lactam analogue (e.g. Clavulanic acid) that has little antibacterial activity *per se*, but can inhibit Lactamase sufficiently for a simultaneously prescribed conventional β-lactam antibiotic to be effective. It remains to be seen whether such inhibition is reliable enough in serious infection for routine exploitation.

Apart from these approaches, the use of β-lactam antibiotics in conjunction with non-β-lactam antibiotics continues to attract attention. There have been assertions of theoretical mutual incompatibilities amongst pairs of antibiotics (e.g. cloxacillin and fusidic acid); such speculation has not been substantiated in man, and it would be reasonable to combine with a β-lactam antibiotic: erythromycin, tetracycline, fusidic acid, rifampicin, aminoglycoside. The limitations to the use of these combinations are likely to be toxicity and cost rather than therapeutic failures or selection of resistant organisms.

Other plasmid-mediated resistances

Most resistance to tetracycline, neomycin, chloramphenicol, streptomycin or erythromycin is mediated by plasmid genes. Evidence over the past two years suggests that plasmid-mediated gentamicin resistance is beginning to spread, like neomycin resistance [7] through the staphylococcal population [8], [9], [10], [11]. There is preliminary evidence for other plasmid traits, e.g. amikacin resistance. Fusidic acid resistance is rarely, if ever, determined by plasmid genes unless they are part of a penicillinase plasmid [12].

These resistance traits are usually determined by separate small plasmids in contrast to multiple resistance of the Enterobacteriaceae that is often coded by a single large plasmid.

For immediate therapy, there is little difference whether a cell harbours several small, or one large plasmid. However, many isolates of staphylococci are resistant to a single antibiotic class.

Mutation to antibiotic resistance in *Staphylococcus aureus*

Much of our earlier view of antibiotic resistance in *Staphylococcus aureus* was coloured by our experience of the ease with which organisms, including staphylococci, became resistant to streptomycin. Subsequently, where the detailed mechanism of streptomycin resistance has been elucidated in the Enterobacteriaceae, it has been found to be enzymic - that is resistant clinical isolates possess a membrane protein that in some way inactivates the drug - generally by adenylylation or phosphorylation. Many resistant laboratory mutants possess abnormal ribosomal proteins. It might be inferred, therefore, that clinical strains of *Staphylococcus aureus* resistant to streptomycin will be expected to possess similar enzymic properties. Although this is partly true, some clinical isolates have acquired Streptomycin resistance by mutation [13], [14].

This evidence is as follows:

1. The level of resistance to streptomycin (MIC > 10 mg/ml) is higher than that typical of enzymic mechanisms and is characteristic of the ribosomal type.
2. The resistance is stable and non-inducible.
3. The genes determining the resistance are chromosomal.
4. *In vitro* protein synthesis from resistant lysates is indifferent to the action of streptomycin.

Other resistant traits may also have arisen by mutation (Fig. 2.1) since stable high level resistance to other antibiotics can be selected *in vitro* in a single step. The theoretical possibility that staphylococci can acquire resistance by a series of mutations, each slightly increasing the level of resistance, is in practice unimportant because: (a) such organisms are usually defective and (b) resistant clinical staphylococci rarely show low levels of resistance; (c) where the resistance mechanism (by mutation) has been elucidated it is due to a single cellular change; (d) in high level resistance a single gene is usually involved.

However, a requirement necessary for mutation to a drug being clinically important is that the resistant progeny are pathogenic. Thus whilst staphylococci can mutate *in vitro* to resistance to tetracycline or

MUTATION TO HIGH LEVEL RESISTANCE IN <u>STAPH. AUREUS</u>.	
STREPTOMYCIN	10^{-10}
NOVOBIOCIN	10^{-9}
RIFAMPICIN	10^{-8}
FUSIDIC ACID	10^{-7}
TRIMETHOPRIM^{x}	10^{-10}
LINCOMYCIN^{x}	10^{-8}
MINOCYCLINE^{x}	10^{-9}

^{x}Specific requirements necessary.

Fig. 2.1 - Approximate frequency of mutation to resistance to antibiotics *in vitro* in *Staph. aureus* that has either been shown, or expected to occur *in vivo*.

erythromycin, such variants are highly defective and likely to be non-pathogenic [15].

Generally, mutants selected *in vitro* to the following antibiotics can be considered non-pathogenic. All *penicillins, cephalosporins, tetracycline, neomycin, erythromycin, chloramphenicol, kanamycin, gentamicin, amikacin.* (Lacey, R. W. unpublished observations).

However, with the following group of agents, mutation occurs at frequencies of 10^{-6} — 10^{-11} *in vitro* to high levels (> x 50 fold the MIC of the wild strain), and the mutants have growth rates comparable to those of sensitive organisms, although some have specific nutritional requirements; *streptomycin, novobiocin, fusidic acid, rifampicin, trimethoprim, lincomycin (clindamycin), minocycline.* Thus there is a threat that staphylococci will acquire resistance to these antibiotics *in vivo* by mutation, and indeed for all except rifampicin, this has probably been observed. Each resistance merits consideration.

1. *Mutation to streptomycin* (see above).
2. *Mutation to novobiocin.* Since some organisms have been isolated that contain novobiocin resistance that is chromosomally mediated [16] this has probably occurred *in vivo.*
3. *Fusidic acid resistance.* Soon after this antibiotic was introduced, staphylococci resistant to this agent were detected. Although some of the resistance is plasmid mediated [12], the ease with which this resistance develops during therapy suggests that mutation may also be important. *In vitro* "sensitive cultures" of staphylococci not previously exposed to this antibiotic contain cocci resistant to fusidic acid at frequencies from about 1 in 10^6 to 1 in 10^9. We have recently examined the properties of these resistant clones selected *in vitro* and find that their survival both in nutrient broth or in human blood is often comparable to that of sensitive organisms (Fig. 2.2 and 2.3; Lacey, R. W. and Stokes, A., unpublished observations). This suggests that *in vivo* these mutants are indeed a potential problem. It has not been established conclusively that fusidic acid resistant staphylococci from clinical material possess such a mutation. This difficulty arises because of the high numbers of resistant mutants in a recipient culture used in transduction experiments.
4. *Rifampicin.* Mutation to this resistance occurs at a frequency of 10^{-8} [17]. The failure to isolate such resistant mutants from clinical material so far is probably a reflection on the small use of this antibiotic against staphylococci.
5. *Trimethoprim.* It is well known that if bacteria are plated on media containing trimethoprim and thymine or thymidine, organisms resistant to trimethoprim with an absolute requirement for thymine or thymidine can be recovered [18]. Such mutants including staphylococci, have been isolated from patients treated with trimethoprim (usually as cotrimoxazole) [19]. The significance of these is two-fold; (a) their isolation indicates that considerable levels of thymine or thymidine are present *in vivo*, at least in some patients, (b) the mechanism of resistance is likely to be mutational as occurs *in vitro.*
6. *Lincomycin* (Clindamycin). *In vitro* and *in vivo* staphylococci mutate to high levels of resistance to these antibiotics, but only if such cells are already resistant to erythromycin [15], [20].
7. *Minocycline resistance.* Some tetracycline resistant staphylococci have chromosomal tetracycline and minocycline resistance [21]; it

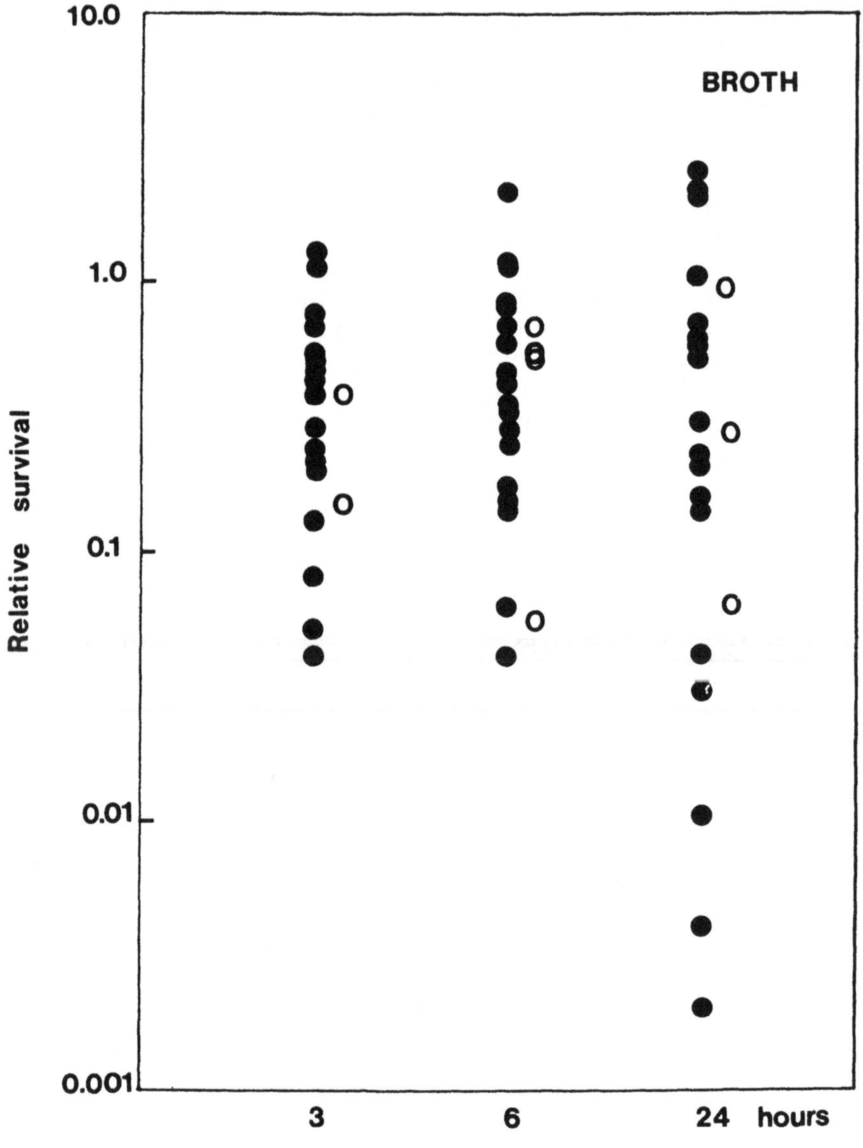

Fig. 2.2 - Relative growth of various fusidic acid cultures relative to their isogenic sensitive counterparts. Closed circles represent chromosomal mutants (MIC > 10 μg fusidic acid/ml). Open circles represent plasmid resistance (MIC ~ 8 μg/ml; the MIC of sensitive cultures ~ 0.2 μg/ml) in nutrient broth.

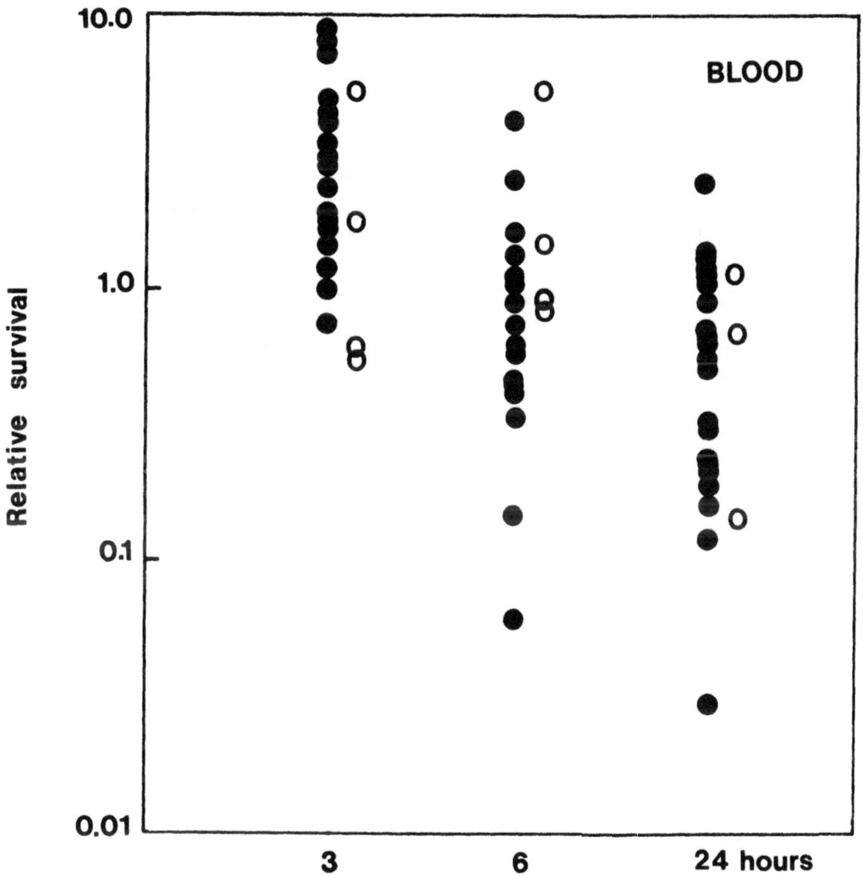

Fig. 2.3 - As Fig. 2.2. In human blood.

is possible that minocycline resistance is mutational in a tetracycline resistant cell, as can be reproduced *in vitro*.

Thus for resistance to streptomycin, novobiocin, fusidic acid, rifampicin, trimethoprim, lincomycin and minocycline, mutation occurs *in vitro* at sufficiently high frequency for it to be anticipated in clinical strains.

The consequences for therapy are obvious; whenever any of these agents are prescribed, consideration should be given to their use as part of combination therapy.

Methicillin resistance

This type of resistance probably occupies a unique category. Resistance arising by mutation or from plasmid carriage is obviously encouraged by the use of certain antibiotics, and attempts to reduce these resistances can be made by optimal deployment of antibacterials, including the use of antibiotics in combination. The incidence of methicillin resistance does not correlate with the use of methicillin or analogues [22]. A detailed analysis of these organisms shows them to be unique in many ways [4].

1. The nature of methicillin resistance is constant in that all show phenotypic heterogeneous resistance, and none produce "methicillinase".
2. All cultures are resistant to streptomycin, and almost all are resistant to tetracycline.
3. Resistance to streptomycin is high (MIC $>$ 10 mg/ml) characteristic of the chromosomal/ ribosomal type.
4. Methicillin-resistant strains are resistant to sulphonamides.
5. Most of the cultures produce enterotoxin B.
6. The bacteriophage typing pattern of methicillin-resistant staphylococci is usually group III [23].

Most of these staphylococci have probably evolved from a single clone characterised by these properties; this type of heterogeneous type of resistance cannot arise in sensitive cultures *in vitro* by mutation.

The clinical significance of these organisms is uncertain; whilst they have on occasion been isolated from a patient's blood following β-lactam therapy, the nature of the resistance (little expressed under physiological conditions) raises the possibility that other properties (e.g. Lactamase production) is responsible for treatment failure. That these organisms are resistant to such a wide range of antibiotics, combination therapy is generally advisable; there is no reason why one agent should not comprisea β-lactam agent (e.g. cloxacillin or cephradine) and the other could be selected from gentamicin, erythromycin, rifampicin, fusidic acid for example. Obviously tetracycline, streptomycin, sulphonamides and penicillinase-susceptible penicillins are appropriate.

Table 2.1 recapitulates the strategies for treating serious staphylococcal infections.

Table 2.1 Strategy for treating serious staphylococcal infections.

Resistance	*Suggested approaches to therapy*
1. Plasmid mediated	The use of available penicillinase-resistant penicillins or cephalosporins singly e.g. (i) cloxacillin; (ii) cephradine.
	The use of a variety of β-lactam antibiotics combined with a lactamase inhibitor (experimental)
(a) β-lactamase only	The use of single improved penicillins and cephalosporins (possibly in the future)
	The use of an available penicillin or cephalosporin (e.g. cloxacillin or cephradine) with a non β-lactam antibiotic. e.g. fusidic acid, erythromycin, rifampicin, trimethoprim.
	The use of the following antibiotics singly: erythromycin, tetracycline, gentamicin (also against fully sensitive organisms and in patients allergic to β-lactams).
b) Other plasmid-mediated resistances, (the cell often also producing β-lactamase)	The use of an available penicillin or cephalosporin (e.g. cloxacillin or cephradine) combined with non-β-lactam antibiotics (e.g. erythromycin, fusidic acid, gentamicin, tetracycline, rifampicin, trimethoprim).
2. Mutational	The use of fusidic acid, rifampicin, trimethoprim, minocycline in combinations with each other, or singly with β-lactam antibiotics.
3. "Methicillin-Resistance" The use of a β-lactam antibiotic (e.g. cloxacillin or cephradine) combined with erythromycin, fusidic acid, gentamicin, rifampicin or trimethoprim.	

Relationship between antibiotic resistance in Staphylococcal strains of animal origin to human strains

Concern has been expressed for many years that the use of antibiotics in animals can select resistant bacteria capable of either colonising man or else donating their transmissible genetic elements to "human" bacteria. Such fears have been reinforced by the practice of adding low concentrations of antibiotics to animal feeds for purposes other than for treating clinical sepsis. Such fears have been largely responsible for prohibiting the use of penicillins and tetracyclines for non-therapeutic purposes.

Macrolide antibiotics, e.g. virginiamycin and tylosin, are commonly used as "growth promoters" in different animal products, and it is important to know whether the use of these agents in animals is liable to prejudice chemotherapy in man. For such a problem to arise in man, the following set of conditions must apparently be met:

a) the use of low levels of macrolides in animals selects resistant organisms;
b) the resistance phenotype itself, or an easily achieved derivative

(e.g. by one step mutation) also expresses resistance to a macrolide antibiotic (i.e. erythromycin) used in human medicine;

c) the animal strains can either be directly pathogenic to man or

d) they are capable of colonising man sufficiently long for the resistance genome in question to become transferred to a "human" strain;

e) such resistant progeny (in 3 and 4) occur at sufficient frequency to significantly elevate the incidence of erythromycin resistance in human strains.

At present, further work is required before it can be stated categorically that these conditions can never be met, but the following all argue against such a train of events.

1. The commonest type of erythromycin resistance in human strains is that described as inducible (or dissociated). In these strains the only macrolide resistance is that to erythromycin; a recent survey showed that 84% of human strains possessed this mechanism [15]. Strains that are resistant to other macrolides, or lincomycins in addition to erythromycin have probably evolved by a further mutation of the inducible type of resistance to erythromycin [20]. Thus the essential resistance mechanism of human strains is different to that in animal strains.
2. Animal and human staphylococci are well known to prefer their respective hosts; it follows that animal strains will be expected to fair poorly in the human.
3. It is unlikely that colonisation of man would occur sufficiently for the genes in question to be transferred to man.
4. Even if such a transfer did occur, the resultant resistant derivatives may not be highly pathogenic, and would be expected to have a disadvantage compared to sensitive organisms, and hence to disappear [24].
5. Macrolide resistance in animal strains is unlikely to be easily selected by sub-therapeutic levels of antibiotic.
6. Even if such transfer of organism (or genes) occurred in man, there is no reason to suspect that the usefulness of the following antibiotics would be threatened: gentamicin, cloxacillin and analogues, cephalosporins, fusidic acid since the genes of these resistances, whether plasmid or chromosmal are not linked to macrolide resistance.

Certainly further work is needed to study this potential problem, but at present there is little to indicate that tylosin or virginiamycin used in animals as growth promotors present a threat to staphylococcal therapy in man. Recent comments [25] about the potential fear of macrolide resistance in animal streptococci prejudicing therapy in man require careful consideration as above, but the risk seems extremely remote.

References

[1] WITTE, W. (1977) *Zbl. Bakt. Hyg., I. Abt. Orig. A.*, **237**, 147.

[2] ASHESHOV, E. H. (1969) *J. Gen. Microbiology*, **59**, 289-301.

[3] IMSANDE, J., and LILLEHOLM, J. L. (1977) *Molecular Gen. Genetics*, **153**, 153-157.

[4] LACEY, R. W. (1975) *Bact. Rev.*, **39**, 1-32.

[5] LACEY, R. W., and LEWIS, E. L. (1975) *J. Med. Microbiol.*, **8**, 337-347.

[6] LACEY, R. W., and STOKES, A. (1977) *J. Clin. Path.*, **30**, 35-39.

[7] LACEY, R. W. (1971) *J. Gen. Microbiol.*, **4**, 73-84.

[8] SOUSSY, C. J., BOUANCHAUD, D. H., FOUACE, J., DUBLANCHET, A., and DUVAL, J. (1975) *Ann. Microbiol.* (Paris), **126B**, 91-94.

[9] SPELLER, D. C. E., STEPHENS, M., RAGHUNATH, D., VIANT, A. C., REEVES, D. S., WILKINSON, P. J., BROUGHALL, J. M., and HOLT, H. A. (1976) *Lancet*, **1**, 464-466.

[10] WYATT, T. D., FERGUSON, W. P., WILSON, T. S., and MCCORMICK, E. (1977), *J. Antimicr. Chemo.*, **3**, 213-217.

[11] DOWDING, J. E. (1977), *Antimicr. Ag. Chemother.*, **11**, 47-50.

[12] LACEY, R. W., and ROSDAHL, V. T. (1974), *J. Med. Microbiol.*, **7**, 1-10.

[13] LACEY, R. W., and CHOPRA, I. (1972), *J. Gen. Microbiol.*, **73**, 175-180.

[14] GRINSTED, J., and LACEY, R. W. (1973), *J. Med. Microbiol.*, **6**, 351-361.

[15] LACEY, R. W. (1977), *Scot. Med. J.*, **22**, 367-374.

[16] LACEY, R. W., and GRINSTED, J. (1973), *J. Med. Microbiol.*, **6**, 511-526.

[17] LACEY, R. W. (1972), *J. Gen. Microbiol.*, **71**, 399-401.

[18] STACEY, K. A. and SIMSON, E. (1965), *J. Bact.*, **90**, 554-555.

[19] STOKES, A., and LACEY, R. W. (1978), *J. Clin. Path.*, **31**, 165-171.

[20] LACEY, R. W., LEWIS, E. L. and HARRISON, J. (1975), *J. Med. Microbiol.*, **8**, 39-46.

[21] ASHESHOV, E. H. (1975), *J. Gen. Microbiol.*, **88**, 132-140.

[22] PARKER, M. T., and HEWITT, J. H. (1970), *Lancet*, **1**, 800-802.

[23] ROSENDAL, K., JESSEN, O., BENTSON, M. W., and BULOW, P. (1977), *Acta Path. Microbiol. Scand.*, **B 85**, 143-152.

[24] LACEY, R. W., and CHOPRA, I. (1975), *J. Med. Microbiol.*, **8**, 137-147.

[25] VAN EMBDEN, J. D. A., SOEDIRMAN, N., and ENGEL, H. W. B. (1978), *Lancet*, **1**, 655-656.

3

Evolution of the polychemotherapeutic approach in the treatment of tuberculosis

C. Grassi

Istituto di Tisiologia e malattie dell'apparato respiratorio, Università di Pavia, Italy

Classification of the antituberculous drugs

In order to examine the evolution of the polychemotherapeutic approach in the treatment of tuberculosis, it is worth briefly recalling the time when the different antituberculous drugs were discovered, as well as their past and present classifications.

As shown in Table 3.1, after 1945 (when Streptomycin became available) from a therapeutic standpoint tuberculosis could be regarded as an infectious disease, since the new drug was specifically active against the *mycobacterium*. During this first period, two other substances exerting a lower antimycobacterial activity (PAS and Thiacetazone) were also discovered.

The discovery of Isoniazid (1952) was clearly more important and marked the beginning of the second period, when other antibiotics and moderately active antimycobacterial chemotherapeutic agents were used (of these, the most active were Ethionamide, Pyrazinamide, Cycloserine, Kanamycin).

The most important antituberculous drug, however, was Rifampicin, discovered in 1965. This year started the third period and witnessed the use of Ethambutol for therapeutic purposes.

Let us now examine past and present classifications of antituberculous agents.

Up to 1965, drugs were subdivided into "major" and "minor", based on an almost purely chronological criterion. The group of "major" drugs included Streptomycin, Isoniazid and PAS, while all other available agents were included on the "minor" group. The discovery of Rifampicin and Ethambutol led however to a thorough revision of these classification criteria, in view of the excellent characteristics of these two new agents (particularly of Rifampicin).

Table 3.1 Antituberculous chemotherapy: historical data.

0 - until 1945:	Pre- antibiotic era (pneumothorax, clima, surgery)
1 - 1945-1952:	Streptomycin (SM) Para-aminosalicylic acid (PAS) Thiacetazone (Tb_1) (Sulfons)
2 - 1952-1965:	Isoniazid (INH) Cycloserine (Terizidone) (CLS-TZD) Pyrazinamide (Morphazinamide) (PZA-MZA) Isonicotinic Thioamides (ET-PT) Kanamycin (KM) Viomycin (VM) Thiocarlide (TC) (Tetracycline)
3 - 1965... :	Rifampicin (RMP) Ethambutol (EMB) Capreomycin (CM) (Rifamycin SV)

Table 3.2 Activity of an antituberculous drug. Criteria of evaluation.

1) Antimycobacterial activity (*in vitro - in vivo*)
2) Intracellular antimycobacterial activity
3) Emergence of resistance
4) Tolerability
5) Interactions with other antituberculous drugs
6) Clinical efficacy

Table 3.3 MIC - Myc. Tub. Hum. H 37 Rv (mcg/ml)

Drug	*In vitro* *	*Intracellular*
Streptomycin	0.5 - 1	25 - 100
PAS	1 - 1.5	100
Isoniazid	0.01 - 0.05	0.01 - 0.05
Viomycin	6.25	100
Cycloserine	10 - 20	20 - 40
Pyrazinamide	250	12.5 - 25
Ethionamide	0.6	0.5
Kanamycin	0.5 - 2	30
Rifampicin	0.2	0.5
Morphazinamide	125	12.5 - 25
Thiocarlide	1 **	1
Ethambutol	0.5 - 2	1 - 2
Prothionamide	0.6	0.5
Capreomycin	2	2 - 3

* Solid medium (Petragnani, Löwenstein-Jensen, IUAT).
**Dubos' medium.

In our opinion, the most useful criteria to assess the effectiveness of antituberculous drugs are those reported in Table 3.2.

An important feature is first of all the degree of antimycobacterial activity *in vitro*, in traditional culture media and in macrophage cultures, to establish endocellular activity.

Table 3.3 reports MICs under both conditions, whereas Table 3.4 [1] lists the drugs exerting the same activity both in culture media and at endocellular level (Isoniazid, Ethionamide, Prothionamide, Ethambutol, Cycloserine, Capreomycin, Rifampicin), those proving less active in monocyte cultures (Streptomycin, Viomycin, Kanamycin, PAS) and finally those with the most marked endocellular activity (Pyrazinamide and Morphazinamide).

A look at these tables clearly reveals the different behavior of the various drugs which, as we will see later on, can explain the usefulness of given combinations.

A large number of investigations have demonstrated the activity of antimycobacterial drugs *in vivo*, responses in the animal having generally been favourable and paving the way to their employement in humans. Although I cannot go into detail, I should like to underscore the importance of concurrent pharmacokinetic studies indicating the bioavailability of given substances in various animal species. A classic example of different pharmacokinetic behaviour in man and animals is provided by Cycloserine, whose bioavailability differs widely in the guinea-pig, in the monkey and in humans [2].

As for the therapeutic efficacy of antimycobacterial agents from this standpoint, we think it interesting to take into account the "number of MICs" recently suggested by the IUAT Commission for Treatment [3], which expresses the ratio between mean blood concentrations and MIC, as shown in Table 3.5.

After establishing the degree of antimycobacterial activity, the possible endocellular activity and the bioavailability of antituberculous drugs, their tolerability should be examined. In this connection, thanks to the high number of studies carried out, it is now possible to exactly define the side-effects of these drugs, summarized in Table 3.6 together with the most commonly employed dosages and administration routes.

The induction of resistance to the various drugs is a more complex problem: *in vitro* investigations are quite numerous and sufficiently suggestive; *in vivo* evaluation is more difficult, since in individual patients the drug attains different concentrations and has therefore varying contact times with the various bacteria present in tubercular lesions. Whilst we know that it is absolutely necessary to administer several drugs simultaneously, in order to avoid the rapid onset of resistance, scanty clinical data are available on the time of onset of resistance to the different drugs.

Table 3.4 Activity of antimycobacterial drugs on human mycobacteria grown on common media and in monocyte-cultures [1].

Same degree of activity in vitro *and in monocyte-cultures*	*Lower degree of activity in monocyte-cultures than* in vitro	*Higher degree of activity in monocyte-cultures than* in vitro
Isoniazid	Streptomycin	Pyrazinamide
Ethionamide	Viomycin	Morphazinamide
Prothionamide	Kanamycin	
Thiocarlide	PAS	
Ethambutol		
Cycloserine		
Capreomycin		
Rifampicin		

At any rate, I think it useful to consider the initial condition of a given mycobacterial population, indicating the proportion of resistant elements in "wild" strains, as shown in Table 3.7 [3]. The "proportion method" of Canetti et al. [5] for antibiogram control in tuberculous patients is based on such an evaluation.

Interactions between antituberculous drugs have also been widely investigated *in vitro* and in laboratory animals. One of the most interesting is the latest study carried out by Dickinson et al. [6]: this work concerns the bactericidal activity of Streptomycin, Isoniazid, Rifampicin, Ethambutol and Pyrazinamide, alone or variously combined.

A final analysis must be made on the basis of results in individual patients.

An overall evaluation of the data obtained on the basis of individual criteria now makes it possible to give a new classification of antituberculous agents.

In 1971, I proposed the classification reported in Table 3.8 [7], distinguishing between primary and secondary drugs, i.e. those to be used in the initial treatment of tuberculosis and those to be employed at later treatment stages or, at any rate, under specific situations.

Subsequently, in 1974 Grosset [8] proposed a "hierarchy" of antituberculous drugs, based on criteria similar to those I had followed, and classified these agents as "major" (Isoniazid and Rifampicin) and "minor" (all the others). More recently, the American Thoracic Society [4] issued a classification including Isoniazid, Ethambutol, Rifampicin and Streptomycin as first-line agents, and all the others as second-line drugs.

It is now a fact that there are drugs to be used in the initial stage of the disease, and others to be employed later in its course: Rifampicin

Table 3.5 "Number of MIC" (Average blood level/MIC) [3].

Drug and dosage		*"Number of MIC"*	
		3 hours	*6 hours*
Rifampicin	600 mg	75	
Isoniazid	450 mg		
slow inact.		100	50
fast inact.		50	10
Streptomycin	1000 mg	30	
Kanamycin	1000 mg	30	
Pyrazinamide	2000 mg	5 - 10	
Ethionamide	500 mg	5	
Ethambutol	1200 mg	3 - 4	
Cycloserine	500 mg	3 - 4	
Viomycin	1000 mg	4	
Capreomycin	1000 mg	8.5	
PAS	4000 mg	20 - 40	
Thiacetazone	150 mg	10 - 15	

Table 3.6 Dosage and side effects of the antituberculous drugs.

Drug	*Daily dosage (g)*	*Route of administration*	*Side effects* *
Isoniazid	0.3 - 0.8	PO or IM or IV	Peripheral neuritis, hepatitis hypersensitivity
Ethambutol	1.5 - 2	PO or IM	Optic neuritis (reversible with discontinuation of drug; very rare at 15 mg/Kg) skin rash
Rifampicin	0.60 - 0.90	PO	Hepatitis, febrile reaction, purpura (rare)
Streptomycin	1	IM or IV	8th nerve damage, nephrotoxicity
Pyrazinamide	1.5	PO	Hyperuricemia, hepatotoxicity
PAS	12 - 15	PO; IV	GI disturbance, hypersensitivity, hepatotoxicity, sodium load
Cycloserine	0.5 - 1	PO	Psychosis, personality changes convulsions, rash
Viomycin	1	IM	Auditory toxicity, nephrotoxicity, vestibular toxicity (rare)
Ethionamide	0.75	PO; IV	GI disturbance, hepatotoxicity hypersensitivity
Prothionamide	0.75	PO	Hepatotoxicity
Kanamycin	1	IM	Auditory toxicity, nephrotoxicity, vestibular toxicity (rare)
Thiacetazone	0.15	PO	Blood dyscrasia and hepatitis
Thiocarlide	6	PO	GI disturbance
Capreomycin	1	IM	8th nerve damage, nephrotoxicity

* According to the statement of ATS [4].

Table 3.7 Primary resistance in "wild" strains of M. Tuberculosis [3].

Rifampicin	1.10^{-8}
Isoniazid Streptomycin Kanamycin PAS Ethambutol	1.10^{-5} to 1.10^{-6}
Ethionamide Pyrazinamide Cycloserina Capreomycin Viomycin Thiacetazone	1.10^{-3}

Table 3.8 Classification of antituberculous drugs [7].

First-line drugs	*Second-line drugs*
Isoniazid Ethambutol Rifampicin (Streptomycin - Capreomycin) (Isonicotinic thioamides)	Cycloserine - Terizidone Pyrazinamide - Morphazinamide Kanamycin PAS (Viomycin) (Thiocarlide) (Thiacetazone)

and Isoniazid unquestionably belong to the first group, Streptomycin and Ethambutol are included with some reserve, whilst the second group comprises all the other substances which, although endowed with favorable properties, are not always effective at the time when the condition is first diagnosed.

Pyrazinamide deserves a place of its own. Its role has been revalued as compared to the past [9], since its marked endocellular activity and its activity at acidic pH can effectively supplement Streptomycin and lead to lesion sterilization.

Drug regimens in the treatment of tuberculosis

After this brief outline of the development and classification of antituberculous drugs, I will now discuss the past and present modes of employment of these agents in the management of the disease, whilst remembering that the therapeutic schemes adopted have been quite different.

Let me now report a few relevant examples, taken either from the literature or from my personal experience.

Whilst during the first period Streptomycin was used as a single agent or together with PAS or Tb_1 to yield a moderately effective combination, after the discovery of Isoniazid a more valid polychemotherapy could be performed, which subsequently included the various newly found secondary drugs. The first attempts of "intermittent" therapy date back to this time [10].

The most effective multidrug combinations were developed during the third period, when Rifampicin and Ethambutol became available: these regimens envisaged the continuous or intermittent administration of antituberculous drugs and suggested the "short course treatment" of tuberculosis.

In Table 3.9 [7] the use of antituberculous drugs in traditional therapy is outlined. During the first cycle (lasting approximately 3 months) Streptomycin, Isoniazid and PAS (or similar combinations) were administered daily. These were partially replaced every 3 months to avoid the emergence of multiple resistance. From a practical standpoint, one of the regimens we adopted for a long time consisted of an initial 3-month course with Isoniazid Streptomycin and PAS, followed by 3 months of Ethionamide Isoniazid and Cycloserine, and 3 further months of Isoniazid Pyrazinamide or Cycloserine and Ethionamide or Kanamycin, and so on for 6-12 months more, according to the sensivity test results. With this regimen, in new cases of tuberculosis we obtained sputum conversion in 100% of patients within 60 to 240 days. In a two-year follow-up after

Table 3.9 Antituberculous chemotherapy - Drug regimens [7].

"Traditional"			*"Intermittent"*
Following treatment	First period of treatment		Following treatment
ET (PT)	SM	INH	INH
CLS (TZD)	INH	EMB	EMB
PZA (MZA)	PAS	RMP	(RMP)
KM		(SM)	(ET-PT)
CM	INH	(CM)	(CLS)
(VM)	EMB	(ET-PT)	(CM)
(TC)	(ET-PT)		
(Tb_1)			
	INH		
	EMB		
	RMP		
	(ET-PT)		

18-24 months of treatment, we observed 9% of relapses, out of 642 patients.

Evaluation of results in chronic tuberculosis is different and more complex. Data are less favorable and not easily comparable, due to differences in disease characteristics between patients: sputum conversion in 62% of the cases (within 6 to 24 months), relapse in 36% of the cases, within 2 years from the end of treatment (lasting 18-24 months) out of 327 patients.

After the introduction of Rifampicin and Ethambutol, the traditional continuous therapy was still carried out, despite changing the initial regimen which, with regard to our patients, was Isoniazid, Ethambutol and Rifampicin or Ethionamide. Also in this case, after 3 months the drugs were partially replaced according to the scheme described above.

Results proved significantly different only with regard to sputum conversion, which was much prompter in the patients receiving Rifampicin from the start (15-75 days); on the contrary, the number of relapses within two years from the end of treatment was not significantly modified (7% of 215 patients treated for 12-24 months).

We have no personal experience of intermittent treatment, which has proved quite promising at both experimental and clinical levels [10], [11], [12], [13].

I should only like to mention that, according to some Authors, intermittent administration could be performed even during the first stage of treatment, whereas others believed this should be preceded by a period of continuous therapy.

The extensive experience reported by Sister Mary Aquinas, et al. [14] at the last International Chemotherapy Congress on 2000 patients treated in Hong Kong, Madras and East Africa sounds quite interesting.

The "short course treatment" of tuberculosis

The "short course" approach to lung tuberculosis is the result of the experience acquired through traditional therapy and more recent continuous or intermittent regimens including Rifampicin and Ethambutol.

The balance of these two periods can already be regarded as positive, and obviously more favorable for the second one, as shown by statistics published throughout the world on the results of specific therapy [15], [16], [17], [18].

Recently short-term treatment has been used. This means to say that daily administration of the appropriate drugs is given for about six months.

In spite of varying duration, the "short-course" management of tuberculosis basically consists of a continuous treatment period, though shorter than with the traditional, or less recent, approach. The rationale for this is well-known and will be a sound one if it is demonstrated that the "short-course" can yield the same favorable results offered by the other schedules.

As an example, let us now analyze the results obtained by some Authors with different therapeutic schemes.

First of all, it should be pointed out that a comparative evaluation of the different therapeutic approaches must include clinical and radiographical results, laboratory tests (particularly the sensitivity tests), and the incidence of relapses during the first two years following the end of treatment.

I report in several tables which follow data from several recent studies [9], [15], [20], [21], [22], [23].

Tables 3.10, 3.11, 3.12 report the work by Fox et al. [9] on bactericidal activity of the multidrug combinations proposed. On the basis of careful checking of experimental data both *in vitro* (routine tests and monocyte cultures) and *in vivo* studies, these authors proposed a reassessment of the relative efficacy of antituberculous agents. The only novel finding is that of Pyrazinamide alone or combined with Streptomycin. This finding is essentially related to the higher activity exhibited by Pyrazinamide both in acid environment and at the endocellular level (as confirmed by us [24] after Mackaness studies [25]) several years ago. The chemotherapeutic agent would be endowed with a stronger activity than believed up to now, due to the concurrent action it may exert as a companion of Streptomycin, or under specific anatomic situations granting unusual protection to mycobacteria.

This explains why the conventional evaluation by Fox et al. [9] is taken as equal to 1 for Isoniazid and Rifampicin, to 0.5 for Streptomycin and Pyrazinamide, and to 0 for Thiacetazone.

From the paper by Fox et al. (three investigations on 1830 patients altogether), which is an overall bacteriologic evaluation of this problem as a whole, it may be concluded that the most effective regimens, to be administered for no more than 6 months, are those which include Rifampicin, since they lead to the lowest number of relapses; as for other treatments, the most favorable results are obtained with those lasting more than 6 months, or including more than three drugs (therefore exceeding the 2.5 index of the SHR group).

The extensive theoretical and practical analysis made by these Authors on the data yielded by the two investigations performed in

Table 3.10 *Bacteriological relapses in the first East African short-course chemotherapy study* [9].

Regimen *	N. of bactericidal drugs	N. of patients	Bacteriological relapses		Bacteriological relapse in months				Relapses with organisms susceptible to INH and SM
			n.	%	7-12	13-18	19-24	25-29	
SHR	2.5	112	2	2	1	1	0	0	2**
SHZ	2	112	12	11	9	3	0	0	9
SHT	1.5	104	23	22	20	2	1	0	22
SH	1.5	112	33	29	28	3	2	0	32
STH/TH	1.5/1	102	4	4	3	0	1	0	3

* SHR = Streptomycin-Isoniazid-Rifampicin (6 months); SHZ = Streptomycin-Isoniazid-Pyrazinamide (6 months); SHT = Streptomycin-Isoniazid-Thiacetazone (6 months); SH = Streptomycin-Isoniazid (6 months); STH/TH = Streptomycin-Isoniazid-Thiacetazone (2 months) then Isoniazid-Thiacetazone (16 months).

** These strains were susceptible also to rifampicin.

Table 3.11 *Bacteriological relapses in the second East African short-course chemotherapy study* [9].

Regimen *	N. of bactericidal drugs	N. of patients	Bacteriological relapses		Bacteriological relapse in months				Relapses with organisms susceptible to INH and SM and RMP
			n.	%	7-12	13-18	19-24	25-29	
SHR	2.5	171	4	2	4	0	0	0	4
HR	2	164	12	7	10	1	0	1	11
SHRZ/TH	3/1	179	13	7	11	1	0	1	13
SHRZ/$S_2H_2Z_2$	3/2 interm.	159	7	4	7	0	0	0	7

* SHR = Streptomycin-Isoniazid-Rifampicin (6 months); HR = Isoniazid-Rifampicin (6 months); SHRZ/TH = Streptomycin-Isoniazid-Rifampicin-Pyrazinamide (2 months) then Isoniazid-Thiacetazone (4 months); SHRZ/$S_2H_2Z_2$ = Streptomycin-Isoniazid-Rifampicin-Pyrazinamide (2 months) then Streptomycin-Isoniazid-Pyrazinamide (twice weekly-4 months).

Table 3.12 ***Bacteriological relapses in the Hong Kong short-course chemotherapy study*** [9].

Regimen *	*Duration (Months)*	*N. of bactericidal drugs*	*N. of patients*	*Bacteriological relapses* ***		*Bacteriological relapse in months*	
				n.	*%*	*1-3*	*4-6*
SHZ	6	2	137	8	6	5	3
$S_3H_3Z_3$	6	2	141	11	8	9	2
$S_2H_2Z_2$	6	2	126	7	6	6	1
SHZ	9**	2	70	2	3	2	0
$S_3H_3Z_3$	9	2	67	3	4	2	1
$S_2H_2Z_2$	9	2	74	2	3	1	1

* 2 = Twice per week; 3 = Three times per week.
** Between 6 and 9 months this series received the 3 times per week schedule.
*** These strains were susceptible both to streptomycin and isoniazid.

Table 3.13 ***French short-course chemotherapy study*** [21].

Groups *	*N. of patients*	*Sputum conversion*		*Persist. sp. conv. at 18th month form the beg. of treatment*		*X ray evaluation*		
		n.	*%*	*n*	*%*	*6*	*12*	*18*
A	55	54	98	46	84	0.54	0.46	0.41
B	62	59	95	58	93	0.63	0.53	0.53
C	60	54	90	47	77	0.62	0.51	0.44

* Daily treatment with isoniazid (450 mg) and rifampicin (600 mg) for 6 (Group A), 9 (Group B) or 12 months (Group C). In addition, each group was treated for three months at the beginning either with streptomycin (1g.) or ethambutol (1g.) daily, according to physician's advice.

East Africa and the one carried out in Hong Kong, suggest important remarks, among which the reintroduction of Pyrazinamide in current therapeutic schemes (especially in developing countries), the lack of untoward toxic effects with short-term treatments, and the negligible influence of previous Streptomycin resistance, as compared to the greater importance of initial INH resistance.

The study by Brouet et al. [21] (see Table 3.13) initially involved 243 patients, subsequently reduced to 177 since 13 of them did not meet the bacteriological criteria suggested for the investigation (they were free from bacilli, or carried bacilli resistant to either Isoniazid or Rifampicin) and other 53 had withdrawn from treatment for various reasons (concomitant diseases, inobservance of protocol, disappearance during treatment or follow-up). The evaluation of these cases leads the French authors to a very optimistic appreciation of the "short-course" approach in the management of tuberculosis. In their view, 6-month regimens including Isoniazid and Rifampicin are quite effective (93%-100% of favorable results) and comparable to traditional 18-months therapy with Streptomycin, Isoniazid and PAS (or Ethambutol). When limited to 6 months or less, other therapeutic schemes are considered both less effective and less tolerated (10-20% of side-effects for traditional regimens, as compared to 2-5% for regimens including Isoniazid and Rifampicin).

The research by Angel [22], [23] (see Table 3.14 and 3.15) which examines in detail a high number of patients treated for periods ranging from 6 to 18 months with Isoniazid and Rifampicin, combined with Ethambutol or Streptomycin, has shown that, irrespective of the regimen adopted and the duration of treatment, the sputum smear becomes negative in 100% of the patients within the sixth month of therapy. In our view, the incidence of relapses in the months following the end of treatment is negligible and, at any rate, never attains significance between groups. However, the Authors conclude that the 9-month course leads to complete curing, which is not assured by the 6-month treatment.

In addition to these appropriate examples of "short-course" therapy, other favorable experiences have been reported [14], [26], [28], [29], [30], [31], [32].

The number of patients treated by us is still too limited to allow personal conclusions to be drawn. Our approach is similar to that of the French authors, i.e. the administration of Isoniazid and Rifampicin for 6-9 months, combined with Ethambutol or Streptomycin and Pyrazinamide during the first 3 months.

Conclusions

Ten years of investigations with the most recently developed agents Rifampicin and Ethambutol, both remarkably active and well tolerated, have allowed a thorough revision of tuberculosis treatment schemes which, during the previous decade, had at any rate yielded excellent, though more delayed, therapeutic results.

No absolute criteria can be proposed to establish *a priori* the duration of treatment, in view of the different anatomico-clinical features which the disease may exhibit when first identified: and it is precisely these characteristics which can condition a relapse in the

Table 3.14 Bacteriological relapses [22], [23].

Group *	N. of patients	Cavity **	Bacteriological relapses inmonths 7-18	13-18	18
EHR 6	98	A	3	–	2
SHR 6	64	A	1	–	1
SHR 6 / EHR 6	12	A	1	–	0
EHR 12	101	A	–	0	0
SHR 12	68	A	–	0	1
SHR 12 / EHR 12	10	A	–	0	0
EHR 9 SHR 9	151	B	–	–	0
EHR 18 SHR 18	155	B	–	–	0

* S = Streptomycin; H = Isoniazid; R = Rifampicin; E = Ethambutol; 6, 9, 12, 18 = months for isoniazid and rifampicin treatment. Ethambutol and streptomycin treatment only 8 weeks at the beginning.

** A = cavitary lesions (diameter ⩽ 2 cm); B = cavitary lesions (diameter > 2 cm).

Table 3.15 Sputum conversion [22], [23].

Group *	N. of patients	Cavity **	Sputum conversion % (months) 1	2	3	4	5	6
EHR 6 EHR 12	ª()	A	39	69	95	97	98	100
SHR 6 SHR 12	141	A	45	77	92	97	99	100
EHR 9 EHR 18	131	B	18	53	85	97	99	100
SHR 9 SHR 18	126	B	25	60	87	98	98	100

* S = Streptomycin; H = Isoniazid; R = Rifampicin; E = Ethambutol; 6, 9, 12, 18 = months for isoniazid and rifampicin treatment. Ethambutol and streptomycin treatment only 8 weeks at the beginning.

** A = cavitary lesions (diameter ⩽ 2 cm); B = cavitary lesions (diameter > 2 cm).

Table 3.16 New drugs regimens (short-course) suggested by IUAT (Commission on Treatment).

Initial treatment (two months)		*Maintenance treatment (four months)*	
Drugs *	*Schedule*	*Drugs*	*Schedule*
INH + RM	daily	INH + RMP	daily or twice per week
INH + RMP + EMB	daily	INH + RMP	daily or twice per week
INH + RMP + SM	daily	INH + RMP	daily or twice per week
INH + SM + PZA	daily or three times per week	INH + SM + PZA	daily or three times per week
INH + RMP + SM + PZA	daily	INH + Tb_1 + INI + SM + PZA	daily twice per week

* INH = Isoniazid; SM = Streptomycin; EMB = Ethambutol; RMP = Rifampicin; PZA = Pyrazinamide; Tb_1 = Thiacetazone.

months following bacteriological cure, as detectable by available means. In fact, whilst prompt sputum conversion is constantly observed by all investigators, the incidence of relapses differs widely during the following months, according to treatment duration and drugs employed.

For these reasons, in our opinion, whilst acknowledging the validity of the assessments carried out by the various Authors, and particularly the possible inclusion of Pyrazinamide in multidrug regimens to be used in developing countries, it should be remembered that in the so-called "developed" countries regimens containing Rifampicin and Isoniazid should be preferentially employed; these two agents should be combined with a third drug (when feasible, Ethambutol, in consideration of its excellent activity and tolerance), whilst keeping Streptomycin as a reserve drug or the agent to be added for especially severe cases, necessitating a four-drug regimen.

Whilst emphasizing that the 9-month treatment is almost sure to lead to complete cure, which is not always the case with the 6-month course, it seems worthwhile underlining once again that it is up to the physician to establish the duration of treatment, that is to say, to decide when drug administration can be discontinued, on the basis of clinical radiological and laboratory (in particular, bacteriological) findings.

The chapter on "short-course" management of tuberculosis is already full of data, but unfortunately still too recent: as suggested by the IUAT Commission for Treatment [3], a series of "régimes courts" (short-course regimens) needs to be submitted to clinical evaluation (Table 3.16).

The chapter on tuberculosis management cannot be regarded as closed, as unfortunately there are still epidemiological conditions so deeply different that no forecast can be made for the eradication of this disease. In this connection, we do not know what to expect from the countries which still form the main "reservoir" of patients, and where serious social problems need to be tackled before medical and therapeutic ones can he solved. To conclude my paper, I can confirm that in our countries the evolution of tuberculosis management has led to far better results than had been obtained up to the beginning of the Sixties, with regard both to rate of cure and reduction of treatment duration.

References

[1] Clini V. and Grassi, C. (1970) *Antibiotica Chemioterapia*, **16**, 20-26.
2] Coletsos, P. J., Bretey, J., Laroche, M. J., et al. (1957) *Ann. Inst. Pasteur*, **93**, 563-580.
[3] Commission du Traitement de l'UICT (1976) *Rev. fr. Mal. Resp.*, **4**, 157-163.
[4] American Thoracic Society (1977) *Am. Rev. Resp. Dis.*, **15**, 185-187.
[5] Canetti, G., Rist, N. and Grosset, J. (1963) *Rev. Tuberc.*, **27**, 217-272.
[6] Dickinson, J. M., Aber, V. R. and Mitchison, D. A. (1977) *Am. Rev. Resp. Dis.*, **116**, 627-635.
[7] Grassi, C. (1971) *Giorn. Ital. Mal. Tor.*, **25**, 301-311.
[8] Grosset, J. (1974) *Proc. 17° Congr. Nat. Tuberculose Clermont-Ferrand*, Masson, Paris, pp. 3-25.
[9] Fox, W., Mitchison, D. A. (1975) *Am. Rev. Resp. Dis.*, **111**, 845-848, 329-353.
[10] Fox, W. (1971) *Postgrad. Med. J.*, **47**, 729-736.
[11] Dickinson, J. M., Mitchison, D. A. (1966) *Tubercle*, **47**, 370-380.
[12] Dickinson, J. M., Mitchison, D. A. (1966) *Tubercle*, **47**, 381-393.
[13] Dickinson, M. J., Ellard, G. A. and Mitchison, D. A. (1968) *Tubercle*, **49**, 351-366.
[14] Aquinas, S. M., Tripathy, S. P. and Darbyshire, J. (1978) in: *Current Chemotherapy* (W. Siegenthaler and R. Lüthy eds.), American Soc. Chemotherapy, Washington, D. C., vol. 1, pp. 43-47.
[15] Fox, W. (1977) *Proc. Roy. Soc. Med.*, **70**, 1-15.
[16] Bates, J. H. (1977) *New Engl. J. Med.*, **297**, 610-611.
[17] Grassi, C. (1977) in: *Aggiornamenti di Chemioterapia* (G. Gialdroni Grassi and C. Grassi eds.), La Goliardica Pavese, Pavia, pp. 1-21.
[18] Angel, J. H. (1977) *J. Antimicrob. Chemother.*, **3**, 290-294.
[19] East African/British Medical Research Council Study (1977) *Am. Rev. Resp. Dis.*, **116**, 3-8.
[20] Gabriel, M. (1975) in: *Chemotherapy* (J. D. Williams and A. M. Geddes eds.), Plenum Press, London, vol. 3, pp. 35-42.
[21] Brouet, G., Canetti, G. and Roussel G. (1977) *Rev. fr. Mal. Resp.*, **5**, 1-62.
[22] Angel, J. H. (1975) in: *Chemotherapy* (J. D. Williams and A. M. Geddes eds.), Plenum Press, London, vol. 3, pp. 25-33.
[23] Angel, J. H., Citron, K., Campbell, I. A. and Somner, A. R. (1976), *Bull. IUAT*, **51**, 57-60.
[24] Daddi, G., Basilico, F. and Grassi, C. (1959) *G. Ital. Tuberc.*, **13**, 123-128.
[25] Mackaness, G. B. (1956) *Am. Rev. Tuberc.*, **74**, 718-728.
[26] Pilheu, J. A. (1977) *Chest*, **71**, 583-586.

[27] Poppe de Figueiredo, F., Alves Brito, A., Laborne Valle, J. H., et al. (1974) *Bull. Un. Int. Tuberc.*, **49**, 411-417.
[28] Koinange Karuga, W. (1974) *Bull. Un. Int. Tuberc.*, **49**, 388-397.
[29] Dubra, F. A. (1976) *Bull. IUAT*, **51**, 61-64.
[30] The Research Committee of the Tuberculosis Association of India, (1976) *Bull. IUAT*, **51**, 65-69.
[31] Gonzalez Montaner, L. J., Leston, J. M., Cetrangolo, A., et al. (1976) *Bull. IUAT*, **51**, 87-89.
[32] Corradini, A., Isekson, I., Deck, C. et al. (1976) *Bull. IUAT*, **51**, 103-105.

4

Combined antimicrobial treatment in tuberculosis and in non-tuberculous infections

D. A. Mitchison

Medical Research Council Unit for Laboratory Studies of Tuberculosis, Department of Bacteriology, Royal Postgraduate Medical School, London W12 OHS, England

The reasons why combined antimicrobial treatment is used both in the treatment of tuberculosis and in non-tuberculous infections are summarized in Table 4.1.

These reasons include:

1. increasing the range of bacterial species which can be affected by the antimicrobial drugs;
2. the prevention of the emergence of drug resistance;
3. synergistic activity of the drugs on bacteria;
4. heterogeneity, either in the conditions surrounding the bacteria or in the bacteria themselves, within the lesions.

I will take each of these reasons in turn indicating whether it applies equally in tuberculosis and in non-tuberculous disease and giving examples.

A wide spectrum of antibacterial activity (Table 4.2) is necessary when the organisms responsible for the infection are unknown. In non-tuberculous disease it is often advisable to start antibacterial treatment before the bacteriological diagnosis has been established and it is then necessary to give two or more antibacterials to inhibit all of the organisms that might be present. For example, leukaemic patients with fever are usually treated immediately with carbenicillin plus gentamicin or tobramycin. In contrast antituberculosis chemotherapy is almost always started after the diagnosis has been made, preferably by bacteriology. There are very rare patients in whom an infection is present that might be tuberculosis or might be non-tuberculous in an inaccessible site such as the pericardium. I have known such patients to have been treated with combinations of

Table 4.1 Use of combined antibacterial therapy.

1	Widens antibacterial spectrum
2	Prevents drug-resistance
3	Bacteriostatic or bactericidal synergism
4	Heterogeneity in lesions

Table 4.2 Wide spectrum activity.

1. Cause of infection unknown	*Non-TB*	common	e.g.	bacteraemia, acute pneumonia
	TB	v. rare	e.g.	TB possible diagnosis
2. Mixed infection in closed lesions	*Non-TB*	common	e.g.	aerobes & anaerobes in abscess
	TB	v. rare	e.g.	*M. tuberculosis* + *M. kansasii*
3. Invasion of open lesions	*Non-TB*	common	e.g.	burns: $AgNO_3$ + sulpha
	TB	v. rare	e.g.	sinus

antibacterials necessary both for tuberculosis and for non-tuberculous disease. This situation should of course be avoided wherever possible by trying a course of non-tuberculous chemotherapy first. Closed lesions such as abscesses, containing several different non-mycobacterial species are relatively common and may need treatment with two or more antibacterials. For instance an abscess containing aerobes and anaerobes might be treated with penicillin plus metronidazole. Tuberculous lesions are seldom colonized by any organism other than the original causative species, *M. tuberculosis* (or *M. bovis*). Very rarely there are mixed infections with more than one species of mycobacteria, such as *M. tuberculosis* plus *M. kansasii*. When non-tuberculous lesions are open, invasion by resistant organisms is common and wide spectrum antibacterial treatment may be given as a preventative. For instance, silver nitrate plus a sulphonamide is used for the treatment of burns. Despite the fact that the cavities of pulmonary tuberculosis communicate via the bronchi to the exterior, invasion by organisms other than tubercle bacilli is a rare event. Examples are the colonization of old tuberculous lesions by other mycobacteria, such as *M. avium*, or by *Aspergillus fumigatus*. A sinus from a tuberculous lesion may also become secondarily infected. Thus, while the use of combined antibacterial therapy to increase the range of bacterial species covered is fairly common in non-tuberculous disease, it is a very rare event in tuberculosis.

An important reason for the use of combined antimicrobial treatment is the prevention of drug resistance (Table 4.3). Resistance may originate in the original flora by overgrowth of mutant bacilli present as a very small proportion of the original sensitive organisms. In non-tuberculous disease this is comparatively rare, an example being the treatment of infections with *Staphylococcus aureus* by rifampicin plus fucidic acid, to prevent the emergence of rifampicin-resistant, mutant staphylococci. However, in tuberculosis the main reason for combined treatment is the prevention of drug resistance arising in this manner. Resistance may also occur as the result of the invasion of open lesions by resistant organisms from the exterior. This has already been dealt with in Table 4.2. I will return to drug resistance in greater detail below.

Antibacterial drugs may be used together in order to obtain synergistic activity (Table 4.4). Synergistic bacteriostatic activity is often considered to be important in the treatment of non-tuberculous infections. Examples are the combination of trimethoprim plus a sulphonamide in respiratory and urinary tract infections and amphotericin B plus 5-fluorocytosine in fungal disease. However, in tuberculosis this type of synergism does not often seem important. It may possibly play a role when combinations of reserve drugs of low activity are used, such as ethionamide, cycloserine and pyrazinamide. The role of bactericidal synergism in many non-tuberculous infec-

Table 4.3 Prevention of drug resistance.

1. Originating in original flora	*Non-TB*	rare	e.g.	rifampicin + fucidic acid in *S. aureus* infections
	TB	Standard practice		
(2. Originating by invasion of open lesions)				

Table 4.4 Synergism.

1. Bacteriostatic	*Non-TB*	Occasionally important	e.g. trimethoprim + sulpha, amphotericin B + 5FC
	TB	Not important	e.g. low activity reserve drugs
2. Bactericidal	*Non-TB*	Uncertain	e.g. meningitis, endocarditis
	TB	Important	Short course chemotherapy

tions is still uncertain. Where host defense mechanisms are effective, as in pneumonia, bactericidal activity is of little significance. When defense mechanisms are less adequate, as in meningococcal meningitis, the response to treatment appears to be as good to chloramphenicol, a bacteriostatic drug, as to penicillin or sulphonamides, which are bactericidal. Even in endocarditis, our dogma tells us that because of ineffective host defenses we must find drug combinations which kill as many as possible of the persisting infecting organisms. Yet there is little hard evidence to correlate the use of synergistic combinations with efficacy in treatment. In tuberculosis, one of the main aim of current research is to develop short course regimens which allow the duration of treatment to be greatly reduced. In short course chemotherapy the bactericidal activity of the drugs in the lesions is of fundamental importance as is bactericidal synergism between them.

In Table 4.5, I have used the term heterogeneity in lesions to cover a variety of situations where organisms may exist in several different types of environment within the lesions or may themselves be physiologically heterogeneous in such a manner as to affect their response to drugs. These conditions may lead to a requirement for combined antibacterial treatment. The first example of heterogeneity is an infection which exists or might exist on both sides of a natural barrier in the body such as that between the blood and the cerebrospinal fluid (CSF), or between the blood and the surface of the bronchial epithelium. These situations occur quite often in non-tuberculous disease and may be a reason for the use of two different antibacterials or for giving the same antibacterial by two different routes. For instance infections in premature babies, where meningitis might be present, are often treated with chloramphenicol (which penetrates well into the CSF) plus gentamicin (which penetrates poorly into the CSF, but is effective against a wide range of organisms elsewhere in the body). In pulmonary infections due to *Ps. aeruginosa*, carbenicillin plus gentamicin may be given parenterally and by inhalation to reach the surface of the bronchi. Gentamicin is also given intraperitoneally as well as systematically when the peritoneum is infected as a result of dialysis. These circumstances are rare in the modern treatment of tuberculosis. Antibacterial drugs are not now given locally as well as systemically, because there is good evidence that they penetrate well into most types of lesion. Nevertheless, a few physicians treat tuberculous meningitis with intrathecal as well as systemic streptomycin and in pleural effusions with drugs locally as well as giving them systemically.

A further reason for giving two drugs is where the micro-environment of the organisms varies considerably within the lesion. An important example of this in the treatment of tuberculosis is in the

Table 4.5 Heterogeneity in lesions.

1. Both sides of barrier - blood-brain, bronchial epithelium etc.	*Non-TB*	occasional	e.g. babies, pulmonary *Ps. aeruginosa*, peritoneal dialysis
	TB	v. rare	e.g. pleural effusion
2. Wide range of environmental pH	*Non-TB*	?	
	TB	Aminoglycoside + pyrazinamide	
3. Physiological heterogeneity in bacteria	*Non-TB*	?	
	TB	Bactericidal activity of rifampicin and pyrazinamide	

use of an aminoglycoside, such as streptomycin, with pyrazinamide. Tubercle bacilli in caseous material are at a neutral pH, where the aminoglycoside is effective. Pyrazinamide is active only at an acid pH when the aminoglycoside is ineffective. Thus a combination of the two drugs covers the complete pH range within which the organisms may exist. I cannot find any example of the use of a combination of antibacterial drugs to span a wide range of environmental pH in the lesions of non-tuberculous disease.

Physiological heterogeneity in bacteria may also be a reason for the efficacy of certain combinations of antibacterials in the treatment of tuberculosis. I will consider this issue in detail under the heading of short course chemotherapy. I will now turn to certain of these issues and consider them in greater detail.

Drug resistance

In tuberculosis, the origin of drug resistance is the occurrence of drug-resistant chromosomal mutants which are present in all populations of drug sensitive tubercle bacilli. They occur in proportions usually ranging from one in 10^5 sensitive organisms to one in 10^7 or 10^8. This statement is true for the major drugs, isoniazid, streptomycin and rifampicin and probably true for all of the other drugs used in the chemotherapy of tuberculosis. If any one of the several drugs is used alone, drug resistant variants are likely to emerge by overgrowth of the original sensitive population by these resistant organisms.

Treatment with two drugs was originally given solely with the aim of preventing the growth of drug resistant mutant organisms, since one drug would prevent the growth of mutant resistance to the other drug and vice versa. An example is the combination of the major drug isoniazid and one of the minor drugs *p*-aminosalicylic acid (PAS) or thiacetazone (Table 4.6). This two-drug combination did not always work for two reasons:

1. strains were sometimes encountered which were resistant to one of the two drugs at the beginning of treatment;
2. even with drug-sensitive organisms the minor drug could not always prevent the growth of mutants resistant to isoniazid.

In consequence treatment was given with three drugs, which protected both against initially resistant strains and also improved efficacy in patients with sensitive strains. Since therapy with three drugs throughout long periods of treatment tended to be toxic, the modern practice of initiating chemotherapy with a period of a few months of a triple drug regime followed by double drug treatment became standard practice throughout the world (Table 4.7).

Triple drug chemotherapy works well in tuberculosis when initial resistance is rare and usually to only one drug, as in the United Kingdom (Table 4.8). Where the prevalence of initial resistance is much higher, as in Hong Kong, resistance is still only to one drug, sometimes to two drugs (streptomycin plus isoniazid) and rarely to three drugs (PAS, streptomycin and isoniazid). Under these circumstances initial treatment with four drugs or even five drugs may result in an almost universally successful response to treatment.

Table 4.6 Prevention of drug resistance by 2-drug combinations (Tuberculosis Chemotherapy Centre, Madras).

Regimen	*Total patients*	*INH-resistant at 12 months (%)*	*Favourable response (%)*
INH	86	47	44
INH + PAS	86	8	91
INH + TB1	65	11	82

Table 4.7 Initial 3-drug regimens in pulmonary TB.

Basic 2-drug regimen	*Supplement for 2-3 months*	*Country*	*Total patients (%)*	*Favourable response (%)*
INH + TB1	—	E. Africa	147	88
INH + TB1	SM	E. Africa	162	96
INH + PAS	SM	UK (MRC)	101	98
INH + EMB	SM	UK (MRC)	91	100
INH + RMP	SM	UK (MRC)	95	99
INH + RMP	EMB	UK (BTTA)	709	?98

Table 4.8 Multiplicity of resistances.

	Plasmid		*Chromosomal*	
	E. coli *urinary (%)*		*Pulmonary TB (%)*	
No of resistances	*Outside hospital*	*Inside hospital*	*UK*	*Hong Kong*
0	68	54	96	60
1	15	6	3	21
2-3	15	9	1	18
4 or more	3	23	0	0

If we now look at non-tuberculous infections the first point to realize is that it is usually determined by resistance factors carried on plasmids within the bacteria. Considering urinary tract infections due to *E. coli* in the United Kingdom in 1974, the number of drugs to which resistance occurred was greater than in the drug resistance of chromosomal origin found in tuberculosis. Where selection pressure for resistance was low, in patients outside hospital, resistance to 2, 3, 4 or more drugs in the infecting strain was fairly common and where the pressure in favour of resistance was even greater, in patients within hospital, the proportion with 4 or more resistances was about a quarter of all of the infections (Table 4.8). Thus combined therapy is much more likely to prevent failure due to initial drug resistance when the resistance is chromosomal in origin as in tuberculosis then when it is plasmid in origin as is usual in non-tuberculous infections.

When drug resistance is chromosomal in origin, even in non-tuberculous infections, there is a further reason why combined chemotherapy may be more effective in tuberculosis. Fig. 4.1 shows an experiment in which cultures of tubercle bacilli were given a pulsed exposure to isoniazid for 1 or 4 days. The drug was added to a culture and then, after removal of the drug by filtration, the organisms were incubated and viable counts done at intervals. It is evident that exposure to isoniazid was followed by a lag in the growth of the culture lasting for about 7 days. In similar experiments lag periods lasting several days occurred after pulsed exposures to almost all of the antibacterial drug used in the chemotherapy of tuberculosis. This means that if patients fail to take drugs for even several days, the organisms in their lesions do not start to divide. Thus small gaps in treatment do not matter and do not promote the emergence of resistance strains. If however there are large and repeated gaps in treatment during which regrowth of bacilli occurs, the chance of a resistant strain emerging is high. This is so, partly because there is some selection of resistant mutants during exposure to the drug and partly because the total population, and therefore the possibility of double

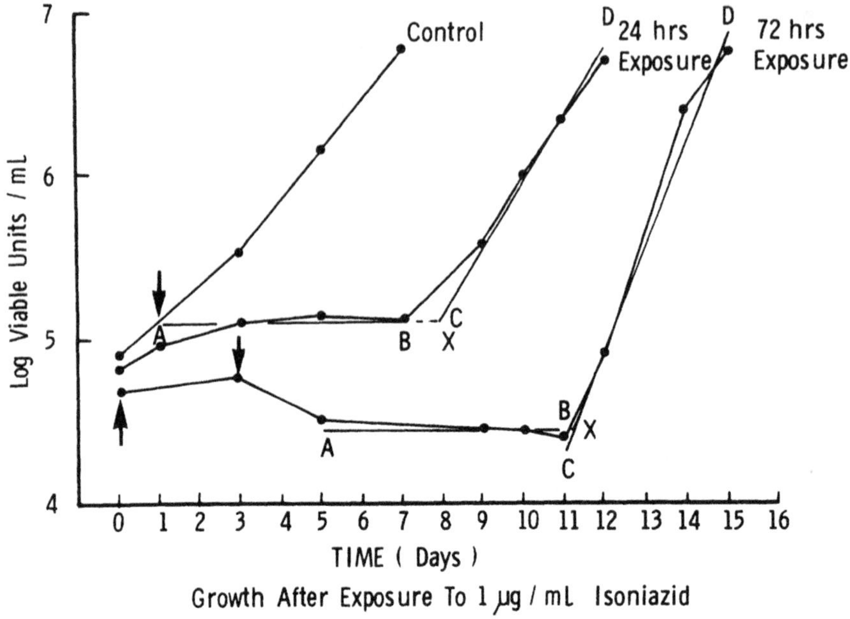

Fig. 4.1 - Lag periods following exposure of *M. tuberculosis* to isoniazid pulses. Cultures of *M. tuberculosis* in 7H9 Tween albumin liquid medium. ↑ = addition of 1 µg/ml isoniazid; ↓ = removal of isoniazid by filtration through a cellulose membrane filter followed by washing and suspension in fresh drug-free 7H9 medium).

mutants, is increased by successive regrowths of thebacterial population. If now we consider the corresponding situation in non-tuberculous infections, experimental evidence suggests that the lag in growth following a pulsed exposure rarely last more than two hours and may even be non-existent. If there is a gap in treatment during which the drug concentrations fall below the MIC, as might happen during the night or because patients fail to take occasional doses of drugs, there would then be periods during which regrowth of organism occurred. Should it be possible for resistance by chromosomal mutation to occur, these are the circumstances which would encourage it.

In summary, the use of combined chemotherapy is less effective in non-tuberculous infections than in tuberculosis partly because drug resistance is usually plasmid-mediated and partly because combined treatment is likely to be less effective should resistance be chromosomal in origin.

Heterogeneity

You will recall that one of the reasons why combined chemotherapy might work was that it provided cover for organisms in a wide range of environmental conditions, and that a wide pH range was covered by a combination of aminoglycoside and pyrazinamide. This situation is illustrated in Fig. 4.2 which shows the estimated size of the bacterial population according to the local pH in the lesion. In human disease most bacilli are in caseous material at a neutral pH. A minority of bacilli are intracellular where, in all probability, some are within phagolysosomes at an acid pH. The figure makes the assumption that streptomycin kills the majority of the bacterial population at a neutral or slightly alkaline pH, whereas

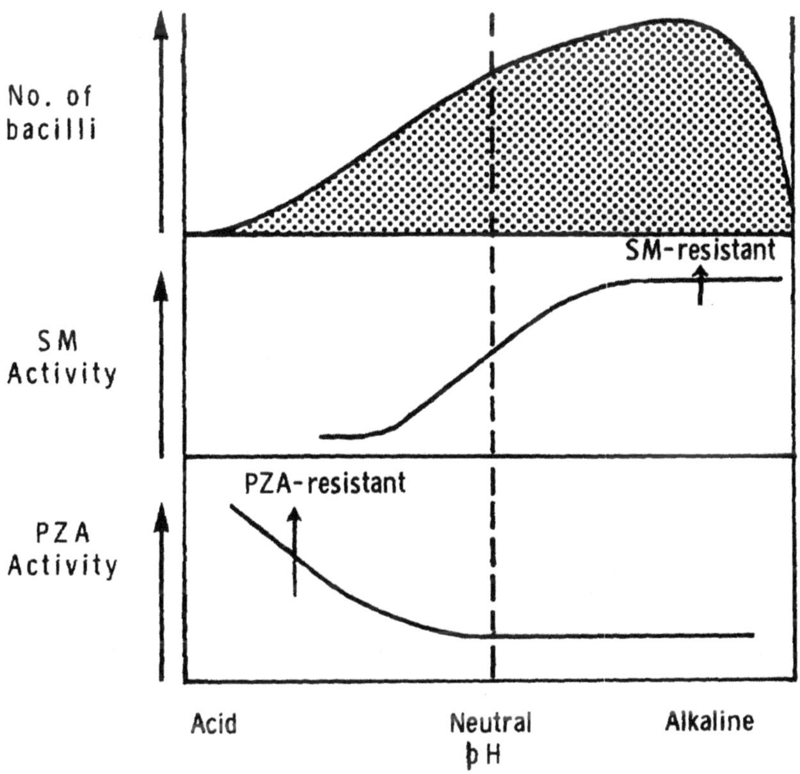

Fig. 4.2 - Bacterial content of tuberculous lesions and the activity of streptomycin and pyrazinamide related to pH.

pyrazinamide kills the minority population at an acid pH. Thus according to this hypothesis when the two drugs are given together they make up one highly efficient drug between them. Given together there is little prevention of drug resistance because streptomycin selects for streptomycin resistant mutants only at a neutral or alkaline pH where their growth is not inhibited by pyrazinamide and *vice versa*. However, if a third drug were added then the triple combination should be highly efficient, yielding results similar to those attained where isoniazid replaces the "single" drug (streptomycin plus pyrazinamide).

Table 4.9 shows the results of clinical studies in East Africa and at the Chemotherapy Tuberculosis Centre, Madras. When streptomycin plus PAS was given only about two-thirds of the patients had a favourable response. Similarly when streptomycin was given with pyrazinamide only slightly more than half the patients had a good response. Although we have no evidence from concomitant studies, these results are probably not much better than would be expected if streptomycin were used alone. However the results obtained by adding on PAS to the "single" drug (streptomycin plus pyrazinamide) was very much better, 92% of the patients having a favourable response. These results are very similar in their efficacy to those obtained using PAS plus isoniazid in the first-line treatment of East African patients. Thus the theoretical concept seems to be borne out by clinical practice.

Short course chemotherapy

Current clinical studies on the development of short course chemotherapy of tuberculosis were based upon the results of a large number of studies of experimental tuberculosis in the mouse carried out originally at Cornell University by McCune and McDermott and more recently by Madam Grumbach, at the Pasteur Institute in Paris,

Table 4.9 Streptomycin, PAS & pyrazinamide for pulmonary TB (Failures of treatment with INH + TB1).

Regimen	*Centre*	*Total patients*	*Culture negative at 6 m. (%)*
SM + PAS	E. Africa	38	58
	Madras	29	69
SM + PZA	E. Africa	72	40
	Madras	57	63
SM + PZA + PAS	E. Africa	60	92

Table 4.10 Conclusions from animal studies.

1	Pyrazinamide and rifampicin are the most potent sterilizing drugs.
2	Isoniazid with pyrazinamide or streptomycin is the most potent sterilizing combination.
3	The addition of streptomycin or ethambutol contributes little or nothing to the sterilizing activity of the combination.

as well as at some other centres. In these experiments mice were injected intravenously, treatment was given for periods ranging from 3 to 18 months, and the number of viable tubercle bacilli in their organs were counted. In situations when the organ cultures became uniformly negative as a result of sterilizing chemotherapy, groups of mice were often left without treatment to see whether relapse would occur; the occurrence of relapse was sometimes encouraged by giving the mice high dosage corticosteroids. It would be impossible to review all of this work here but some of the more important findings are summarized in Table 4.10.

Rifampicin and pyrazinamide were found to be key sterilizing drugs while streptomycin and ethambutol played little or no role in sterilizing the organs. Furthermore combinations of rifampicin or of pyrazinamide with isoniazid were more effective than combinations with drugs other than isoniazid. These findings in animal work have altered the whole course of current research in the chemotherapy of tuberculosis. Instead of concentrating on the prevention of drug resistance, the emphasis has changed to obtaining regimens that are as bactericidal as possible so that the duration of treatment can be cut down. In doing this we hope to be able to obtain greater cooperation from patients, to minimize the risk of relapse should they default from treatment and to obtain more effective and less toxic treatment.

Let us now see what happens when we give rifampicin or pyrazinamide (the sterilizing drugs in the mouse) to human patients. The first three lines of Table 4.11 summarize the results of the first short course chemotherapy study done in East Africa. Three of the regimens are shown, each having been given for 6 months. Cultures taken at 2 months after the start of treatment were more frequently negative when pyrazinamide or rifampicin had been added to the basic regimen of isoniazid plus streptomycin. After the end of the 6-month period of treatment, the patients were followed by regular monthly bacteriology. Relapses occurred during the next two years in 22% of the patients treated with isoniazid plus streptomycin butin only 9% when pyrazinamide was added and in only 2% when rifampicin was added. Thus there is clear evidence that both

Table 4.11 Bactericidal activity in pulmonay TB.

Regimen		Total patient	Culture negative at 2 m. (%)		Relapse rate 7-30 m. (%)
INH + SM		154*	49	**	22
INH + SM + PZA	for 6 m.	150	66		9
INH + SM + RMP		148	69		2
INH + SM + RMP		361*	73		
INH + SM + RMP + PZA		517	83		
INH + SM + PZA		129***	72		
INH + SM + PZA + RMP		261	92		

No effect on addition of SM, TB1 or EMB.
* E. AFrica ** $p < 0.05$ *** Madras.

pyrazinamide and rifampicin were active in sterilizing the lesions in man, similar to their activity in the mouse. The next two lines of Table 4.11 show the effect of adding pyrazinamide to a basic regimen of isoniazid, streptomycin and rifampicin. The addition of pyrazinamide increased the culture negative rate at 2 months from 73% to 83%, a statistically significant difference. The last two lines show that the addition of rifampicin to a regimen of isoniazid, streptomycin and pyrazinamide also increased the culture negativity at 2 months from 72% to 90%, again a significant difference. Thus pyrazinamide and rifampicin both markedly increase the sterilizing activity of antituberculosis regimens and, when given together, their activity is synergistic.

The nature of the synergism between isoniazid, rifampicin and pyrazinamide is illustrated in Fig. 4.3. All 3 drugs penetrate well through the various barriers and into cells. We think that the major part of the bacterial population is actively dividing and is killed by isoniazid and by rifampicin fairly early in treatment (Fig. 4.3 A). However the rate at which isoniazid kills tubercle bacilli depends upon their growth rate, so that it is much less bactericidal against semi-dormant bacilli and inactive against fully dormant bacilli. Rifampicin is also inactive against dormant bacilli. Thus there are two smaller parts of the bacterial population which are not so easily killed by isoniazid. One of these contains organisms which are almost completely inhibited by an acid environment (Fig. 4.3 C). Because growth is so slow, neither isoniazid nor rifampicin kills them; however they are particularly susceptible to the action of pyrazinamide. If pyrazinamide were not given, this sub-population might well start to grow after chemotherapy and cause clinical relapse. The second sub-population consists of bacilli that are dormant most of the time but occasionally have short periods of metabolic activity lasting a few hours (Fig. 4.3 B). There is not time during these brief periods of metabolism for isoniazid to kill the organisms, since there is a lag of at least a day before isoniazid begins to be bactericidal. However rifampicin starts to be bactericidal much more quickly than isoniazid, and so manages to kill this sub-population of organisms. One cannot explain the greater sterilizing activity of rifampicin, as compared to isoniazid, by an increased rate of bactericidal activity, since isoniazid and rifampicin kill bacilli at about the same rate; they only differ in the speed with which this activity starts. Neither can one argue that rifampicin penetrates better, since isoniazid itself penetrates excellently into all manner of tuberculous lesions. Thus we are left with the concept of physiological heterogeneity in the bacterial population within the lesions of tuberculosis. Short course treatment is most effective when several drugs are given to cope with each type of

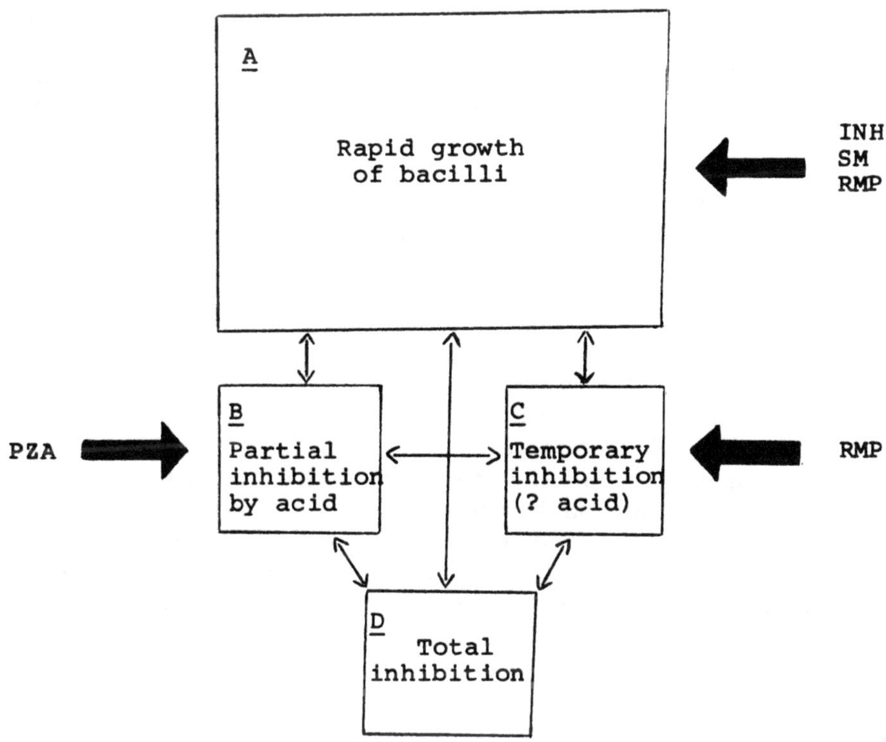

Fig. 4.3 - Population synergism in short course chemotherapy of tuberculosis. (Concept that bacillary subpopulations within lesions differ in their physiological state and micro-environment, and therefore in their susceptibility to the sterilizing activity of antibacterial drugs).

Table 4.12 Special characteristics of the chemotherapy of tuberculosis.

1	Laboratory studies confirmed by large scale clinical trials.
2	Diagnosis established before chemotherapy.
3	Infection of lesions with other organisms very rare.
4	Drug resistance mediated by chromosomes not plasmids.
5	Lag period of several days after pulsed drug exposure.
6	Bactericidal more important than bacteriostatic synergism.

physiological state which the organisms may have. Thus we have a new type of synergism.

The special characteristics of combined antibacterial treatment of tuberculosis are set out in Table 4.12. As we have seen there have been several important issues in which laboratory and clinical investigations have proceeded in step, reinforcing the validity of the hypotheses and demonstrating their practical importance. These issues include the early demonstration that double drug treatment prevails the emergence of resistance, the synergistic activity of pyrazinamide and streptomycin in conventional reserve drug regimes and the development of short course chemotherapy. Pulmonary tuberculosis is a good chemotherapeutic model to study because changes in the bacterial population occur slowly during treatment, because samples of this population can easily be obtained in the sputum and because there are plenty of patients with the disease.

There are several other ways in which tuberculosis differs from many non-tuberculous infections. The diagnosis is almost always established before chemotherapy is started, usually on firm bacteriological evidence, whereas non-tuberculous antibacterials are frequently started before the bacteriological diagnosis is made or when it is incomplete. Then in tuberculosis mixed infections or invasion from the exterior are rare, drug resistance is chromosomal in origin and bacilli do not grow during gaps of several days in antibacterial activity. All these factors profoundly modify the influence of combined therapy on the emergence of resistance. Finally, almost all of the emphasis on synergistic drug activity in tuberculosis is concerned with bactericidal synergism, whereas bacteriostatic synergism is often considered of greater importance in non-tuberculous infections.

5

Rationale and results of combined antineoplastic chemotherapy

S. Monfardini

Division of Clinical Oncology F - Istituto Nazionale Tumori (Via Venezian, 1 - Milano - Italy)

Summary

Recent progress in the chemotherapy of human cancer has resulted mainly from the discovery of new antitumor drugs and from the application of combination chemotherapy. While single agent chemotherapy has developed from predictive evidence of drug efficacy in experimental systems, combination chemotherapy regimens have not evolved to clinical use from basic experimental data but from success obtained in empirical studies. In fact to improve the results previously achieved with single agent chemotherapy, medical oncologists have been developing some broad guidelines concerning the mechanism of action, type of toxicity and clinical spectrum of activity of drug candidates for combination. More recently cytokinetic considerations have also been used in an attempt to improve the neoplastic cell kill.

The clinical experience acquired over the last 10 years has shown a definite superiority of combination over single-agent chemotherapy in most haematological malignancies, but these achievements have only partially applied to solid tumours. In the search for combination regimens active on solid tumours, previously known to be poorly sensitive to chemotherapy, the area of clinical empiricism should be restricted by evaluating only those regimes which have proved effective in appropriate *in vitro* tests.

Introduction

Recent progress in the chemotherapy of human cancer has resulted mainly from the discovery of new antitumor drugs and from the application of combination chemotherapy.

While single drug chemotherapy developed from predictive evidence of drug efficacy in experimental systems, combination chemotherapy regimens have not evolved to clinical use from basic experimental data but from success obtained in empirical studies. The lack of detailed knowledge on several factors such as specific mechanism of action of individual agents, development of drug resistance at the cell level and proliferation rates of neoplastic cell populations, have probably been the main reasons for the partially empirical selection of agents for use in combination chemotherapy [1].

To achieve further results in combining antitumor agents there is presently a need to evaluate experimental systems that can provide predictive data. This new approach requires however a close cooperation between the experimentalist and the medical oncologist in order to establish a true interaction of experimental and clinical studies.

Rationale of combination chemotherapy

In comparison with single-agent chemotherapy, combination chemotherapy can be defined as the association of two or more drugs in the attempt to increase the number of neoplastic cells killed and to delay the onset of drug resistance, improving thereby the quality and duration of clinical responses. However it must be pointed out that drug combinations are not necessarily more effective than single agent chemotherapy. In fact considering the combination of two drugs several theoretical results are possible [2]:

1. The activity is superior to additive effects of both agents;
2. the activity is simply additive;
3. the activity is superior to either single agent alone, but less than additive;
4. the activity is the same as either single agent alone;
5. the combination is less active than either single agent alone.

In the absence of any discriminating factor Lloyd has estimated that 2,211 two-drug combinations and 47,905 three-drug combinations could be empirically derived from an available pool of 67 drugs [3]. A disease-oriented view of two-drug combinations using the most active drugs [2] has shown how few of the potentially effective two drug combinations (which could be devised with 9 or 10 most active drugs) have actually been tested in solid tumors and hematologic malignancies. Rarely the number of combinations tested exceeded 20% of the possibilities.

Other variables in drug combinations besides the number of possible combinations, are the diverse dosage of drugs, the different drug schedules and the sequence of drug administration.

It is evident that it is not possible to test all or even a reduced fraction of the available drug combinations. Therefore it has become important to develop certain principles of combination chemotherapy to aid in the selection of potentially synergistic drug combinations, [4]. In the attempt to improve the results previously achieved with single agent chemotherapy, medical oncologists have been developing the following broad guidelines for the selection of agents to be used in combination.

1. The choice of candidate drug for use in combination should be restricted to those drugs which are definitely active when used alone against the particular tumor in question.
2. The drugs to be chosen should have minimal overlapping toxicity so that full doses of each component can be used with tolerable host toxicity.
3. The drugs to be combined should have a different mode of action.

More recently some cytokinetic considerations have been also used in the attempt to improve the log kill of neoplastic cells [5]. Combinations of drugs active in different phases of the cell cycle have been designed since in most tumors few cells are in the sensitive phase when an effective blood concentration of a phase dependent agent is present.

Effective combination of drugs may be obtained by using one agent which blocks cells in one portion of the cycle. If a second drug is given which kills the cells maximally during the stage immediately following the block, the arrest in the cell proliferation caused by the first drug may result in a partial synchronisation of the neoplastic population which increases the degree of cell kill by the second therapeutic agent. This approach may prove to be of clinical importance.

Since the usefulness of combination chemotherapy has not been fully assessed in all types of tumors a critical approach to the use of poly-drug therapy should be a constant guideline in designing new combinations [6]. With single-agent chemotherapy the utility of each agent can be adequately assessed and resistance develops to only one drug a time. The disadvantage is however that developement of resistance to a single agent may be more rapid than to any of the several agents used simultaneously. The advantage of combination chemotherapy consists mainly of the possible summation of therapeutic effects and the possible delay of drug resistance but at the same time toxic liability may be increased by therapeutically inactive compounds and simultaneous development of resistance to several drugs would limit future therapeutic options.

Results of combination chemotherapy

The clinical experience acquired over the last 15 years has shown a definite superiority of combination over single agent chemotherapy in some tumors and hematological malignancies, whilst in some other solid tumors there is only partial evidence in favor of a combined drug modality. Furthermore only in a few solid tumors is there definite evidence of no superiority of combination chemotherapy in comparison with single agents (Table 5.1).

Due to the magnitude of the subject which encompasses the whole field of medical oncology, this lecture is intended to present only a short review of the progress achieved for neoplasias in which combination chemotherapy has been proven definetely superior to single-agent chemotherapy.

Hodgkin's disease. Single-agent chemotherapy in Hodgkin's disease has been replaced in the past few years by combination chemotherapy because of its proven superiority in the incidence of complete remissions, duration of response and survival. Table 5.2 shows that for the MOPP (nitrogen mustard, vincristine, procarbazine and prednisone) and ABVD (adriamycin, bleomycin, vinblastine, imidazolcarboxamide) combinations the percentage of complete responses is superior to the addition of the activity of the single drugs used for these combinations. The achievement of complete remission is the most important factor affecting the prognosis since about 65% of complete remitters treated with MOPP continue to remain free of disease after 5 and 10 years [7].

Non-Hodgkin's lymphomas. With single-agent chemotherapy, even in previously untreated patients the overall complete remission rate has usually been less than 20%. As in Hodgkin's disease the concepts of intensive combination chemotherapy have recently also been applied to non-Hodgkin's lymphomas. The most widely used combinations have been CVP (cyclophosphamide, vincristine, prednisone) C-MOPP (cyclophosphamide, vincristine, prednisone, procarbazine), HOP (adriamycin, vincristine, prednisone), CHOP (cyclophosphamide, adriamycin, vincristine, prednisone), ABP (adriamycin, bleomycin, prednisone) and BACOP (bleomycin, adriamycin, cyclophosphamide, vincristine, prednisone). Complete remission has been obtained in 50 to 80% of patients with nodular patterns and in 40 to 60% of patients with diffuse patterns [8]. All multiple drug regimens, regardless of the number and types of drugs, doses and sequence, produced a remarkable improvement in the duration of complete remission, when retrospectively compared to single agent chemotherapy.

Acute lymphocytic leukemia. Most of the improvements in treatment to date, at least in the clinical area, have occurred largely through an empirical approach. However sufficient information has now been accumulated to encourage one to regard acute lymphocytic

Table 5.1 Combination chemotherapy versus single agent chemotherapy in the major tumor types.

Combination Versus single agent Chemotherapy	*Tumor types*
Definite superiority	Hodgkin's disease Non-Hodgkin's lymphomas Acute lymphocytic leukemia Acute myelocytic leukemia Multiple myeloma Breast carcinoma Testicular carcinoma Small cell carcinoma of the lung Soft tissue sarcomas
Possible superiority	Gastric carcinoma Colon carcinoma Ovarian carcinoma Lung carcinoma (other than small cell) Head and neck carcinoma Neuroblastoma Osteosarcoma Ewing sarcoma Chronic lymphocytic leukemia
No superiority	Melanoma Cervix carcinoma

Table 5.2 Clinical activity of single agents compared to that of principal combination chemotherapy regimens most effective in Hodgkin's disease.

Drugs		*Complete remission*
Nitrogen mustard		15%
Vinblastine		30%
Vincristine		15%
Adriamycin		10%
Bleomycin		10%
Bis-Chloroethyl-Nitrosourea		10%
Chloroethyl-Cyclohexyl-Nitrosourea		20%
Prednisone		5%
Procarbazine		20%
Imidazole carboxamide		10%
MOPP	Nitrogen mustard Vincristine Procarbazine Prednisone	70-80%
ABVD	Adriamycin Bleomycin Vinblastine DTIC	75%

leukemia as a potentially curable disease.

Nine distinct drugs have been proven beneficial in arresting and reducing the acute leukemia cell populations. Prednisone, vincristine, adriamycin, daunomycin, 6-mercaptopurine, methotrexate, cytarabine, l-asparaginase, cyclophosphamide administered as single agents induce remission in no more than 50% of patients. A comparison in terms of complete remission obtained with various combination regimens reveals that in almost every instance the complete remission rate is as great or greater for the combination than would be predicted from the additive effect of each drug alone [9]. The most widely used treatment for remission induction in previously untreated patients with acute lymphocytic leukemia is the combination of vincristine and prednisone. Remissions are generally induced in 90% of children. By adding daunomycin to the combination of vincristine and prednisone a complete remission has been obtained in 100% of children and in 75% of adults [10].

Since the increased survival of the remitting patient with acute lymphocytic leukemia is virtually equivalent to the total time spent in remission, the question has been raised whether combinations of drugs do prolong the duration of remission to a greater extent than single drug administration. Since no better way has yet been found of eradicating all leukemic cells during induction, probably the optimal therapeutic strategies at the present are:

1. to reduce the leukemic population as much as possible consolidating the remission during the initial intensive polydrug treatment regimen;
2. to incorporate in the maintenance regimen periodic administration of drugs acting in different phases of the cell cycle.

Acute myelocytic leukemia. Acute myelocytic leukemia is much less sensitive to a number of agents than acute lymphocytic leukemia. However combinations of: vincristine, prednisolone, methotrexate; 6-mercaptopurine and MeGAG; 6-mercaptopurine, MeGAG and cytarabine; cytarabine and 6-thioguanine; and cytarabine and daunomycin appear in general to be more effective remission inducers than any of their component drugs given at similar doses and schedules [9].

Patient responding to induction have a longer survival in comparison with non-responders. The maintenance of a remission in acute myelocytic leukemia is still an unsolved problem. Since there are not enough truly effective drugs to allow one to design a variety of complementary regimens, to decrease the risk of selecting resistant cell lines, generally the same drugs as those used for induction are employed.

Multiple myeloma. Melphalan has been extensively used as a single agent to treat multiple myeloma; a significant difference in survival has been observed between responders and non-responders (41 versus 9 months) [11]. Other drugs effective in myeloma are prednisone, adriamycin, procarbazine and the nitrosoureas. Overall response rate to single agents has varied from 10 to 49%. The use of various combinations such as melphalan and prednisone; melphalan, prednisone and procarbazine; melphalan, prednisone, vincristine, cyclophosphamide and bis-chloroethyl-nitrosourea has provided better results since the response rate has varied from 48 to 90% [10].

Breast carcinoma. A wide variety of single agents received clinical trials in breast cancer. Significant tumor regression were noted in 10 to 33% of patients. The most commonly effective agents are cyclophosphamide, methotrexate, 5-fluorouracil, vincristine and adriamycin. Cumulative collected data on combination of cytoxan, methotrexate, 5-fluorouracil, vincristine and prednisone have shown fairly comparable results for five and four-drug combinations, while the results with three drugs appear to be nearly as effective [12]. With these combinations significant responses have been seen in 35 to 65% of patients. It has been reported that response rates for adriamycin containing combinations are higher than for non-adriamycin containing combinations when both had the same number of drugs.

Combination chemotherapy has improved the overall response rate relative to single agents, however complete response rates are still seen in less than 25% of patients. The median duration of responses is relatively short (8 months) but survival gain can be demonstrated for responders versus non responders.

Testicular carcinoma. Many agents have a substantial activity in advanced testicular carcinoma, notably actinomycin D, mithramicin, vinblastine, bleomycin, adriamycin and cis-diammine-dichloroplatinum. Objective responses were seen in a percentage varying from 20 to 50%. One of the major significant achievements of these single-agents studies was not only that a complete remission could be obtained, but that approximately half of these complete remissions were permanent cures [13].

One of the most widely used combinations has been that of vinblastine plus continuous bleomycin infusion. This two drug combination appears to represent a truly synergistic combination, for the complete remission rate was higher than would be predicted by the use of these two drugs given alone. In fact complete remissions were obtained in 39-52% of patients with advanced disease [14], [15]. A further substantial progress has been achieved with the addition of cis-diammine-dichloroplatinum since this triple drug combination was recently reported as producing complete remission in 74% of patients [16].

Small cell carcinoma of the lung. Small cell bronchogenic carcinoma is the most responsive type of lung cancer to chemotherapy. Cyclophosphamide, methotrexate, mechlorethamine, adriamycin, CCNU, procarbazine, hexamethylmelamine, VP-16 have established activity superior to 20% in previously untreated patients [18]. However the responses to single agents are short lived, lasting a median of 3 months.

An increasing number of combination chemotherapy regimens have been tested in the past few years. The overall response has been around 60%, the complete response rate was 23% and median survival 7 months [17]. The drugs used in these studies were cyclophosphamide, vincristine, nitrosoureas, methotrexate, adriamycin, prednisone, VP-16, bleomycin and imidazole carboxamide. The available data suggest that response rates and median survival increase respectively with two, three, and four drug combinations. As with other tumors showing a marked sensitivity to chemotherapy complete responders live longer than partial and non-responders; however, few patients are cured.

Soft tissue sarcomas. Single-agent chemotherapy of metastatic soft tissue sarcomas in adults has given poor results. Most of the responses reported for vincristine (52%), cyclophosphamide (47%) and actinomycin D (24%) were obtained in studies done on children. In broad phase II studies adriamycin and imidazole carboxamide both proved to have definite activity when given as single agents. Later on it was shown that the effects of adriamycin and imidazole carboxamide were additive since their combination produced a total response rate of 47% [18]. The most effective combined chemotherapeutic regimen for soft tissue sarcomas is however formed by cyclophosphamide, vincristine, adriamycin and imidazole carboxamide. Significant responses have in part been obtained with this combination in 52% to 59% of patients with advanced soft tissue sarcomas.

Comment

Combination chemotherapy has had a definite impact against neoplastic diseases for which a number of active drugs with diverse mechanisms of action, different toxicity and capability of inducing complete remission when used alone are available. For tumors which show a marked sensitivity to combination chemotherapy patients achieving complete response survive significantly longer than non-responders and partial responders.

The most consistent progress of combination chemotherapy has been against hematological malignancies, especially acute lymphocytic leukemia, Hodgkin's disease and diffuse histiocytic lymphoma, but these achievements have only partially translated to the common solid tumors. Unfortunately for the more common solid tumors the supply of effective drugs is more limited, however only a few of the potentially useful combinations have actually been tested. In a number of solid tumors there has actually only been partial evidence of possible superiority of combination chemotherapy over single-agent chemotherapy, but single conventional agents chemotherapy is no more expected to provide better results than combination chemotherapy. For this reason randomized studies testing non cross-resistant combinations should be developed. In the search for combination regimens active on solid tumors previously known to be poorly sensitive to chemotherapy, the area of "clinical empiricism" should be restricted by evaluating experimental systems which can provide predictive data to be used in the rational design of combination chemotherapy regimens. Furthermore a better understanding of cell kinetics of human solid tumors would facilitate pharmacological studies, so that the available drugs could be used more efficaciously.

References

[1] Griswold, D. P., Dykes, D. J., Kelly, C. A., Roberts, B. J. and Dominick, C. A. (1974) "Approaches to combination chemotherapy in rat, mouse and hamster tumors", *Cancer Chem. Rep.*, **2**, 99-108.

[2] Carter, S. K. (1974) "Planning combined therapy. The interaction of experimental and clinical studies", *Cancer Chem. Rep.*, **2**, 3-11.

[3] Lloyd, H. H. (1974) "Combination chemotherapy: considerations for design and analysis", *Cancer Chem. Rep.*, **2**, 157-166.

[4] Capizzi, R. L., Keiser, L. W. and Sartorelli, A. C. (1977) "Combination chemotherapy. Theory and practice", *Seminars in Oncology*, **4**, 227-253.

[5] Valeriote, F. A. (1977) "The role of cell kinetics in cancer chemotherapy", *Seminars in Oncology*, **4**, 217-226.

[6] Young, C. W. (1969) "Formal discussion: cautionary consideration of combination chemotherapy in the treatment of human malignancies", *Cancer Res.*, **29**, 2281-2283.

[7] Bonadonna, G. (1978) *Hodgkin's disease: chemotherapy. Current therapy*, W. B. Saunders Company, Phyladelphia, London, Toronto, pp. 293-301.

[8] Bonadonna, G. and Monfardini, S. (1974) "Chemotherapy of non-Hodgkin's lymphomas", *Cancer Treat. Reviews*, **1**, 167-181.

[9] Henderson, E. S. (1969) "Treatment of acute leukemia", *Seminars in Hematology*, **6**, 271-319.

[10] Clarysse, A., Kenis, Y. and Mathé, G. (1976) *Cancer Chemotherapy. Its role in*

the treatment strategy of hematologic malignancies and solid tumors, Springer, Berlin - Heidelberg-New York.

[11] ALEXANIAN, R., BERGSAGEL, D. E., MIGLIORE, G. J., VAUGHIN, W. K., HOWE, C.D. (1968) "Melphalan therapy for plasma cell myeloma", *Blood*, **31**, 1-9.

[12] CARBONE, P. P., BAUER, M., BAND, P. and TORMEY, D. (1977) "Chemotherapy of disseminated breast cancer", *Cancer*, **39**, 2916-2922.

[13] KENNEDY, B. J. (1970) "Mithramycin therapy in advanced testicular neoplasms", *Cancer*, **26**, 755-756.

[14] SAMUELS, M. L., HOHNSON, E. E., HOLOYE, P. Y. (1975) "Continuous intravenous bleomycin therapy with vinblastine in stage III testicular neoplasia", *Cancer Chem. Rep.*, **59**, 563-570.

[15] SAMUELS, M. L., LANZOTTI, V. J., HOLOYE, P. Y. and HOWE, C. D. (1977) "Stage III testicular cancer: complete response by substage to velben plus continuous bleomycin infusion (VB-3)", *Proc. Am. Ass. Cancer Res.*, Denver, May 18-21 (Abstr. 584).

[16] EINHORN, L. H. and DONOHUE, J. (1977) "Cis-Diamminedichloroplatinum, vinblastine and bleomycin combination chemotherapy in disseminated testicular cancer", *Ann. Int. Med.*, **87**, 293-298.

[17] BUNN, P. A., COHEN, M. H., IHDE, D. C., FOSSIECK, B. E. Jr., MATTHEWS, M. and MINNA, J. D. (1977) "Advances in small cell bronchogenic carcinoma", *Cancer Treat. Rep.*, **61**, 333-342.

[18] PINEDO, H. M. and KENIS, Y. (1977) "Chemotherapy of advanced soft tissue sarcomas in adults", *Cancer Treat. Rev.*, **4**, 67-86.

6

Quantitative evaluation of antibacterial activity *in vitro* of associated chemoantibiotics

E. Gori

Istituto di Farmacologia, Facoltà di Scienze, Via Vanvitelli 32, Milano, Italy

A. Visconti, U. Alieri

Laboratorio di Microbiologia, Ospedale S. Carlo, Via Pio II 3, Milano, Italy

M. Sangiovanni, F. Scrollini

Laboratorio Ricerche Dr. L. Zambeletti, Baranzate di Bollate (Mi), Italy

M. Basagni, P. Beggi

Istituto di Matematica, Facoltà di Scienze, Via Saldini 50, Milano, Italy

Summary

The procedures used at the present time in assessing the interaction between two antibacterial agents are reviewed briefly and evaluated critically.

A new procedure is then suggested, for which the data of the antibacterial activity *in vitro* of many pairs of chemoantibiotics tested either in solid medium (measuring the inhibition zone diameters in a deliberately modified multiplate agar diffusion-test) or in liquid medium (measuring by turbidimetry the intensity of the microbial growth at a pre-settled time) have been utilized.

In order to circumvent the three-dimensional representation, a diagram was adopted based on the general principle of "equivalence". According to this principle concentrations of the associated agents may be expressed simply as the sum of the concentration of the first agent plus the concentration of the second agent converted into the equi-effectual concentration of the first.

Thanks to this device, the additive interaction may be visualized as a linear regression coincident with the linear regression of the first agent, and the super - or infra - additive interactions may be visualized as curvilinear regressions starting from the linear regression of the first agent, attaining a positive or negative maximum and then slowly declining in order to coincide asymptotically with the linear regression.

Such curvilinear regressions may be explained by a mathematical equation based on the hypothesis that the maximum of the positive or negative interaction may be attained only when the two antibacterial agents are associated according to a definite ratio.

A statistical approach able to estimate from the experimental data the parameters of such a mathematical model is presented.

Among the many advantages usually listed for chemoantibiotic associations, the synergism of the antibacterial effect is paradoxically confined to the last place, even if the synergism continues to be claimed (also quite recently by O.M.S. [1]) as the basic rationale for each combined therapy.

There are several reasons for this position:

1. it is doubtful whether antibacterial synergism as well as antagonism are clinically relevant;
2. there are few well-established examples of synergism and antagonism;
3. there is no clear-cut agreement on the meaning of this term and, more broadly, of all terms concerning drug interaction, with the result that since 1971 we ourselves have felt the need to propose a new classification and nomenclature [2];
4. in general, the quantitative assessment of antibacterial interaction is somewhat crude and the methods in use have been substantially the same for about twenty years and in most cases confined to *in vitro* [3].

With regard to the quantitative assessment, some critical remarks seem worth considering.

It is well known that the most widely applied technique is isobolic representation, which is obtained by putting for each agent on the corresponding axis placed at the right angle the fractional values of a pre-selected activity unit and then joining them up.

Such an activity unit may in turn be:

1. the minimal bacteriostatic (MIC) or the minimal bactericidal concentration (MBC) assessed in liquid or solid medium [4-18];
2. the concentration capable of ensuring a definite rate of microbial generation, which may be expressed by a constant K [19-22];
3. the CI_{50}, that is the concentration of the antimicrobial agent required to cause a definite % inhibition of growth (turbidimetrically determined) [23-28];
4. the ED_{50}, that is the dose (expressed as mg/kg) capable of ensuring 50% of survival of experimentally infected mice [29], [30].

It may also be shown that the many indices proposed in order to evaluate the antibacterial interaction [31-35] are not only substantially coincident with each other [3], but may in fact be considered as derived from the isobolic principle, of which they are a satisfactory short-cut, provided that the limits of such indices are evaluated (*).

Also the so often used diffusion test, obtained by placing two paper strips each containing a single antibiotic at right angles to each other on an agar plate (with or without the successive transfer on cellophane) gives inhibition zones, the profiles of which we consider may be interpreted as isoboles (**) and may be explained by a corresponding exponential equation [38], [39], [40]. From the practical point of view, such isoboles cannot be used quantitatively, since the exact concentrations of interacting antibiotics are not measurable (***).

Although the isoboles, in direct or indirect form, are so widely applied, they may be submitted to more than one criticism.

1. It is rather difficult to establish *a priori* the exact proportions of two substances to be put in a mixture in order to produce the preselected end-point effect. It is obviously much easier to measure the effect induced by different mixtures of the two substances.
2. It is rather cumbersome to obtain in only one experiment all the experimental data concerning the effects of the single as well as of the associated substances. Therefore the isoboles are generally based on so called "historical data", that is on data collected from a series of successive experiments. However, such successive data are not strictly comparable, due to large variations from day to day.
3. Isoboles are usually made up of very few experimental points, so that in practice they will not be smooth curves.
4. Such experimental points are generally single (and not mean) values, so that even the straight line corresponding to the additivity isobole cannot be tested for linearity. Furthermore, in the case of super- or infra-additive interaction, it may be mathematically proven that isobole is a rectangular hyperbola [42], [43] and, in fact, this hyperbola has been sometimes fitted [36], but it has never been submitted to statistical analysis, even if a method of estimation is available [44].
5. The total effect chosen as activity unit is very often a 100% effect (such as MIC) which, in fact, is not statistically sound.

(*) If the chosen activity unit is the MIC and the modulus adopted for serial diluition is 2, the sum of the concentrations of both antibiotics assuring the MIC must be less than 50% or greater than 200% in order to conclude respectively for the existence of synergism or antagonism [36].

(**) In order to strengthen this similarity, Chabbert [37] suggests adopting the log transformation of the axis, but the result is, in our opinion, rather disappointing, because the differences between addition and synergism became elusive.

(***) The last criticism may be applied *a fortiori* to all the other less used diffusion tests, such as "strip gradient plates" [41] and so on.

6. It is customary to subdivide mechanically the two axes of isobolic diagram in % fractions of the pre-selected activity unit, but this procedure may be accepted only provided that the single log dose-effect regressions of the two associated drugs are linear and parallel. In fact it may be shown that when the two log dose-effect regression lines are not parallel, to the same fraction of the activity unit of each drug does not correspond the same intensity of effect, so that summing up two fractions in order to attain the activity unit cannot assure the full effect corresponding to such an activity unit (*). In other words, in this condition the isoboles may be still drawn, provided that the mathematical fractions of each activity unit are converted into equi-effectual fractions. Such a conversion requires the knowledge of the dose-effect regression lines, which is so often lacking in chemotherapy.
7. The last criticism of isobole design is that the interaction is investigated only at a pre-chosen level of effect, all the other levels of the effect being unexplored. With regard to this, it may be observed that, even if, from the theoretical point of view, isoboles of the same set may be predicted to be all superimposable [43] and an amazing variety of sets of isoboles (identified by a complicated nomenclature) has been presented since 1921 by Loewe [47], no convincing collection of experimental data is available up to now, in order to confirm the predicted designs.

For a long time statistical techniques much more sophisticated than isoboles have been devised in order to evaluate quantitatively the joint action by a series of authoritative workers, such as Bliss [48], Finney [49] and, more recently, Hewlett and Plackett [50], [51].

However, such techniques have never been utilized in the antibacterial field, except in the most simple form of "similar action", for some not even very convincing experiments *in vitro* [52] and *in vivo* [53].

Also the technique of so called "independent action" has been applied to ED_{50} values determined for antibacterial effects *in vitro* [54] and for antimitotic chemotherapy *in vivo* [55], but it has never been submitted to the very cumbersome statistical treatment suggested by Hewlett and Plackett [50], [51].

On the other hand, all these techniques are restricted to quantal responses and to a pair of agents associated according to a fixed ratio.

Some statistical procedures have also been devised in order to

(*) This criticism, put forward by us in 1972 [3], agrees substantially with the criticism put forward by Weaver and co-workers [45] and recently requoted by Tedeschi [46], according to which adding the doses of two drugs having a similar action does not necessarily mean adding their effects.

explore the entire field of joint effects of two agents associated according to all possible ratios.

The best one is the so-called "steepest ascent method" of Box and Wilson [56], [57], [58], which, in fact, has been applied, even if incompletely, to antitumoral chemotherapy [59], [60], but it has never been applied to antibacterial chemotherapy.

Other interesting skilled techniques, such as the so called "simplex" design, with its "lattice" and "centroid" variants [61], [62], [63], devised in order to explore the factor space, are still awaiting an extensive pharmacological and chemotherapeutic application.

In order to remove most of the above-quoted criticisms, we have elaborated since 1964 [64] and successively modified [2], [65] an experimental as well as a mathematical-statistical procedure, which is suitable for any pharmacological combination, but which, in fact, finds in the antibacterial test *in vitro* all the conditions needed to be successfully applied and verified.

The experimental pattern of such a procedure has been devised in order to satisfy the following conditions:

1. the effects induced by single as well as by joint concentrations have to be measured all at the same time, that is under exactly the same experimental conditions;
2. for every joint effect the number of replications has to be high enough to allow the interpolating equations to be submitted to statistical analysis;
3. the range of the effects has to extend from 0% to 100% for single as well as for joint agents;
4. the grid of joint concentrations has to be, on the one hand, very extended and, on the other hand, very close in order to cover the whole of the yield surface to be explored.

In practice, a modified checker-board design is adopted. The design consists of a series of increasing concentrations of a first as well as of a second antibacterial agent, plus a series of increasing concentrations of the first agent, associated to one or more than one fixed concentrations of the second agent. This factorial design may also be inverted.

For the sake of simplicity the antibacterial effect chosen was the bacteriostatic one.

In turn, this bacteriostatic effect has been evaluated by two different techniques modified in order to satisfy the above-quoted conditions.

The first technique has been that of one-center agar diffusion test, by which the two antibacterial agents are put together in the same hole. As a matter of fact, this test has been always discarded on the grounds of the belief, authoritatively stated by Gunnison and Jawetz since 1952 [66], that the inhibition zone of the faster diffusing antibiotic cannot be modified by the slower diffusing antibiotic. However, we have shown that (except, perhaps, for some large antibiotics, such as bacitracin) incidental differences in the diffusion speed of different antibiotics may be easily overcome by an over-night period of pre-incubation at 4 °C, so that the diameters of the inhibition zones of associated antibiotics are always significantly different from those of single agents [3], as predicted, on the other hand, by the theoretical formula of Humphrey and Lightbown [67].

The details of this technique are omitted here, as they have been described in more than one of our previous papers [2], [3], [64], [65]. It is only noteworthy that the experiment corresponding to the factorial design above quoted was always accomplished at one time, by using a set of rectangular glass trays for the solid medium; that the stainless steel cylinders for the solutions were randomly allocated in such trays; that each concentration was tested at least six times; that the diameters of the inhibition zones were measured to the nearest 0,5 mm.

More recently it has been possible to measure the bacteriostatic effect in a liquid medium by using the Autobac 1 System of the Pfizer Diagnostic Division. The procedure has been adapted to our experimental pattern and details of this technique will be published elsewhere.

It is only noteworthy here that, thanks to this system, it is possible to test simultaneously 360 single or associated concentrations of the antibacterial agents; the time of incubation may be restricted to only 4 hours; the turbidity may be read automatically as difference (Δ) versus controls; a high number of replica for each tested concentration may be obtained all closely alike, provided that the antibacterial solution as well as the inoculum are stored at -20 °C and that a strictly randomized allocation of the solutions into the system is applied.

It may also be stated that for solid as well as for liquid tests the inoculum was always picked by us from one colony; that the mixtures of the antibacterial solutions were always pre-tested for possible evidence of physico-chemical interactions and pH variations. Finally preliminary experiments were carried out to assess the range of concentrations to be tested.

The second and more relevant part of our procedure was concerned with the graphic representation of data and their related mathematical interpretation and statistical evaluation.

The first, even if incidental, problem to be faced has been which type of effect-concentration regression for a single agent is the most suitable to interpolate the data expressed as mm of diameters by the diffusion test or as % transmittance by the liquid test. In both cases it is very often possible to draw a linear or a slightly parabolic regression of the data on the log of the used concentrations. Only in some cases must linearization be restricted to the central values or may be obtained by using some special metametrization procedure such as the square of data or the more complicated ones recently suggested by Kavanagh [68] (*).

The second and again incidental problem concerns the fact that in a diffusion test it is impossible to attain by definition a maximum effect, whereas in a liquid medium it is always possible to attain the 100% effect, so that, in the last case, any joint effect, even if largely super-additive, will never be able to exceed this ceiling and in fact will seem truncated.

The fundamental problem to be solved was how to circumvent the isobolic representation and attain a planar but clear-cut plotting of data.

Apart from the isobolic representation, two procedures have been adopted up to now for such a planar drawing in the case of mixture of drugs.

The first procedure is usually adopted when the drugs are associated according to one fixed ratio (for instance 2: 1): in this case the effects are plotted against the increasing concentrations or doses of the mixture assumed as total [48], [49].

The second procedure is adopted when a series of increasing concentrations or doses of a first agent is associated to one fixed concentration of a second agent. In this case the measured effects are plotted only against the increasing concentrations of the first agent [70].

As a matter of fact, in the first, but especially in the second case, as pointed out by Gero [71], it is practically impossible to differentiate, at least at first sight, the simple additive effect from a true synergistic effect.

On our part we have devised since 1964 [64] a new procedure, named by us "equi-effectual", based on the metametric transformation of the concentrations of one of two drugs into the equi-effectual concentrations of the other.

In fact, if two drugs are "equivalent" according to the nomenclature of Gallego [36], the expected effect of their mixture will be exactly equal to the effect obtained with that concentration of one component, chosen as reference, which results from the sum of the concentration of the reference component plus the concentration of the second component transformed into the equieffectual concen-

(*) In no case did the probit transformantion of transmittance data, as suggested by Treffers [69] and by Maccacaro [52], yield a linear plot.

tration of the reference component. This transformation is obviously easier if the two dose-effect regression lines of the two agents are parallel, but it is also possible to make this transformation at each level, if the parallelism is lacking.

In practice, the mean experimental data obtained with the associated concentrations of the two antibacterial agents are to be plotted against the sum of the two concentrations expressed only as a concentration of the reference agent.

Thanks to this device all the mean experimental data will lie, in the case of additivity, on the straight regression line of the reference agent, as exemplified in Fig. 6.1.

However, when a super or infra-additive interaction intervenes, as exemplified in Figs 6.2, 6.3, 6.4 and 6.5, there will be a curvilinear regression, which starts from the straight regression line of the reference agent, attains a maximum respectively above or under this line and then slowly declines in order to attain again asymptotically the straight regression line.

Many other examples could be presented, from which it is possible to deduce that these curvilinear regressions are all substantially alike, whatever bacteriological test adopted, whatever pair of antibacterial agents tested and, most relevant, whatever type of interaction takes place.

At first, we searched for an empirical formula interpolating this kind of curvilinear regressions and in fact the interpolation was successfully achieved in 1965 [65] by using an exponential parabolic equation. However, the fact remains that such an empirical formula, even if successfully applicable, does not explain the form of the regression.

Such a form suggests that the maximum positive or negative interaction is attained only at a definite ratio between the concentrations of two antibacterial agents (not necessarily 1: 1), under and above which the number of the molecules of one or of the other agent are in excess so that they cannot interact but simply add or subtract their single effects.

The mathematical models satisfying this hypothesis may be written as follows:

$$E_{IF} = a_I + b_I \ln \ (pI + \mathrm{qF}\rho)\, K + (1 - p)\, I + (1 - q)\, F\rho$$

where:

E_{IF} is the total effect;

a_I, b_I are the parameters of the straight line relating the effect of agent I, chosen as reference agent, to the ln of the corresponding concentrations;

I is the increasing concentrations of the reference agent;

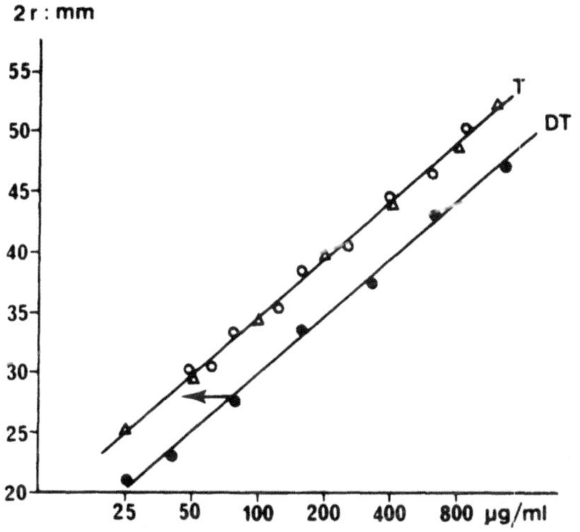

Fig. 6.1 - Mean experimental data (marks) and calculated linear regressions relating the antibacterial effect respectively to the log of concentrations of Demethylchlortetracycline (DT: ●), of Tetracycline (T: △) and of one fixed concentration of DT associated to increasing concentrations of T (o). Effect on bacterial growth of *Sarcina lutea ATCC 9341* measured in one center agar diffusion test as diameter (2r: mm) of the inhibition zone.

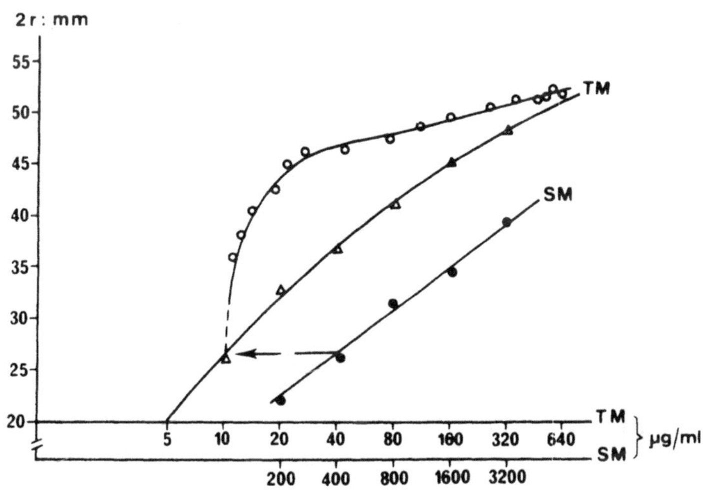

Fig. 6.2 - Mean experimental data (marks) and calculated linear regressions relating the antibacterial effect respectively to the log of concentrations of Sulphamethoxypiridazine (SM: ●) and of Trimethoprim (TM: △). Mean experimental data (o) and calculated curvilinear super-additive regression relating the antibacterial effect to the log of one fixed concentration of SM associated to increasing concentrations of TM. Effect on bacterial growth of *Staphylococcus aureus ATCC 6438P* measured in one center agar diffusion test as diameter (2r: mm) of the inhibition zone.

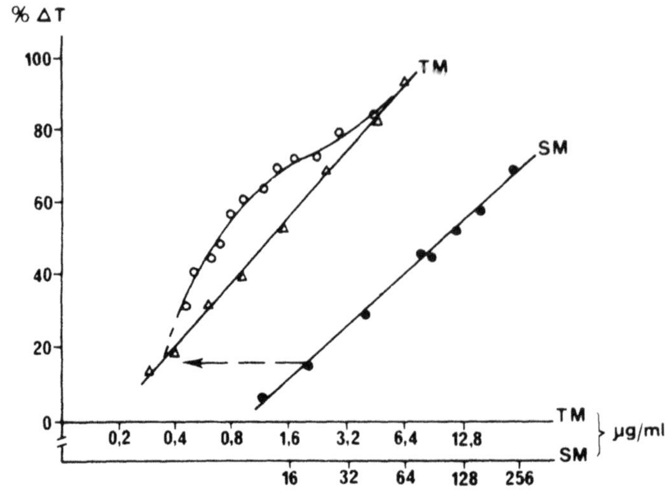

Fig. 6.3 - Mean experimental data (marks) and calculated linear regressions relating the antibacterial effect respectively to the log of concentrations of Sulphamethoxypiridazine (SM: ●) and of Trimethoprim (TM: △). Mean experimental data (o) and calculated curvilinear super-additive regression relating the antibacterial effect to the log of one fixed concentration of SM associated to increasing concentrations of TM. Effect on bacterial growth of *Klebsiella pneumoniae SCH 147/77* measured on a liquid medium as △% of transmittance versus control.

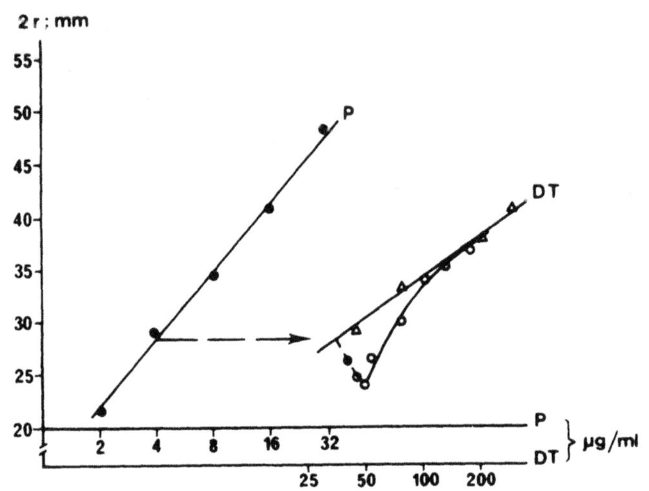

Fig. 6.4 - Mean experimental data (marks) and calculated linear regressions relating the antibacterial effect respectively to the log of concentrations of Penicillin (P: ●) and of Demethylchlortetracycline (DT: △). Mean experimental data (o) and calculated curvilinear infra-additive regression relating the antibacterial effect to the log of one fixed concentration of P associated to increasing concentrations of DT. Effect on bacterial growth of *Sarcina lutea ATCC 9341* measured in one center agar diffusion test as diameter (2r: mm) of the inhibition zone.

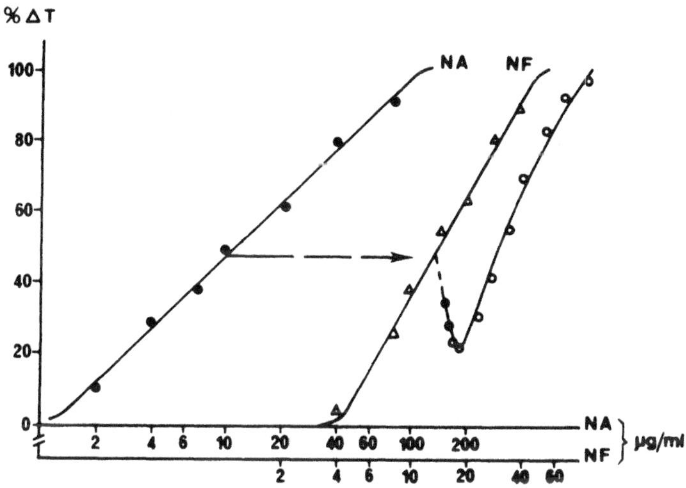

Fig. 6.5 - Mean experimental data (marks) and calcuted linear regressions relating the antibacterial effect respectively to the log of concentrations of Nalidixic acid (NA: ●) and of Nitrofurantoin (NF: △). Mean experimental data (o) and calculated curvilinear infra-additive regression relating the antibacterial effect to the log of one fixed concentration of NA associated to increasing concentrations NF. Effect on bacterial growth of *Escherichia coli SCH 120/77* measured on a liquid medium as Δ% of transmittance versus control.

F is the fixed concentration of agent F converted into the equieffectual concentration of agent I by the factor "ρ";
$(1 - p)\, I$ is the non-interacting fraction of I;
$(1 - q)F$ is the non-interacting fraction of F;
pI and qF are the interacting fractions of I and F;
K is the factor by which the sum of the interacting fractions of I and F must be multiplied in order to give the interactive component of the total effect E_{IF}. K is > 1 in case of super-additive interaction, is < 1 in case of infra-additive interaction.

In order to test whether this mathematical model really fits the experimental data, the following procedure was devised.

The above-quoted equation is transformed into the following two-branched function, where the first branch corresponds to the first part of the curvilinear regression and the second branch corresponds to the second part:

$$E_{IF} \begin{cases} a_I + b_I \ln (\alpha I + F\rho) \text{ when } F\rho > p/q \cdot I \\ a_I + b_I \ln (I + \beta F) \text{ when } F\rho < p/q \cdot pI \end{cases}$$

The parameters a_I, b_I, α, β and their respective errors are simultaneously estimated according to an iterative method quite like that suggested by Snedecor and Cochran [72] for the solution of an exponential asymptotic regression.

The two coupled regression lines are tested, as usual, by a variance analysis, which is reiterated in order to achieve the optimization for the best fitting of the α and β values.

In turn, α and β are used in order to solve the following system:

$$\alpha = K + K \cdot p/q - p/q$$

$$\beta = K + \frac{K}{p/q} - \frac{1}{p/q}$$

Finally the parameters K and p/q are isolated and their respective errors calculated solving a system given by the two expressions of the total differential of α and β.

Such parameters are able to define adequately the interaction, in the sense that K is expressing the mean super- or infra-additive intensity of the interaction and p/q indicate the mean ratio of the equiactive concentrations of the two antibacterial agents at which the interaction intervenes.

From the calculated regression it is also possible to determine the p/q ratio (which may be also expressed as ratio of μg/ml, or of mM) at which the interaction attains its positive or negative maximum.

In any case the proposed equation has been shown to fit perfectly all the above presented regressions (Table 6.1).

However, it must be observed that, according to our fully extended design, we can draw for each tested pair of antibacterial agents not one, but a set of many curvilinear regressions, as exemplified in Figs 6.6, 6.7, 6.8 and 6.9.

Therefore, a final problem is emerging: that is whether or not the K and p/q calculated for each single curvilinear regression of the same set may be assumed to be all superimposable, as theoretically predictable.

In order to solve this problem we formerly presented in 1971 [2] a bi-exponential general equation developed from the single exponential parabolic equation above quoted, but this general equation was affected by the same criticism of the single, that is the meaning of its parameters is empirical and the number of these parameters is so high that any kind of regression may in fact be interpolated.

More correctly, the above quoted two-branched equation used for the single curvilinear regressions may be used in order to determine simultaneously all K and p/q of the set of the curvilinear regressions, provided that, subsequently, K and p/q so calculated are tested for homogeneity against their variance by the usual "t" test. In fact the parameters so calculated are shown to be not significantly

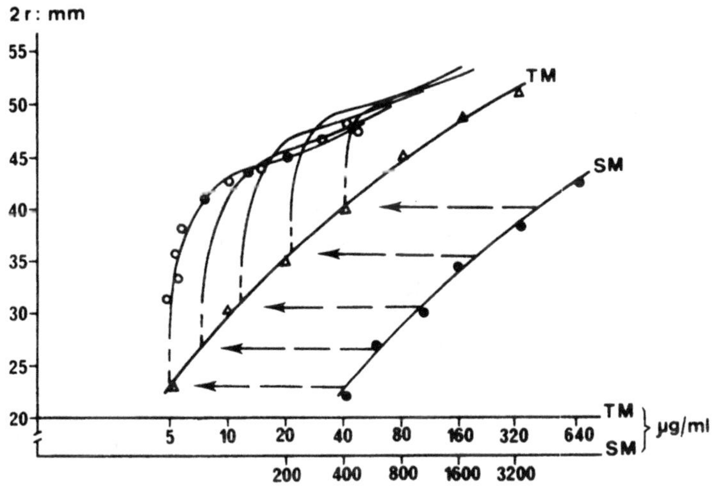

Fig. 6.6 - Mean experimental data (marks) and calculated parabolic regressions relating the antibacterial effect respectively to the log of concentrations of Sulphamethoxypiridazine (SM: ●) and of Trimethoprim (TM: △). Set of curvilinear super-additive regressions relating the antibacterial effect to the log of more than one fixed concentration of SM associated to increasing concentrations of TM. Only the mean experimental data of the first curvilinear regression (o) are plotted. Effect on bacterial growth of *Staphylococcus aureus ATCC 6538P* measured in one center agar diffusion test as diameter (2r: mm) of the inhibition zone.

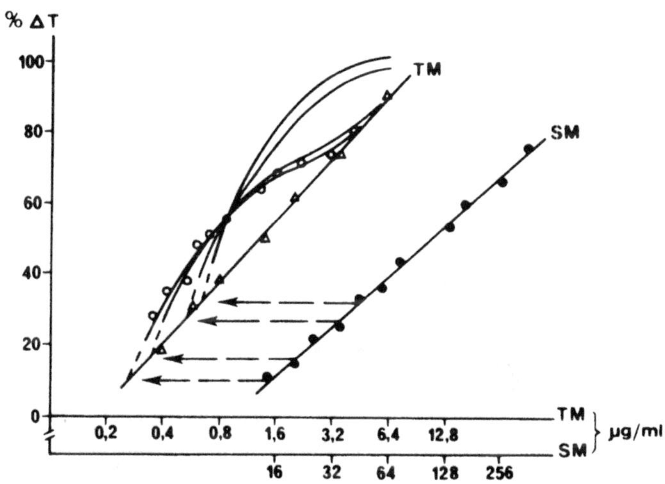

Fig. 6.7 - Mean experimental data (marks) and calculated linear regressions relating the antibacterial effect respectively to the log of concentrations of Sulphamethoxypiridazine (SM: ●) and of Trimethoprim (TM: △). Set of curvilinear super-additive regressions relating the antibacterial effect to the log of more than one fixed concentration of SM associated to increasing concentrations of TM. Only the mean experimental data of the first curvilinear regression (o) are plotted. Effect on bacterial growth of *Klebsiella pneumoniae SCH 147/77* measured on a liquid medium as Δ% of transmittance versus control.

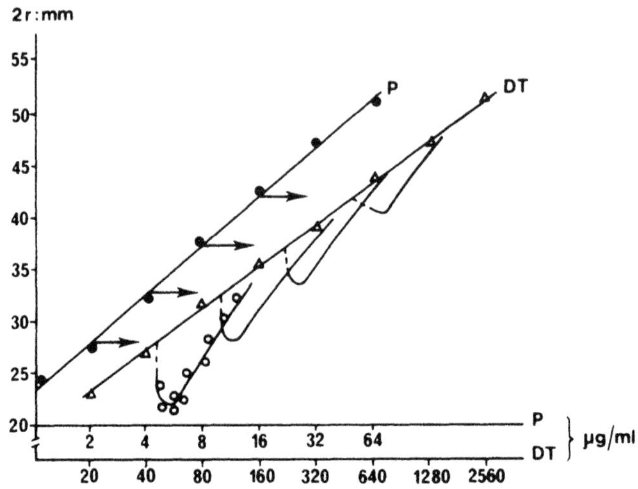

Fig. 6.8 - Mean experimental data (marks) and calculated linear regressions relating the antibacterial effect respectively to the log of concentrations of Penicillin (P: ●) and of Demethylchlortetracycline (DT: △). Set of curvilinear infra-additive regressions relating the antibacterial effect to the log of more than one fixed concentration of P associated to increasing concentrations of DT. Only the mean experimental data of the first curvilinear regression (o) are plotted. Effect on bacterial growth of *Sarcina lutea ATCC 9341* measured in one center agar diffusion test as diameter (2r: mm) of the inhibition zone.

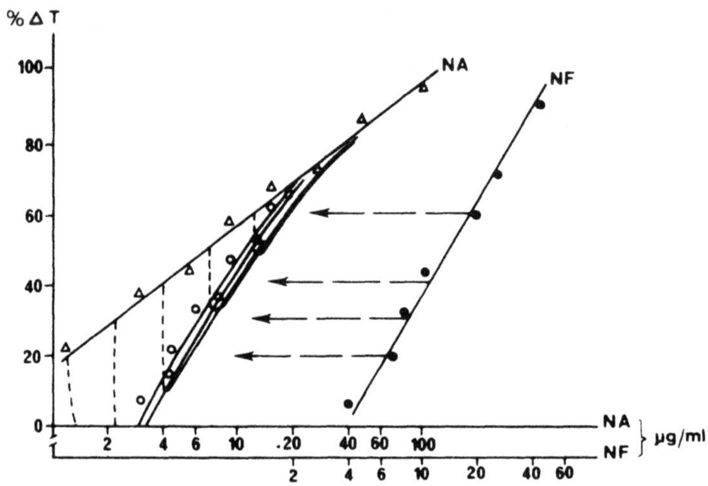

Fig. 6.9 - Mean experimental data (marks) and calculated linear regressions relating the antibacterial effect respectively to the log of concentrations of Nitrofurantoin (NF: ●) and of Nalidixic acid (NA: △). Set of curvilinear infra-additive regressions relating the antibacterial effect to the log of more than one fixed concentration of NF associated to increasing concentrations of NA. Only the mean experimental data of the first curvilinear regression (o) are plotted. Effect on bacterial growth of *Escherichia coli SCH 120/77* measured on a liquid medium as Δ% of transmittance versus control.

Table 6.1 K *and* p/q *(with corresponding standard errors) calculated for single curvilinear regressions presented in Figs 6.2, 6.3, 6.4 and 6.5.*

Figure	*Chemo-antibiotics*		*Type of interaction*	*K*	*p/q*
	Fixed (p)	*Increasing* (q)			
6.2	Sulphamethoxypiridazine	Trimethoprim	Super-additive	25.17 ± 3.86	0.131 ± 0.009
6.3	Sulphamethoxypiridazine	Trimethoprim	Super-additive	10.46 ± 2.26	0.055 ± 0.015
6.4	Penicillin	Demethylchlortetracycline	Infra-additive	0.43 ± 0.09	1.288 ± 0.080
6.5	Nalidixic Acid	Nitrofurantoin	Infra-additive	0.46 ± 0.01	0.459 ± 0.012

Table 6.2 *Mean* K *and* p/q *(with corresponding standard errors) calculated for the sets of curvilinear regressions presented in Figs 6.6, 6.7, 6.8 and 6.9.*

Figure	*Chemo-antibiotics*		*Type of interaction*	*K*	*p/q*
	Fixed (p)	*Increasing* (q)			
6.6	Sulphamethoxypiridazine	Trimethoprim	Super-additive	12.25 ± 1.41	0.639 ± 0.017
6.7	Sulphamethoxypiridazine	Trimethoprim	Super-additive	5.93 ± 0.40	0.383 ± 0.009
6.8	Penicillin	Demethylchlortetracycline	Infra-additive	0.52 ± 0.06	2.060 ± 0.129
6.9	Nitrofurantoin	Nalidixic acid	Infra-additive	0.77 ± 0.02	0.200 ± 0.011

different from each other for all the above presented examples (Tab. 6.2).

On account of the complexity of the equation and the tediousness of reiterative procedures, a computer program is now being prepared according to which many other sets of curvilinear regressions, part obtained, part being carried out, will be processed.

When all these results are available, it will be possible to assess quantitatively the interaction of a long series of old and new pairs of antibacterial agents tested against the same or different microorganisms and their strains.

It will be also possible, on the one hand, to ascertain the existence in the specific field of antibacterial chemotherapy of the types of interaction usually indicated as similar, synergistic and independent in the classical biometric papers [48], [49], [50], on the other hand to identify, as foreseeable from some initial data, other yet not described types of interaction.

Finally, if it is successful, the procedure will be curtailed in order to adapt it to a routine practice in clinical antibacterial chemotherapy.

References

[1] O.M.S. (1977) *Série de Rapports techniques 615*, Genève.

[2] Fedeli, L., Meneghini, L., Sangiovanni, M., Scrollini, F. and Gori, E. (1971) in: *Toxicological problems of drug combination* (De C. Baker, S. B. ed.), Excerpta Medica, Amsterdam, pp. 231-245.

[3] Gori, E. and Sangiovanni, M. (1972) in: *Le associazioni antibatteriche* (T. Berti, ed.), Piccin Editore, Padova, pp. 63-98.

[4] Nichols, A. C. (1948) *Proc. Soc. Exp. Biol. Med.*, **69**, 477-478.

[5] Despois, R. and Pinnert, S. (1950) *Compt. Rend. Soc. Biol.*, **144**, 23-25.

[6] Zipf, H. F. and Halbeisen, T. (1954) *Arzneimittel-Forsch.*, **4**, 463-467.

[7] Hobby, G. L., Celmer, W. D., Lenert, T. F., Pikula-Wrabec, D., Domikian, M. A., Daly, J. and Sarocco, G. (1957) *Antibiotics Annual 1956-1957*, Med. Encycl. Inc., New York, pp. 15-33.

[8] Klein, P. (1957) *Bacteriologische Grundlagen der chemotherapeutischen Laboratoriumspraxis*, Springer-Verlag, Berlin, pp. 26-35.

[9] Welch, H., Wright, W., Reedy, R. J. and Wintermere, D. (1958) *Antibiotics Annual 1957-1958*, Med. Encycl. Inc., New York, pp. 738-744.

[10] Laplassotte, J. and Ferrabouc, L. (1963) *Thérapie*, **18**, 915-920.

[11] Sabath, L. D., Steinhawer, C. B. and Finland, M. (1963) *New Engl. J. Med.*, **268**, 284-286.

[12] Bushby, S. R. M. and Hitchings, G. H. (1968) *Brit. J. Pharmacol.*, **33**, 72-88.

[13] Schmid, E. E. (1969) *Arzneimittel-Forsch.*, **19**, 1007-1011.

[14] Kerry, D. W., Hamilton-Miller, J. M. T. and Brumfitt, W. (1975) *J. Antimicrobial Chemother.*, **1**, 417-427.

[15] Weinstein, R. J., Young, L. S. and Hewitt, W. L. (1975) *J. Lab. Clin. Med.*, **86**, 853-862.

[16] D'Alessandri, R. M., Mc Neely, D. J. and Kluge, R. M. (1976) *Antimicrob. Ag. Chemother.*, **10**, 889-892.
[17] Kohlmann, F. W. and Sous, M. (1976) *Arzneimittel-Forsch.*, **26**, 613-618.
[18] Parsley, T. L., Provonchee, R. B., Glicksman, C. and Zinner, S. H. (1977) *Antimicrob. Ag. Chemoter.*, **12**, 349-352.
[19] Garrett, E. R. (1966) *Arzneimittel-Forsch.*, **16**, 1364-1369.
[20] Garrett, E. R. and Won, C. M. (1973) *Antimicrob. Ag. Chemother.*, **4**, 626-633.
[21] Nolte, H. (1973) *Arzneimittel-Forsch.*, **23**, 1654-1657.
[22] Calabi, O. (1975) *Experientia*, **31**, 479-482.
[23] Elion, G., Singer, S. and Hitchings, G. H. (1954) *J. Biol. Chem.*, **208**, 477-488.
[24] Ciack, J. and Hahn, F. E. (1958) *J. Bacteriol.*, **75**, 125-129.
[25] Chabbert, Y. (1963) *L'antibiogramme: Sensibilité et Résistance des Bactéries aux Antibiotiques*, De La Tourelle-Saint-Mondé.
[26] Checcacci, L. (1971) *Minerva Med.*, **62**, 521-532.
[27] De Lajudie, P. (1971) *Pathol. Biol.*, **19**, 607-612.
[28] Duval, J., Mathieu-Saint-Laurent, P. and Soussy, C. (1971) *Pathol. Biol.*, **19**, 637-648.
[29] Miller, A. K. and Verwey, W. F. (1954) *Antibiot. Chemotherapy*, **4**, 169-172.
[30] Böhni, E. (1969) *Schweiz. Med. Wochschr.*, **99**, 1505-1510.
[31] Donovick, R., Lapedes, D. and Pansy, F. (1947) *J. Bacteriol.*, **54**, 425-444.
[32] Zwart Voorspuij, A. J. Z. and Bokma, L. H. (1958) *Antibiot. Chemotherapy*, **8**, 171-179.
[33] Dettori, R. (1962) *Giorn. Microbiol.*, **10**, 49-64.
[34] Kaplan, D. and Koch, W. (1962) *Nature*, **195**, 958-959.
[35] Cimmino, A., Filadoro, F. and Renzini, G. (1964) *Antibiotica*, **2**, 19-32.
[36] Gallego, A. (1958) *Proc. IV Int. Congr. Biochemistry Symp. V*, Pergamon Press, pp. 134-165.
[37] Patte, J. C., Hirsch, H. and Chabbert, Y. (1958) *Ann. Inst. Pasteur*, **94**, 621-635.
[38] Dye, W. E. (1956) *Antibiotics Annual 1955-1956*, Med. Encycl. Inc., New York, pp. 374-382.
[39] Knapp, W. and Jacherts, D. (1955) *Zbl. Bakt.*, **162**, 474-478.
[40] Maccacaro, G. A. (1961) *Progr. Ind. Microbiol.*, **3**, 175-210.
[41] Scherr, G. H. and Bechtle, R. M. (1959) *Antibiotics Annual 1958-1959*, Med. Encycl. Inc., New York, pp. 855-864.
[42] Hewlett, P. S. (1969) *Biometrics*, **25**, 477-487.
[43] Ashford, J. R. and Cobby, J. M. (1974) *Biometrics*, **30**, 11-31.
[44] Hey, E. N. and Hey, M. H. (1960) *Biometrics*, **16**, 606-617.
[45] Weaver, L. C., Swinyard, E. A., Woodbyry, L. A. and Goodman, L. S. (1955) *J. Pharmacol. Exptl. Therap.*, **113**, 359-370.
[46] Tedeschi, H. (1976) *Ann. N.Y. Acad. Sci.*, **281**, 243.
[47] Loewe, S. (1959) *Arzneimittel-Forsch.*, **9**, 449-456.
[48] Bliss, C. I. (1939) *Ann. appl. Biol.*, **26**, 585-615.
[49] Finney, D. J. (1952) *Probits Analysis: a Statistical treatment of the Sigmoid Response Curve*, 2nd ed., Cambridge University Press, London.
[50] Hewlett, P. S. and Plackett, R. L. (1960) in: *Quantitative Methods in Pharmacology* (H. De Jonge, ed.), North-Holland Publishing Company, Amsterdam, pp. 328-336.
[51] Plackett, R. L. and Hewlett, P. S. (1967) *Biometrics*, **23**, 27-44.
[52] Maccacaro, G. A. (1956) *Giorn. Microbiol.*, **2**, 463-484.
[53] Eyles, D. E. and Coleman, N. (1955) *Antibiot. Chemotherapy*, **5**, 529-539.
[54] Arioli, V., Berti, M., Carniti, G., Rossi, E. and Silvestri, L. G. (1977) *J. Antimicr. Chemother.*, **3**, 87-94.
[55] Rall, D. P. (1965) *Antimicrobial Agents and Chemotherapy* (G. L. Hobby ed.), American Soc. Microbiol., Ann Arbor, pp. 1086-1095.
[56] Box, G. E. and Wilson, K. B. (1951) *J. Roy. Stat. Soc.*, **13**, 1-45.

[57] Box, G. E. (1958) *Ann. N.Y. Acad. Sci.*, **76**, 921-924.
[58] Davies, O. L. (1960) *The design and analysis of industrial experiments*, Oliver and Boyd, London, pp. 495-551.
[59] Addelman, S., Gaylor, D. W. and Bohrer, R. E. (1966) *Biometrics*, **22**, 730-746.
[60] De Carli, L. (1957) *La Ricerca Scientifica Supplemento*, **27**, 185-196.
[61] Schaffé, H. (1963) *J. Roy. Statist. Soc.*, **25**, 235-263.
[62] Gorman, J. W. and Hinman, J. E. (1962) *Technometrics*, **4**, 463-487.
[63] Claringbold, P. J. (1955) *Biometrics*, **11**, 174-185.
[64] Gori, E., Sangiovanni, M. and Floris, E. (1964) *Proc. 3rd International Congress on Chemotherapy* (H. P. Kuemmerle, and P. Preziosi, eds.), Thieme Verlag, Stuttgart, pp. 695-699.
[65] Gori, E., Sangiovanni, M., Floris, E. and Pagan, M. (1965) in: *Antimicrobial Agents and Chemotherapy* (G. L. Hobby, ed.), American Society for Microbiology, Ann Arbor, pp. 273-277.
[66] Gunnison, J. B. and Jawetz, E. (1953) *J. Lab. Clin. Med.*, **42**, 163-164.
[67] Humphrey, J. K. and Lightbown, J. H. (1952) *J. Gen. Microbiol.*, **7**, 129.
[68] Kavanagh, F. (1977) *J. Pharm. Sci*, **66**, 1520-1525.
[69] Treffers, H. P. and Muschel, L. K. (1954) *J. Exptl. Med.*, **99**, 157-165.
[70] Ariens, E. J., Simonis, M. and Van Rossum, J. M. (1964) *Molecular Pharmacology*, Vol. I (E. J. Ariens, ed.), Academic Press, New York.
[71] Gero, A. (1965) *Mol. Pharmacol.*, **1**, 312-313.
[72] Snedecor, G. W. and Cochran, W. G. (1967) *Statistical Methods*, VI ed., Iowa State University Press, Ames, USA, p.p. 465-471.

7

Experimental approach to the pharmacokinetic study of combined antimicrobial therapy

M. Ghione

Montedison Divisione Farmaceutica - 20146 Milano

C. Della Bruna, A. Sanfilippo, G. Schioppacassi

Farmitalia - 20146 Milano

Summary

The basic concept justifying the simultaneous administration of two antibacterial agents in the treatment of infectious diseases relies upon the existence of a synergic interaction between the drugs.

The appraisal of this synergism cannot be based only on the results observed in steady state experimental conditions, but ought to consider the kinetic properties of the compounds and their interaction in disease states, as well as their influence on the bacterial pathogenetic factors.

An experimental approach to clarify the aspects of these not yet studied interactions has been carried out by determining the kinetic parameters of ampicillin and dicloxacillin administered (as single agents or in combination) to groups of mice either healthy or injected i.v. with non lethal doses of viable or UV irradiated suspensions of *E. coli* B strains producer, or non-producer, of β-lactamase. The research demonstrated that the processes of drug absorption, distribution and excretion are influenced by body function alterations induced by the infectious disease and that the pathologic phenomena are in their turn selectively modified by the activity of the drugs. These observations are of relevance for a proper scheduling of antibiotics or any other drug, not only in overt infectious diseases but even in the morbid states where production or absorption of pharmacodynamically active bacterial metabolites (chiefly endotoxin) can occur.

Introduction

Combinations of antibiotics are extensively employed in clinical practice. Only a few of such drug combinations have a sound theoretical and experimental background [1], [2], [3]. Among them, the combinations between aminopenicillins and isoxazolylpenicillins are in the front rank.

The aminopenicillins are widely used antibiotics by both oral and parenteral routes. Owing to the presence of an amino group in the α-position of the side chain, these compounds possess, besides the well known tolerance of the penicillins, the clinically advantageous characteristics of stability in an acid environment and activity against gram-negative schizomycetes.

The clinical use of aminopenicillins is seriously limited by the presence of bacterial strains turned resistant either because of intrinsic mechanisms such as decreased permeability or, more frequently, through the production of β-lactamase. To circumvent this obstacle, combinations of β-lactamase resistant penicillins (chiefly isoxazolylpenicillins) and aminopenicillins have been studied *in vitro* and *in vivo* [4]. The rationale of these combinations lies in the role played by the β-lactamase resistant penicillins in killing the strong enzyme producer gram-positive microorganisms and in binding the enzyme itself [5]. With the help of these sheltering mechanisms the aminopenicillin present in the combination can escape hydrolysis and perform its inhibiting action on the gram-negative microorganisms.

The two compounds can display an optimal synergic action if and when optimal concentrations of both are contemporaneously present in the infectious foci. The design of a successful treatment schedule requires, therefore, a knowledge of the kinetic behaviour of the drugs. To this effect the kinetic parameters obtained in healthy animals or healthy human volunteers [7] cannot be considered adequate because antibiotics are usually not administered to healthy people but rather to patients suffering from an infectious disease process. The morbid state sustained by infectious processes is, by definition, a derangement of the homeostasis. It is highly plausible that this derangement can affect absorption, distribution, metabolism and excretion of an antibiotic. Moreover, these phenomena are likely to be modified by the presence of the parasites and/or some of their metabolites e.g. β-lactamase.

In spite of the obvious interest in these problems, the literature reports very scanty data on drug kinetics in infectious diseases. Therefore we tried to gather experimental information by studying the change of kinetic parameters of an aminopenicillin (Ampicillin) and an isoxazolylpenicillin (Dicloxacillin) in groups of either healthy mice or those injected with viable or UV irradiated suspensions of gram-negative microorganism strains, producers or non-producers of β-lactamase, and treated with Ampicillin or Dicloxacillin as single agents or with their combination.

Materials and methods

Groups of male CD1 mice (COBS, SPF, obtained from Charles River, Italy) were inoculated i.v. with 0,2 ml of a bacterial suspension ($2.5\ 10^6$ germs/mouse) from an overnight culture in Tryptose Broth either of *E. coli* BP^- (non-producer of β-lactamase), or of its mutant *E. coli* BP^+ (strong producer of the enzyme), or with suspension of the latter strain irradiated with UV rays until 99% of the microorganisms were killed but without modifying the β-lactamase content of the culture. Groups of animals were kept untreated and carefully observed for 10 days.

Two hours after inoculation, different groups of mice were treated subcutaneously with 100 mg/kg (in aqueous solution) of Ampicillin (A) or Dicloxacillin (D) (henceforth referred to as groups IA or ID) or with 200 mg/kg of the combination of Dicloxacillin and Ampicillin (i.e. 100 mg/kg of A plus 100 mg/kg of D, henceforth referred to as groups IIA or IID).

At zero time and after 10, 20, 30, 60 and 120 min, six animals from each experimental group were sacrificed by an overdose of ether. Blood was collected with heparin and centrifuged. Specimens of pooled plasma samples were then assayed for A (in groups IA and IIA) on. *S. abortivoequina* and for D (in groups ID and IID) on *S. aureus* 153 P^+ (β-lactamase producer), by means of the agar diffusion method against fresh standards containing known concentrations of the antibiotics in pooled normal mouse plasma.

For the determination of A, 2.5% Na benzoate was added to the plasma samples of the mice injected with the β-lactamase producing *E. coli* BP^+ (UV irradiated or not) in order to stop the enzyme activity.

A few basic pharmacokinetic parameters including the biological half-life time (t 1/2), the hypothetical concentration at time O ($c^{\circ}p$) and the peak value have been determined by graphical interpolation or extrapolation. The formula Dose/$c^{\circ}p$ was used to calculate the apparent volume of distribution (app. Vd).

Results

In healthy animals neither the blood level curves nor the pharmacokinetic parameters of A (or D) were significantly modified by the contemporaneous administration of D (or A). The curves (Fig. 7.1) are characterized by a rapid ascendent limb (corresponding to the absorption phase), reaching within 20 min a peak value higher for D than for A, and by a descendent limb (elimination phase) steeper for A than for D. The blood levels approached the threshold of detectability, with the analytical method we employed, 60 min after administration of A and 120 min after administration of D. The t 1/2 value of A, administered either as a single drug or in combination, was shorter and the app. Vd values were lower than the matched values of D. The c°_p value was higher for A than for D (See Fig. 7.5, 7.6, 7.7).

In animals infected with *E. coli* BP^- (non-producer of β-lactamase) the antibiotic blood level curves showed a partially modified shape (Fig. 7.2).

The absorption phase was rather slow in all the experimental groups. Peak values were reached 30-40 min after injection of the drugs. The D elimination rate was virtually unmodified, but c°_p values were increased and the app. Vd values decreased. In comparison with the healthy controls the elimination of A was prolonged, the c°_p values were decreased and the app. Vd values increased particularly in animals treated with A alone (See Fig. 7.5, 7.6, 7.7).

In the animal groups infected with *E. coli* BP^+, producer of β-lactamase, a drastic modification of the shape of the drug blood level curves was seen (Fig. 7.3). The absorption was delayed, the peak values were lower and the elimination phases were definitely longer than in the corresponding control groups or in animals infected with the β-lactamase non-producer mutant strain. A notable difference in the drug blood level curves was seen between the subgroups of animals treated either with the single antibiotics or their combination (Fig. 7.3). Accordingly, the changes of the pharmacokinetic parameters were impressive. The t 1/2 values of I A or I D groups were longer and the Vd values higher than the matched healthy control values. In the animals treated with the drug combination (groups II A and II D) the kinetic parameters were less modified and the values never reached the level of the healthy control groups (See Fig. 7.5, 7.6, 7.7).

The blood level curves of the animals infected i.v. with UV irradiated bacterial suspension of *E. coli* BP^+ showed more resemblance to the groups infected with *E. coli* BP^- and, less closely, to the healthy control groups if analyzed in terms of curve shape (Fig. 7.4). However, the values of the pharmacokinetic parameters were clearly different from the control and were like the corresponding values of the animal groups infected with *E. coli* BP^+ suspension (Fig. 7.5, 7.6, 7.7). In the animal groups treated with the drug combination (II A and II D) in comparison with those injected with the single antibiotics (I A and I D) higher c°_p and lower app. Vd values were observed.

No death occurred in the infected untreated animal groups and no apparent sign of disease was observed. The behaviour and food intake were the same as in the healthy untreated animals.

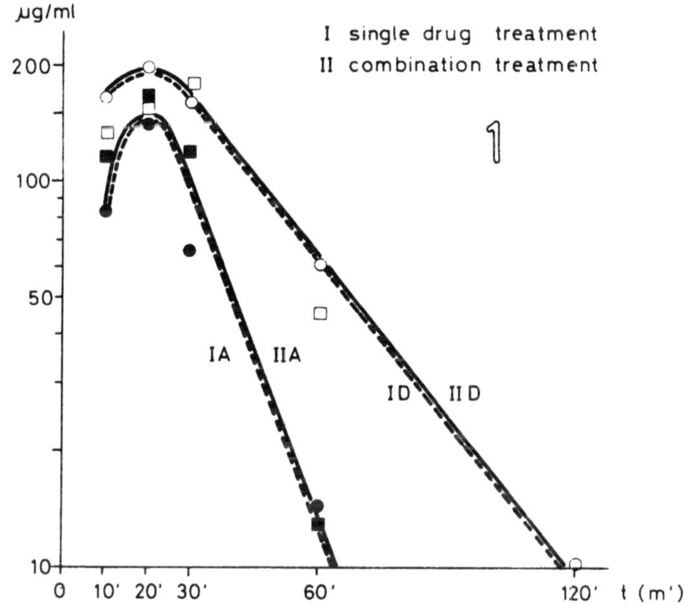

Fig. 7.1 - Blood level of Ampicillin (A) or Dicloxacillin (D) administered s.c., either as a single agent (I) or in combination (II) to healthy mice.

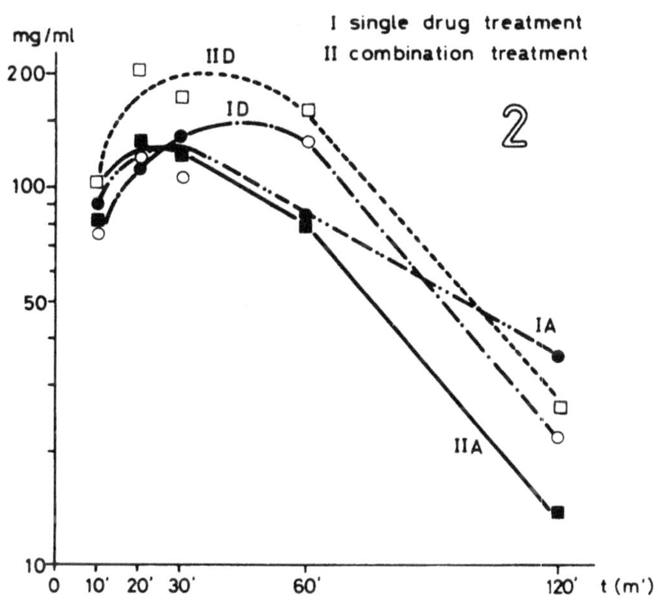

Fig. 7.2 - Blood level of Ampicillin (A) or Dicloxacillin (D) administered s.c., either as a single agent (I) or in combination (II) to mice injected i.v. two hours before the treatment, with a suspension of viable bacteria (*E. coli* BP^-, strain non producer of β-lactamase).

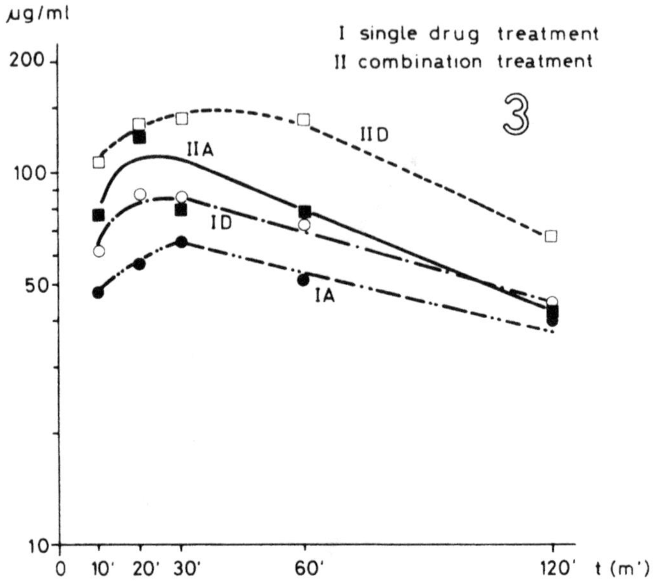

Fig. 7.3 - Blood level of Ampicillin (A) or Dicloxacillin (D) administered s.c., either as a single agent (I) or (II) to mice injected i.v. two hours before the treatment with a suspension of viable bacteria (*E. coli* BP+, producer of β-lactamase).

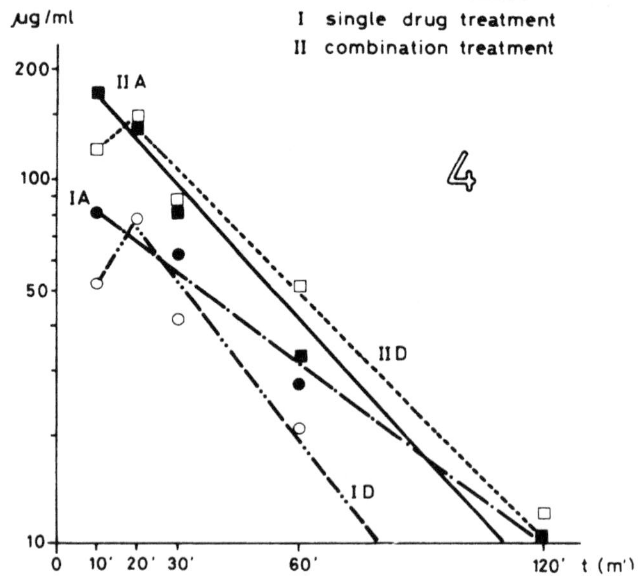

Fig. 7.4 - Blood level of Ampicillin (A) or Dicloxacillin (D) administered s.c., either as a single agent (I) or in combination (II) to mice injected i.v., two hours before the treatment with a suspension of UV irradiated bacteria (*E. coli* BP+ producer of β-lactamase).

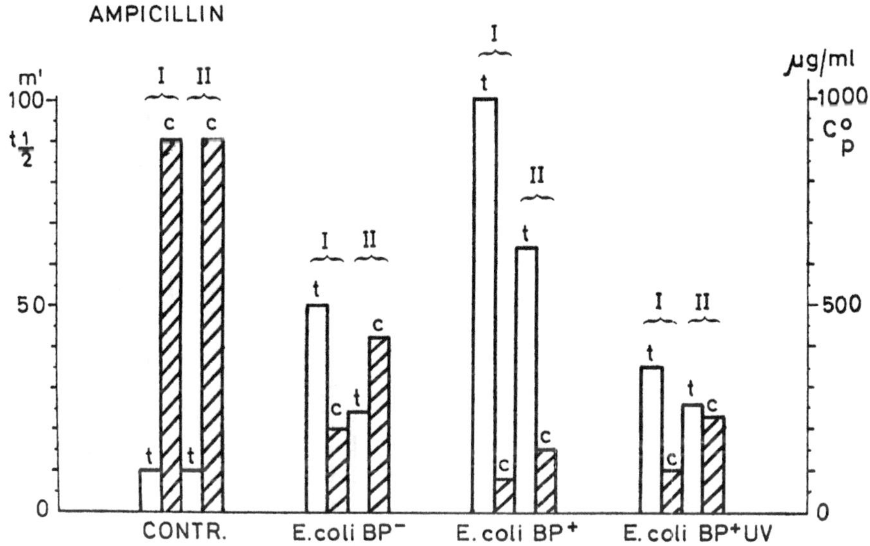

Fig. 7.5 - Values of the biological half-life time (t 1/2) and of the hypothetical concentration at t_0 (c^o_p) of Ampicillin administered s.c., as a single drug (I) or combined with Dicloxacillin (II), to groups of either healthy mice (Contr.) or those injected i.v. with different strains of *E. coli* B non-producer (P^-) or producer (P^+) of β-lactamase, either viable or UV irradiated (UV).

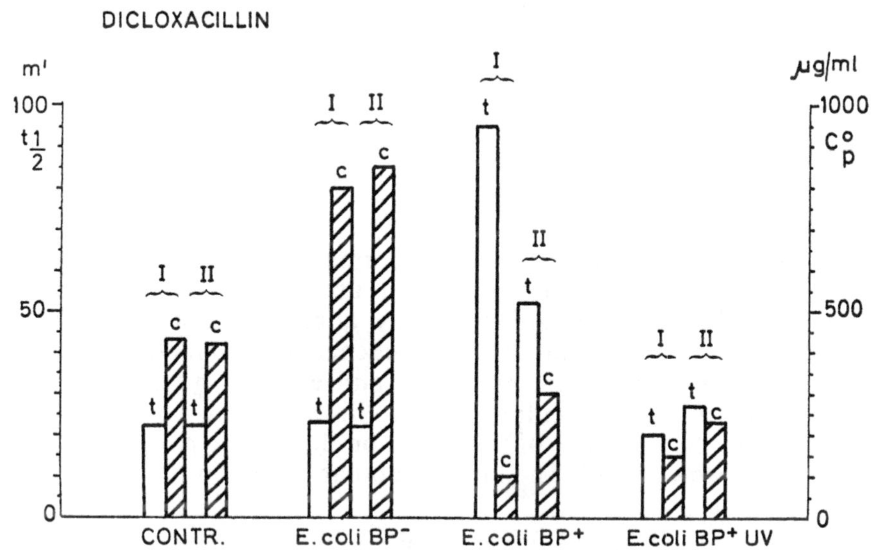

Fig. 7.6 - Values of the biological half-life time (t 1/2) and of the hypothetical concentration at t_0 (c^o_p) of Dicloxacillin administered s.c., as a single drug (I) or combined with Ampicillin (II), to groups of either healthy mice (Contr.) or those injected i.v. with different strains of *E. coli* B non-producer (P^-) or producer (P^+) of β-lactamase, either viable or UV irradiated (UV).

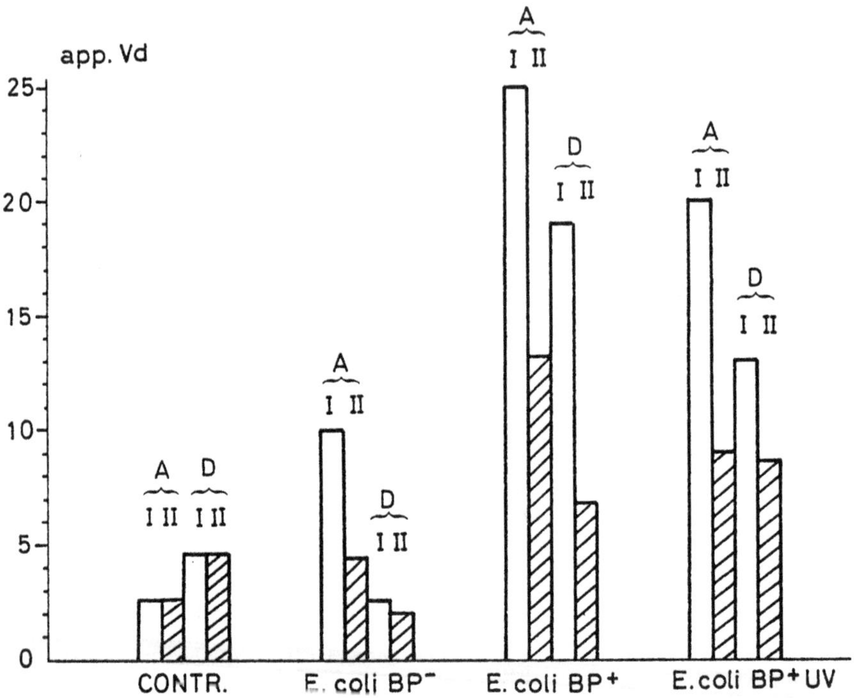

Fig. 7.7 - Values of the apparent volume of distribution of Ampicillin (A) or Dicloxacillin (D) administered s.c., as a single agent (I) or in combination (II), to groups of mice either healthy (Contr.) or injected i.v. with different strains of *E. coli* B non-producer (P^-) or producer (P^+) of β-lactamase, either viable or UV irradiated (UV).

Discussion

The blood level curves of each experimental group fit the scheme of an open one compartment model. In such a model drug absorption and distribution after subcutaneous administration depend on vascular permeability and drug physicochemical characteristics.

Excretion depends on metabolic and emunctory functions [8]. The semisynthetic penicillins used in this study are poorly metabolized, and the influence of this factor on the kinetic parameters can therefore be neglected. The kidney tubular excretion plays the most relevant role [9].

In healthy mice, in spite of the high dosage administration (100 mg/kg), the kinetic parameters of one drug (either A or D) were not modified by the contemporaneous presence of the other (Fig. 7.1). The normal mice emunctory function did not appear saturated by this drug. Moreover, in all experimental groups the elimination rates of the compounds followed a first order kinetic, i.e. the elimination of the drugs depends upon their concentrations in the blood. This means that a capacity limited excretion mechanism is never exceeded in our experimental models.

The blood stream invasion by *E. coli* B, even in non-lethal and apparently well tolerated doses, gives rise, at least temporarily, to an endotoxemic and septicemic state.

The speed of the onset and the extent of the body function alterations which rapidly followed the bacterial challenge are attributable to the effect of the endotoxin. Some complex self-amplifying mechanisms (like the complement and the coagulation systems) are activated. Blood and vessel-lining cells and platelets are altered. Highly active mediators of the inflammation (prostaglandin, histamine, etc.), are discharged into the blood stream and interstitial fluids. The vascular permeability is increased. The excretory functions are impaired. These alterations accounts for the changes of the t 1/2 and app. Vd values.

The drug-dependent factors will now be considered. The aminopenicillin (A), a highly hydrosoluble and poorly protein-bound antibiotic inhibits *E. coli* B, but is easily hydrolized by the β-lactamase produced by the P^+ mutant. The isoxazolylpenicillin (D), a lipophilic and highly protein bound derivative, is devoid of any activity against *E. coli* B but can bind the β-lactamase.

The administration of the antibiotics, either as single drugs or combined, acts in different ways. The administration of A, a cidal and lytic antibiotic, elicits at least two phenomena that can be considered antithetical in terms of their influence on the body functions. There is, on the one hand, a loss of germ viability by several log units within a fraction of the bacterial generation time ($<$ 20-30 m' for *E. coli*) and on the other hand a massive bacterial lysis with liberation of bacterial antigens, chiefly of endotoxin. In a short-term observation period the former event will tend to restore the homeostasis and the latter to enhance its alterations.

No attempt will be made to examine further the peculiarities of these factors in a paper limited to the study of the kinetic phenomena. They have been mentioned only to recall that infectious disease processes and therapy interact at different levels. Sometimes opposing effects can exert a strong modulatory influence on a drug's kinetic profile. Therefore it is very difficult, if not impossible, to correlate the observed kinetic parameter changes with a single factor.

Drug-dependent drug kinetic alterations can be seen by comparing the values of the kinetic parameters in the animal groups submitted either to the single drug or to the drug combination treatment and challenged with the same bacterial preparation. The presence of D consistently reduces the extent of the A kinetic aberrations induced by the bacterial challenge.

The t 1/2 and the apparent Vd values of the II A set of data are uniformly reduced at the ~ 50% of the matched I A values. The actual peak values of A are substantially reduced in the animal groups injected with (either viable or irradiated) suspensions of the penicillinase-producing strain and are consistently enhanced by the contemporaneous administration of D. These findings are consistent with the interpretation that the penicillinase-binding properties of the latter compound are efficient in protecting the former one from enzymatic hydrolysis.

The D kinetic alterations are correspondingly influenced by the presence of A in the animal groups injected with either viable or UV irradiated bacterial suspension of the penicillinase-producing strain. The enzyme binds D irreversibly as a non-hydrolyzable substrate. The complex is microbiologically inactive. For this reason the presence of high amounts of the enzyme causes an evident decrease of the D peak levels determined by microbiological assay. The D levels are reinstated by combination with A which in its turn, being protected by A against the hydrolysis, can kill the enzyme producer bacteria.

Obviously in the animal group injected with viable, actively metabolizing and enzyme producing bacteria, the influence displayed by the presence of A can be more clearly observed than in the groups injected with irradiated bacteria.

Conclusion

The efficacy of a drug depends upon its intrinsic activity and kinetic properties. The latter can be modified by a disease process. This phenomenon has been extensively studied, with particular interest to derangements of the cardio-vascular system, modifications of the metabolic and emunctory functions, changes in the binding properties of blood and tissue proteins induced by drug treatment [10] or associated with non-infectious disease [11], [12], [13], [14], fasting [15] and age [16].

Very little attention has been devoted to bacterial infectious diseases and antibiotics with regard to interaction displayed on the drug kinetic by the parasite as well as by the altered host function.

A few studies on the antibiotic level in inflammatory exudates [17], [18] and the time-honoured cognition that inflammation enhances the penetration of some antibiotics into normally secluded areas (like CNS) summarize the total knowledge in this field.

According to our data, gram-negative sublethal bacterial infections can cause extensive changes of drug kinetics due to the binding of antibiotic to bacterial structures or metabolites and to the release of biologically-active bacterial substances, chiefly endotoxin. Endotoxin interferes with hepatic excretory function [19] but this consideration can be omitted in this context owing to the ancillary role played by hepatic excretion in penicillin kinetics.

In the course of non-infectious disease states (traumatic, hemodynamic, metabolic, tumors, etc.) overt or subclinical infectious processes can supervene, or endotoxin elaborated by the bacteria normally present in the intestinal tract can be absorbed. In these conditions, drug kinetic changes occur which modify to a certain extent the expected efficacy of the drugs. In the management of these cases, treatment with antibiotics is mandatory in addition to the specific therapy. The administration of a combination of aminopenicillin with isoxazolylpenicillin offers a fair chance of counteracting the bacterial infection [20] and repressing the drug kinetic alterations. The reaction to endotoxin is genetically conditioned [21] and modulated by the neurohormonal interference.

Further experimental work is required to take all these variables into consideration with respect to proper drug scheduling. As already suggested for non-infectious diseases, new model systems [22] and a better understanding of the role of neurohormones and their antagonists in infections [23] are required to accomplish this task.

References

[1] Jawetz, E. (1968) *Ann. Rev. Pharmacol.*, **8**, 151-170.
[2] Segre, G. (1977) *Farmaci*, **1**, 15-20.
[3] Ullmann, E. (1977) *Deutsche Med. Wchnschr.*, 1781-1783.
[4] Acred, P. and Sutherland, R. (1966) *Antimicrob. Ag. Chemother.* **6**, 53-61.
[5] Hamilton-Miller, J. M. T. and Smith, J. T. (1964) *Nature*, **201**, 999-1003.
[6] Brusch, J. L., Barza, M., Brown, R. B., Berseron, M. G. and Weinstein, L. (1976) *Infection*, 4 suppl., **2**, 82-90.
[7] Scalvini, A., Del Monte, G., Schioppacassi, G. and Morvillo, E. (1971) *Minerva Med.*, **62**, 2747-2750.
[8] Notari, R. E. (1971) *Biopharmaceutics and Pharmacokinetics*, Marcel Dekker, New York.
[9] Barza, M. and Weinstein, L. (1976) *Clin. Pharmacokinetics.* **1**, 297-308.
[10] Warren, R. D. and Bender, R. A. (1977) *Cancer Treatm. Rep.*, **61**, 1231-1241.
[11] Craig, W. A. (1976) *Infection*, 4 suppl., **2**, 137-141.
[12] Kato, R. (1977) *Xenobiotica*, **7**, 25-92.
[13] Gibson, T. P. and Nelson, H. A. (1977) *Clin. Pharmacokinetics*, **2**, 403-426.
[14] Lavigne, J. G. (1977) *Sem. Hôp. Paris*, **53**, 2399-2401.
[15] Reidemberg, M. M. (1977) *Clin. Pharmacol. Therap.*, **22**, 729-734.
[16] Morselli, P. L. (1976) *Clin. Pharmacokinetics*, **1**, 81-98.
[17] Ellis, B. W., Stanbridge, R. D., Sikorski, J. M. and Dudley, H. A. F. (1975) *J. Antimicrob. Chemother.*, **1**, 291-296.
[18] Raeburn, J. A. (1976) *Infection*, 4 suppl., **2**, 149-151.

[19] Utili, R., Abernathy, C. O., Aron, S. A. and Zimmerman, H. J. (1977) *Experientia*, **33**, 1631-1632.
[20] Schioppacassi, G., Curcio, L., Perrotta, P. (1975) *Giorn. Ital. Chemiot.*, **22**, 95-98.
[21] Watson, J., Largen, M. and Mc Adam, K. P. W. J. (1978) *J. Exp. Med.*, **147**, 39-49.
[22] Benet, L. Z. (Ed.) (1976) *The effect of disease states on drug pharmacokinetics*, American Pharmaceutical Association, Washington, DC.
[23] Smith, L. M., Kennedy, L. R., Regme-Karlson, M. H., Johnson, V. L. and Burmeister, L. F. (1977) *Am. J. Clin. Nutr.*, **30**, 1285-1288.

8

Pharmacokinetic aspects of combined antimicrobial therapy

F. Follath, S. Vozeh *

Division of Clinical Pharmacology, Department of Medicine, Kantonsspital Basel, Switzerland

Summary

The simultaneous use of multiple therapeutic agents carries the risk of drug interactions, which may be harmful to the patient. Chemical incompatibility, diminished gastrointestinal absorption, displacement from protein binding sites, alteration of hepatic drug metabolism and competition for renal excretion are the principal mechanisms responsible. Several antibiotics (rifampicin, isoniazid, tetracyclines, penicillines and aminoglycosides) have been shown to participate in such reactions with different drugs. There are, however, only few well documented examples of pharmacokinetic interactions among individual antimicrobial agents.

Practical problems with combined antimicrobial treatment may also arise in cases with renal and hepatic damage. Since the elimination rate of the different components will be affected to a variable degree, the dosage must be individually adapted for each drug.

Introduction

The simultaneous use of multiple therapeutic agents is always connected with a certain probability, that different components of a drug regimen will influence each other's pharmacological activity. Antibiotics are no exception to this rule and may interact with other antimicrobial and non-antimicrobial drugs. As interactions often result in diminished drug effects or increased toxicity, possible benefits of a combined treatment should be carefully balanced against its potential risks in each case.

A virtually endless list of clinically more or less relevant interactions has been reported in the literature, but it is readily apparent,

* Authors supported by the Swiss National Research Foundation.

that a great majority of such reactions occur by a limited number of basic mechanisms. An understanding of these mechanisms enables the physician to prevent, interpret and treat unwanted drug interactions. To a limited extent the likelihood of interactions can also be predicted for newly introduced therapeutic agents.

Concurrent administration of two or more drugs may alter the time course of drug levels in the organism (pharmacokinetic interactions) and modify pharmacological effects (pharmacodynamic interactions) by the following mechanisms:

1. Direct chemical and physical interactions.
2. Changes in the rate and extent of absorption.
3. Displacement from macromolecular binding sites.
4. Induction or inhibition of drug metabolising enzymes.
5. Competition for tubular and biliary excretion.
6. Interactions at receptor or tissue sites.

Direct interactions

Incompatibilities among antibiotics and other drugs may result in a marked loss of activity even before they are administered to the patient. Amphotericin B, cephalothin, chloramphenicol, erythromycin, kanamycin, penicillin and tetracycline participate in numerous pharmaceutical interactions and should therefore not be mixed in infusion fluids [1].

In this category of interactions the inactivation of aminoglycosides by penicillins is particularly interesting. Mc Laughlin and Rees were the first to demonstrate a carbenicillin induced inactivation of gentamicin *in vitro*, in animals and in a patient with renal failure [2]. Confirming these observations, Noone et al. [3] showed a rapid decline of microbiologically measured gentamicin levels in dextrose and saline solutions when carbenicillin or ampicillin were added, whereas inactivation in serum reached only 15% in 8 hours. A recent study indicates that carbenicillin and ticarcillin also interact with tobramycin, sisomicin and to a lesser degree with amikacin [4]. It is evident from these findings, that aminoglycosides and penicillins should never be mixed in infusion fluids. Furthermore, in patients with renal insufficiency sustained carbenicillin levels may result in decreased therapeutic efficacy by in vivo inactivation of gentamicin [5]. In such situations the adequacy of dosage should be monitored by repeated determinations of serum concentrations of the aminoglycoside. On the other hand gentamicin-carbenicillin interaction is unlikely to occur in patients with normal kidney function.

Drug-induced changes in gastrointestinal absorption

Absorption of drugs from the gastrointestinal tract is influenced by dissolution rates of the ingested preparations, local pH, gastric emptying time, gut motility and blood flow. The rate and extent of antibiotic absorption may also be significantly altered by several drugs. The formation of insoluble complexes between tetracyclines and iron, calcium, magnesium and aluminium salts is well known [6]. Suboptimal serum levels of penicillin V may result from simultaneous treatment with neomycin [7] cholestyramin, metoclopramide and bicarbonate [8]. Another example of practical importance is the reduction of rifampicin (RMP) biavailability by concurrent oral administration of paraaminosalicylic acid [9]. The exact mechanism of this interaction is not known, but peak concentrations of RMP and areas under the serum concentration curve remain unchanged when PAS is given intravenously or orally 12 hours before the RMP dose. Fortunately this combination of drugs is only exceptionally required in a modern antituberculous drug regimen. Rifampicin, isoniazid and ethambutol do not influence their mutual absorption.

As these observations illustrate, interactions at the gastrointestinal level might be responsible for an unsatisfactory antimicrobial effect.

Displacement from macromolecular binding

Most drugs circulate in the blood reversibly bound to a macromolecule, usually albumin. The extent of binding depends on drug and protein concentrations, the number of available binding sites and specific affinity constants. When a second drug competes for the same binding sites displacement takes place and the fraction of pharmacologically active free molecules increases. At the same time, however, distribution and elimination of the affected drug may also change, which tends to lower the unbound drug level to its previous value. For this reason significant interactions by displacement should only be expected among highly protein bound drugs with small distribution volumes [10]. Enhanced drug effects will usually occur during the first days of combined treatment and will then dissipate with continued administration of the interacting substances.

Displacement from protein binding plays an important role in adverse effects of oral anticoagulants and antidiabetics. Among the antimicrobial agents only sulfonamides (sulfaphenazole, sulfisoxazole, sulfamethazine) were reported to cause hypoglycemia in patients on tolbutamide [11] or chlorpropramide [12]. To our knowledge clinically relevant displacing reactions between different antibiotics are not described.

Metabolic interactions

Metabolic transformation of drugs by hepatic and possibly other microsomal enzymes represents a principal route of elimination for many pharmacologically active substances. Not surprisingly, drug induced changes of enzyme activity frequently form the basis of clinically relevant interactions. The following reasons can be given to explain this tendency:

a) drug metabolizing enzymes are unspecific;
b) the number of enzyme molecules, i.e. metabolizing capacity, is limited;
c) enzyme activity is enhanced by repeated administration of certain drugs or other foreign compounds.

While enzyme induction accelerates drug inactivation and elimination, competition for the same enzyme leads to drug accumulation or even saturation of the metabolic process. As a rule, any lipid soluble agent may be considered as a candidate for enzymatic interactions.

Antibiotics participate in several metabolic interactions. For example RMP is a potent enzyme inducer causing an increase of cytochrome P-450 in microsomes and a proliferation of the endoplasmatic reticulum in liver cells. Acocella et al. [13] have shown that this drug stimulates its own metabolic degradation, as evidenced by a shortened half life after repeated administration. Self-induction of RMP metabolism is evident even in patients with liver disease. A major effect of RMP on hexobarbital and tolbutamidekinetics was demonstrated by Zilly et al. [14]. Half lives of both drugs were reduced under simultaneous RMP administration to less than 50% of control values without changes in distribution volumes. Other drugs similarly affected by RMP are anticoagulants, corticosteroids and oral contraceptives. An increase in cortisone requirement under antituberculous treatment in patients with adrenocortical insufficiency can be especially dangerous if this interaction is not recognised [15].

The problem of enzyme induction needs special consideration with respect to the combined use of RMP and isoniazid (INH). Clinical reports indicate a higher rate of hepatic toxicity with this combination, than with either drug alone [16], [17]. INH hepatitis has recently been explained by the production of a toxic metabolite from acetylhydrazine [18]. In animal experiments hepatic necrosis could be potentiated by phenobarbital and prevented by enzyme inhibitors, such as cobalt chloride [19]. Thestimulation of microsomal P-450 activity by RMP could explain a more extended liver damage. At present, there are no pharmacokinetic data in humans to confirm or invalidatc this assumption. Acocella [13] and Llorens [17] found no modifications of INH kinetics by concomitant RMP treatment during 7 days, neither in adults nor in children. Half lives of INH and elimination of acetyl-INH in the urine remained unchanged. However, these negative findings do not exclude the possibility of an increased production of the toxic acetyl hydrazine derivative, since the analytical methods utilised by these Authors were not suitable to indentify such metabolites. Since the acetylation itself is not affected by inducing agents [19], major changes of INH half lives should not be expected. Another explanation for the increased toxicity of INH combined with RMP could be a simple summation of side effects, affecting both slow and rapid acetylators. In view of its great practical importance, the question of combined antituberculous drug toxicity requires further elucidation.

The aformentioned interactions illustrate the role of RMP as an enzyme inducing agent. In contrast to this, INH may act as an inhibitor of phenytoin metabolism and produce serious toxic reactions in patients on combined antituberculous and antiepileptic treatment [20]. A final relevant example of altered antibiotic metabolism is the acceleration of doxycyclin (DXC) elimination by antiepileptic drugs. In a Finnish study [21] DXC half life was reduced by phenytoin and carbamazepin from 13.3 hours to 7. 1 hours. Patients receiving these drugs should therefore have two daily doses of DXC, to insure adequate serum levels.

Competition for tubular and biliary excretion

Renal elimination of drugs is determined by three mechanisms: Glomerular filtration, tubular reabsorption and active tubular secretion. Glomerular filtration is unlikely to be directly altered by simultaneous administration of two or more drugs. Tubular reabsorption mainly consists of passive diffusion of drug molecules through cellular membranes and is thus not subject to primary interactions. However, the reabsorption of weak acids or bases, categories to which most drugs belong, is also influenced by urinary pH values. Since only unionized molecules are reabsorbed, acidification or alkalinization of urine will change the elimination rate of such drugs. Among the antimicrobial agents sulfonamides [22] and nalidixic acid [23] are examples of pH dependent urinary excretion.

Some of the weak acids and bases are transported from the blood across the renal tubuli by an active process against a concentration gradient. This secretory mechanism is a saturable process and may therefore be influenced by substances competing for the same macromolecular carrier sites. The reduced excretion rate of penicillin by probenecid is well known, but also ampicillin, carbenicillin and cephalosporins may be affected by uricosuric agents. A similar influence on an active transport process may also occur in the liver: rifampicin elimination can be slowed bv probenecid [24].

Interactions at receptor sites

Synergism between two antibiotics can be regarded as a drug interaction at receptor or tissue sites. The clinical advantage of certain antibiotic combinations is well recognised, especially in patients with granulocytopenia or in cases of severe infection due to unidentified pathogens. Aminoglycosides plus carbenicillin or cephalosporins constitute the regimens most frequently employed. In recent years, however, numerous reports called attention to the potential risks of the aminoglycoside-cephalosporin combination. A high incidence of acute tubular necrosis in patients receiving gentamicin with cephalotin was observed in retrospective [25], [26] and in a few prospective studies [27], [28]. Increased creatinine or BUN levels occured in 20-34% of patients, as compared to only 2-6% when penicillins plus aminoglycosides were used [27], [28], [29]. In view of this apparent toxic interaction the question arises, whether the simultaneous use of aminoglycosides and cephalosporins should be abandoned or at least limited to exceptional cases. A critical evaluation of the experimental and clinical data casts some doubt on the widely accepted assumption of nephrotoxic potentiation. In most clinical cases renal damage occurred in old, severely ill and septic patients with initially elevated serum creatinine concentrations. Regardless of this fact, both gentamicin and cephalotin were given in a “usual” or even very high dosage. Since the elimination rates of these antibiotics depend on glomerular filtration, abnormal drug accumulation was probably present in a high proportion of patients. A recent controlled trial indicates that gentamicin alone may induce nephrotoxicity in 11% of patients despite careful adaptation of the dose regimen to creatinine clearance values [30]. Cephalotin itself has also a toxic potential [31]. An additional adverse factor is often the simultaneous administration of furosemide with the antibiotics. Tilstone et al. [32] demonstrated a significant reduction of inulin, gentamicin and cephaloridine clearance by this potent diuretic. Thus, an increased risk of the aminoglycoside-cephalosporin combination need not be postulated to explain the reported adverse effects.

Recent animal experiments provide no evidence of toxic potentiation [33]; on the contrary, they show a protective effect of cephalosporins against kidney damage from gentamicin [34], [35]. Data on urinary enzyme elimination in humans also point to a certain protective effect of cephalothin and cephazolin against tubular impairment by aminoglycosides [36]. The Boston surveillance study equally failed to confirm an increased risk of the aminoglycoside-cephalosporine combination [37].

In conclusion, the present data do not prove a potentiation of gentamicin toxicity by cephalosporins. On the other hand, there is some indication that the combination of gentamicin or tobramycin with carbenicillin, ticarcillin or methicillin will reduce the incidence of renal damage. This protective interaction could be due to an influence on tubular handling and renal accumulation of aminoglycosides.

Combined antimicrobial therapy in renal disease

Pharmacokinetic principles must be especially observed when combined drug regimens are used in patients with reduced renal function. The elimination rate of different antibiotics in renal failure is affected to a highly variable degree [36]. In order to ensure adequate serum levels but avoid toxic accumulation, the dosage has to be individually adapted for each drug. For example, a patient with a creatinine clearance of 10 ml/min will have a serum half life of about 6 hours for carbenicillin and 20 hours for gentamicin. In order to achieve similar serum concentrations as in cases with normal renal function, one half of the usual maintenance dose should be administered at intervals equal to these $t_{1/2}$ values. A dosage modification in renal failure will also be required with the currently used triple combination of antituberculous drugs. Rifampicin elimination is independent of creatinine clearance, whereas INH (in slow acetylators) and ethambutol are partially excreted unchanged by the kidney. Maintenance doses should therefore be slightly reduced for these two agents.

The choice of the optimal dosage regimen depends on the antibiotics employed and on the disease to be treated. The use of a generally applicable dosage rule [36], will greatly facilitate combined drug treatment in patients with renal disease.

References

[1] KING J. C. (1973) *Guide to parenteral admixtures*, Cutter Laboratories, St. Louis.

[2] MC LAUGHLIN J. E. and REEVES D. S. (1971) "Clinical and laboratory evidence for inactivation of gentamicin by carbenicillin; *Lancet*, **1**, 261-264.

[3] NOONE P. and PATTISON J. R. (1971) "Therapeutic implications of interaction of gentamicin and penicillins", *Lancet*, **II**, 575-578.

[4] HOLT H. A., BROUGHALL J. M., MC CARTHY M. and REEVES D. S. (1976) "Interactions between aminoglycoside antibiotics and carbenicillin or ticarcillin", *Infection*, **4**, 107-109.

[5] DAVIES M., MORGAN J. R. and ANAND CH. (1975) "Interactions of carbenicillin and ticarcillin with gentamicin", *Antimicrob. Ag. Chemoter.*, **4**, 431-434.

[6] NEUVONEN P. J., GOTHONI G., HACKMAN R. and BJÖRKSTEN K. (1970) "Interference of iron with the absorption of tetracyclines in man", *Brit. Med. J.*, **4**, 532-534.

[7] CHENG S. H. and WHITE A. (1962) "Effect of orally administered neomycin on the absorption of penicillin V", *New Engl. J. Med.*, **267**, 1296-1297.

[8] BLANC M.H., BERTHOUD S., RUDHARDT M., HERRERA A., LUPINSKY I., BOURRIT C., BOURRIT B. and FABRE J. (1973) "Pénicilline V et interactions médicamenteuses", *Schweiz. Rundschau Med.*, **62**, 861-867.

[9] BOMAN G. (1974) "Serum concentration and half-life of rifampicin after simultaneous oral administration of aminosalicylic acid or isoniazid», Eur. J. Clin. Pharmacol., **7**, 217-225.

[10] DETTLI L. (1974) "Pharmacokinetic aspects of drug interactions", in *Clinical effects of interaction between drugs*, Excerpta Medica, Amsterdam, pp 39-68.

[11] CHRISTENSEN L. K., HANSEN M. S. and KRISTENSEN M. (1963) "Sulfaphenazole-induced hypoglycemic attacks in tolbutamide/treated diabetics", *Lancet*, **II**, 1298-1301.

[12] TUCKER G. H. and Hirsch J. I. (1972) "Sulfonamide-sulfonylurea interaction", *New Engl. J. Med.*, **286**, 110-111.

[13] ACOCELLA G., BONOLLO L., GARIMOLDI M., MAINARDI M., TENCONI L. T. and NICOLIS F. B. (1972) "Kinetics of rifampicin and isoniazid administered alone and in combination to normal subjects and patients with liver disease", *Gut*, **13**, 47-53.

[14] ZILLY W., BREIMER D. D. and RICHTER E. (1975) "Induction of drug metabolism in man after rifampicin treatment measured by increased hexobarbital and tolbutamid clearance", *Europ. J. Clin. Pharmacol.*, **9**, 219-227.

[15] EDWARDS O. M., CORTENAY-EVANS R. S., GALLEY J. M., HUNTER J. and TAIT A. D. (1974) "Changes in cortisol metabolism following rifampicin therapy", *Lancet*, **II**, 549-551.

[16] LEES A. W., ALLAN G. W., SMITH J., TYRELL W. F. and FALLON R. J. (1971) "Toxicity from rifampicin plus isoniazid and rifampicin plus ethambutol therapy", *Tubercle*, **52**, 182.

[17] LLORENS J. SERRANO R. J. and SANCHEZ R. (1978) "Pharmacodynamic interference between rifampicin and isoniazid", *Chemotherapy*, **24**, 97-103.

[18] MITCHELL J. R., THORGEIRSSON U. P., BLACK M., TIMBRELL J. A., SNODGRASS W. R., POTTER W. Z., JOLLOW D. J. and KEISER H. R. (1975) "Increased incidence of isoniazid hepatitis in rapid acetylators: possible relation to hydrazine metabolites", *Clin. Pharmacol. Ther.*, **18**, 70-79.

[19] NELSON S. D., MITCHELL R. J., TIMBRELL J. A., SNODGRASS W. and CORCORAN G. B. (1976) "Isoniazid and iproniazid: Activation of metabolites to toxic intermediates in man and rat", *Science*, **193**, 901-903.

[20] GEERING J. M., RUCH W. and DETTLI L. (1974) "Diphenylhydantoin-Intoxikation durch Diphenyl-hydantoin-isoniazid-Interaction", *Schweiz. Med. Wschr.*, **104**, 1224-1228.

[21] Neuvonen P. J., Penttilä O., Lehtovaara R. and Aho K. (1975) "Effect of antiepileptic drugs on the elimination of various tetracycline derivatives", *Europ. J. Clin. Pharmacol.*, **9**, 147-154.
[22] Dettli L. and Spring P. (1966) "Diurnal variations in the elimination rate of sulfonamide in man", *Helv. med. Acta*, **33**, 291.
[23] Mc Chesney E. W., Froelich E. J., Lesher G. Y., Crain A. V. R. and Rosi D. (1964) "Absorption, excretion and metabolism of a new antibacterial agent, nalidixic acid", *Toxicol. Appl. Pharmacol.*, **6**, 292-309.
[24] Kenwright S. and Levi A. J. (1973) "Impairment of hepatic uptake of rifamycin antibiotics by probenecid, and its therapeutic implications", *Lancet*, **II**, 1401-1405.
[25] Opitz A., Herrmann I., v. Herrath D., Schaefer K. (1971) "Akute Niereninsuffizienz nach Gentamicin-Cephalosporin-Kombinationstherapie", *Med. Welt*, **22**, 434-438.
[26] Fillastre J. P., Laumonier R., Humbert G., Dubois D., Metayer J., Delpech A., Leroy J. and Robert M. (1973): "Acute renal failure associated with combined gentamicin and cephalotin therapy", *Brit. med. J.*, **2**, 396-397.
[27] Klastersky J., Hensgens C. and Debusscher L. (1975) "Empiric therapy for cancer patients: Comparative study of ticarcillin-tobramycin, ticarcillin--cephalothin, and cephalothin-tobramycin", *Antimicrob., Ag. Chemother*, **7**, 640-645.
[28] EORTC International Antimicrobial Project Group (1978) "Three antibiotic regimens in the treatment of infection in febrile granulocytopenic patients with cancer", *J. Infect. Dis.*, **137**, 14-29.
[29] Wade J. C., Petty B. G., Smith C. R., Lipsky J. J., Conrad G. L., Ellner G. and Lietman P. S. (1978) "Nephrotoxicity of gentamicin or tobramycin with methicillin or cephalothin in sepsis of unknown etiology", *Proceedings of the 10th Internat. Congr. Chemotherapy*, American Society for Microbiology, Washington, D. C., vol. II, 971-972.
[30] Smith C. R., Baughman K. L., Edwards C. Q., Rogers J. F. and Lietman P. S. (1977) "Controlled comparison of amikacin and gentamicin", *New Engl. J. Med.*, **296**, 349-353.
[31] Carling P. C., Idelson B. A., Casano A., Alexander E. A. and Mc Cabe W. R. (1975) "Nephrotoxicity associated with cephalothin administration", *Arch. Int. Med.*, **135**, 797-801.
[32] Tilstone W. G., Semple P. F., Lawson D. H. and Boyle J. A. (1977) "Effects of furosemide on glomerular filtration rate and clearance of practolol, digoxin, cephaloxidine and gentamicin", *Clin. Pharmacol. Therapeut.*, **22**, 389-394.
[33] Harrison W. O., Silverblatt F. J. and Turck M. (1975) "Gentamicin nephrotoxicity: failure of three cephalosporins to potentiate injury in rats", *Antimicrob. Ag. Chemother.*, **8**, 209-215.
[34] Dellinger P., Murphy T., Pinn V., Barza M. and Weinstein L. (1976) "Protective effect of cephalothin against gentamicin-induced nephrotoxicity in rats", *Antimicrob. Ag. Chemother.*, **9**, 172-178.
[35] Luft F. C., Patel V., Moo Nahm Yum and Kleit S. A. (1976) "Nephrotoxicity of cephalosporin-gentamicin combinations in rats", *Antimicrob. Ag. Chemother.*, **9**, 831-839.
[36] Mondorf A. W., Breier J., Zegelman M. and Müller H. (1978) "Die Wirkung von Aminoglykosiden auf den proximalen Tubulus des Menschen", *Advances in Clinical Pharmacology*, Urban & Schwarzenberg, München, vol. 15, pp. 101-110.
[37] Fanning W. L., Gump D. and Jick H. (1976) "Gentamicin and cephalothin-associated rises in blood urea nitrogen", *Antimicrob. Ag. Chemother.*, **10**, 80-87.
[38] Dettli L. (1977) "Elimination kinetics and dosage adjustment of drugs in patients with kidney disease", in *Progress in Pharmacology*, Fischer, Stuttgart & New York, vol. 1.

9

New developments in combined chemotherapy of non-tuberculous infections

W. Brumfitt, J. M. T. Hamilton-Miller

Dept. of Medical Microbiology, Royal Free Hospital, (University of London), Hampstead, London NW3 2QG, England

Summary

The reasons for the introduction of combined chemotherapy in the early years of antimicrobial treatment are discussed. Reasons are given as to why such therapy originally fell into disrepute.

More recently the concept of combined chemotherapy has been re-examined, for a number of reasons which are critically reviewed.

In general, antibiotic combinations are not given as a single pharmaceutical preparation containing both drugs. Such combinations, of course, prevent flexibility in the use of the individual components but may have the advantage of providing the optimal combination in a more acceptable form to both doctor and patient.

In our own work we have chosen the combination of rifampicin and trimethoprim (ratio 3.75:1) as a model to study the problem. The *in vitro* and pharmacokinetic findings are discussed, and justification made for a proper clinical trial to assess the value of the combination. Non-tuberculous urinary infections (of all kinds) which have been shown to be difficult to eradicate, would appear to be an excellent subject for such a study.

Historical Introduction

Although in the early 1950's there were several different antimicrobial agents in clinical use (benzylpenicillin, tetracyclines, sulphonamides, streptomycin, neomycin, bacitracin, erythromycin and chloramphenicol) adequate single-drug therapy was not available for all bacterial infections. Clinicians therefore quite properly resorted to combination therapy. Such an approach, however, became unnecessary as many new antibiotics were discovered during the second part of the 1950's. Further, there had been gross misuse of certain combinations both as a result of bad clinical judgement and lack of accurate as well as rapid laboratory testing. Another factor was the commercial interests of some companies who resorted to misleading promotion. In this way the concept of combination antibiotic treatment fell into disrepute for reasons described vividly by Dowling [1] in 1965.

Some years ago the combination of tetracycline and novobiocin were marketed commercially as a single preparation and it was claimed that they were synergistic [2], [3]. However, a careful investigation by Hirsch and Finland [4] showed that this claim was erroneous.

The Current Situation

The introduction of co-trimoxazole (Septrin, Bactrim), a fixed combination of trimethoprim and sulphamethoxazole at the ratio of 1:5, in the late 1960's has been justified commercially on the grounds of synergy and suppression of resistance. However, it is not entirely clear, even 10 years later, to what extent these phenomena occur under clinical conditions. There is no doubt that co-trimoxazole is highly effective for treating a variety of infections, but only in the case of gonorrhoea is clinical synergy without doubt. For other infections some workers have found trimethoprim alone to be equally effective [5], [6].

The use of non-fixed combinations has now become a respectable, indeed virtually mandatory, practice in hospitals. The most common justification for combined therapy is to give the widest possible "cover" against pathogens (although this justification is often shown to have been an excuse without support from clinical information let alone laboratory findings). Combinations are necessary for a variety of reasons: the marked slowing down of the introduction of broad spectrum antibiotics, increased acquisition of resistance to existing drugs and advances in immunosuppression and cytotoxic therapy (particularly in leukaemias).

A major misuse of antibiotic combinations is their systemic use in prophylaxis.

Synergistic activity is another common justification for using combined antibiotics. Although this is an attractive concept, as we have discussed elsewhere [7], the demonstration of synergy under clinical conditions is rather rare. Klastersky's work [8] shows clearly that the use of antibiotic combinations, especially synergistic ones, is associated with a better clinical response in cancer patients.

It seems that in the minds of some workers a compelling reason for using combinations is for the prevention of resistance. However, study of literature shows that most of the good clinical evidence in support of this idea comes from the various carefully controlled trials of the antimicrobial therapy of tuberculosis. Study of these data shows clearly that multiple drug therapy is mandatory to prevent the emergence of resistance [9]. In sharp contrast, however, infections due to the Group A β-haemolytic streptococcus have been treated with penicillin alone since its introduction, yet not even a slightly resistant variant of this organism, resulting from treatment, has been found. Thus, had combined therapy been used it might have been judged to be a great success. This shows the importance of assessing every situation on its merits.

The present situation is that combinations, in form of two separate drugs, are being used with increasing frequency to treat a number of infections. The idea of fixed combinations is opposed strongly by some members of the medical profession as well as drug regulating bodies. Furthermore, the acceptance or otherwise of the various fixed combinations varies from country to country. Examples of fixed combinations available in the United Kingdom are: sulphamethoxazole/trimethoprim (co-trimoxazole), penicillin/streptomycin, flucloxacillin/ampicillin, tetracycline/novobiocin and tetracycline/nystatin. All have been criticised on various grounds, but of those cited the most logical would seem to be a combination of the antifolate agents namely sulphamethoxazole/trimethoprim. As already pointed out, in a number of infections one or other of the components of the latter combination may be equally effective by itself, with less risk of toxicity and side-effects.

Desirable properties of components of a combination

In an effort to contribute to this topic, we have endeavoured to delineate what characteristics two antibiotics should have to form an ideal combination. The drugs chosen for combination must fulfil all or most of the following criteria (in approximate order of importance).

1. They should have as wide a range of antimicrobial activity as possible, and should be complementary.
2. Primary bacterial resistance to the drugs chosen for combination should be rare, and there must be no cross-resistance.
3. They must be compatible pharmacokinetically, in terms of distribution, half-lives and routes of excretion.
4. They must also be compatible pharmaceutically, and should be suitable for both oral and parenteral administration, preferably once or (at most) twice a day.
5. Synergistic activity is desirable, but an additive effect is acceptable. This contingency is made more likely if the two drugs have modes of action which are related.

In addition, it may be convenient for descriptive purposes if one is able to regard one as the primary (or "killer") drug, and the other as the secondary (or "back-up") compound which among other things should assist in preventing the development of resistance.

The number of possible combinations of two drugs which can be put together from the antibiotics at present available is vast [7], so any approach of this sort must be empirical. As we decided to follow the advice of one of the doyens of this subject, Professor Ernest Jawetz, and test rather than predict interactions against many pathogenic species of bacteria, we settled upon a single combination [10]. We chose rifampicin, an antibiotic inhibitory at therapeutically attainable concentrations for almost all pathogens as the "killer" drug, with trimethoprim as the "back-up" compound. Our *in vitro* work (see below) took more than one year to carry out. Hence the reason for selecting the pair of drugs which have been judged to be the most effective combination. Even this combination is not ideal, as trimethoprim is poorly active against *Pseudomonas aeruginosa* and anaerobes, and recent reports suggest that resistance to trimethoprim may be increasing-although at present this is not our personal experience [11]. However, on theoretical grounds, trimethoprim seemed overall the most suitable "back-up" drug. Our finding of synergy between the two compounds was encouraging and we noted that both drugs interfere with different steps in the biosynthesis by bacteria of nucleic acids. We also took into account the results of an earlier small scale study [12].

For anaerobic bacteria, we suggest, using the same criteria, that erythromycin would be an appropriate "back-up" compound. Alternatively, a nitro-compound may be suitable.

Studies on rifampicin/trimethoprim

In vitro findings. We undertook, starting in 1973, a comprehensive study to investigate, quantitatively, the interaction of rifampicin and trimethoprim in vitro against a large variety of commonly isolated pathogenic bacteria of medical importance. It soon became obvious that, where synergy did occur, it was not on the same scale as that between trimethoprim and sulphamethoxazole; therefore, to be sure of accuracy, we had to develop a more discriminating technique for its detection. Therefore, we considered it mandatory to use drug concentration steps smaller than those employed in the conventional doubling dilution technique. We also defined synergy by rigorous criteria [13], [14]. All our results (including previously unpublished material) obtained using 363 unselected strains are summarized in Table 9.1. Synergy was observed in 41% of the strains tested. Antag-

Table 9.1 Combined in vitro *activity of rifampicin and trimethoprim. Data from:* [13], [14], [15] *and unpublished experiments.*

Synergy often observed		*Synergy not uncommon*		*Synergy rare or additive only*	*Antagonism**
Strept. pyogenes *Strept. faecalis*	44/44**	*Enterobacter spp.*	3/13	*E. coli*	*Staph. aureus*
		Kleb. aerogenes	14/40	*Salm. typhimurium*	*N. gonorrhoeae*
Proteus mirabilis *Pr. morganii* *Pr. rettgeri* *Pr. vulgaris*	52/72	*Ps. aeruginosa* *Bact. fragilis* *H. influenzae*	9/20 7/19 3/14	*Prov. stuartii*	
Serratia marcescens	15/15				

* Only when rifampicin < 0.02 μg/ml.
** Strains for which synergy observed/strains tested.

onism was observed only with two species; *Staphylococcus aureus* and *Neisseria gonorrhoeae*, but only at extremely low concentrations of rifampicin (< 0.02 μg/ml) so it is unlikely that this would have any clinical relevance.

Two other groups of workers from London have subsequently substantiated, and to some extent extended, our findings. Grüneberg & Emmerson [16] tested selected strains of bacteria isolated from urinary tract infections, including some resistant to trimethoprim or rifampicin, or both. They used a different methodology (doubling dilutions) and an alternative definition of synergy, so their results cannot be compared directly with ours-but only 16% of the strains investigated showed synergy. Farrell et al [17] also tested a selected group of 61 rather unusual strains in that they were all resistant to gentamicin and found synergy in 43%.

As a result of the work described above, the following species must be added to those appearing in Table 9.1 for which synergy has been reported: *Staph. epidermidis*, *Citrobacter freundi*, *Flavobacterium spp.*, *Acinetobacter spp.* and *Alcaligenes spp.*

Grüneberg and Emmerson [16] pointed out that the extent of synergy most commonly observed in these studies (where ΣFIC lies between 0.6 and 0.7) has been defined by a Study Group [18] as "slight or no potentiation". However, we wish to point out that, under these conditions, inhibition is occurring with each drug present only at one-third its individual MIC which many workers would regard as being of importance. However, we would emphasize that too much stress should not be laid upon the results of *in vitro* synergy experiments carried out by the technique described above. In our opinion, the absence of antagonism is of much greater significance. We would also stress that we have never been able to demonstrate synergy by short-term killing curves, turbidimetric measurements or disc tests.

One report of *in vivo* synergy has been made [19]. However, we are reluctant to attribute too much importance to this type of experiment — carried out in the mouse — as the pharmacokinetics of these drugs are known to differ in man and in the mouse (e.g. the half-life of trimethoprim is 1 hour in the mouse and 8-10 hours in man). [20], [21].

Suppression of Resistance Emerging. An *in vitro* finding which we regard as of much greater significance than that of synergy was made by Kerry et al [13]. The emergence of resistance to rifampicin was dramatically reduced in the presence of sub-inhibitory concentrations of trimethoprim. This has since been confirmed by Arioli et al [19]. It is also clear that when trimethoprim was present at inhibitory levels, rifampicin resistance cannot emerge.

Results of studies on the effect of the single drugs and of the combination on resistance emergence in gut flora organisms would be very helpful in establishing the validity of these *in vitro* findings to the *in vivo* situation. We regard such experiments as of the highest priority. The results would obviously have a great effect on the prescribing recommendations of those concerned with the control of infection in the hospital environment.

Pharmacokinetic considerations. The findings described above dictated the design of pharmacokinetic experiments. The object of these was two-fold:

1. to determine whether rifampicin and trimethoprim interfere with the handling of one another in man;
2. to see how much latitude in dosage schedules was possible while avoiding a "trimethoprim gap", which is that situation where rifampicin is present alone. Clearly, such a "trimethoprim-gap" would encourage rifampicin-resistance to emerge.

Acocella and Scotti [22] used 600 mg rifampicin + 160 mg trimethoprim (a ratio of 3.75 : 1), given once daily, and Hamilton-Miller & Brumfitt [23] tested 300 mg rifampicin + 160 mg trimethoprim (a ratio of 1.88 : 1), the mixture being taken 12-hourly. Healthy volunteers were used and all drugs were given as capsules (separately and in combination). In view of the long half-life of trimethoprim it is clearly important that pharmacokinetic studies in people with *normal renal function* should be continued for at least 12 hours.

Only minor evidence of pharmacokinetic interference was found between the two drugs: the half-life of trimethoprim fell to 70-80% of its usual value (8-10 hours) after 7 days' treatment with the combination. Both groups of workers suggested that this finding may be connected with the well-known inducing effect of rifampicin on liver enzymes; further experimental studies will be required to quantitate any changes in biliary excretion of the two drugs occurring during the course of a 7 day regimen.

The feared "trimethoprim-gap" (i.e. the situation when trimethoprim is not present in the serum in effective concentrations) did not occur on either dosage schedule. Once daily dosage resulted in sub-optimal levels of rifampicin during the second 12-hour period, and on the twice daily regimen the rifampicin dosage was also too small. We therefore suggest the use of a different combination, in a total daily dosage of 900 mg rifampicin + 240 mg trimethoprim. The combination has now been prepared in tablet form. The proposed dosage scheme for clinical trial work in Great Britain will be one tablet (300 mg rifampicin and 80 mg trimethoprim) in the morning and 2 tablets (600 mg rifampicin and 160 mg trimethoprim) in the evening. The combination is to be given the name "Rifaprim" (Dow-Lepetit). A multi-centre controlled trial is now in progress for the treatment of urinary tract infections, which are selected because of their failure to respond to conventional treatment.

All the available evidence suggests that there will be no toxicity or side-effects using this type of schedule for a 7 day period.

Lack of risk of emergence of rifampicin-resistant *Mycobacterium tuberculosis*

As expected, the proposed use of rifampicin for non-tuberculous conditions has met with some opposition but this appears to be of the emotive kind and has no scientific background. Rifampicin is a very useful drug in the treatment of tuberculosis and has been responsible for the dramatic shortening in the total length required for treatment in this disease [24]. Hence, many chest physicians are extremely reluctant for it to be used outside tuberculosis [e.g. 25], for fear that rifampicin-resistant *M. tuberculosis* strains could emerge. This question was put squarely in a Leading Article in Lancet [26]. This editorial was deliberately provocative and encouraged the "reserve the drug for tuberculosis" lobby to state their case. It is significant, however, that no response was forthcoming in the letters to the Lancet. A single letter [27] appeared on the subject agreeing with the editorial but there was no dissent. Any lingering fears which chest physicians might still have had should have been dispelled by a recent paper by Acocella, Brumfitt and Hamilton-Miller [28]. Here it was shown that the incidence of strains of *M. tuberculosis* showing primary resistance to rifampicin is no greater in countries where rifampicin is used freely and often alone, for non-tuberculous disease than it is in countries where rifampicin use is strictly controlled. There are good grounds for believing that rifampicin-resistant *M. tuberculosis* strains arise when rifampicin is improperly used for treating tuberculosis; however, this is secondary resistance. Again, no dissension was expressed to this second paper and no figures disproving the thesis have appeared since.

Thus, it must be concluded that the case has been made beyond reasonable doubt that the risk of selecting rifampicin-resistant strains of *M. tuberculosis* by using rifampicin for extra-tuberculous indications for 7 days is very small. On the other hand, it is vital when "Rifaprim" becomes more generally available that a running check be made on the sensitivity of *M. tuberculosis* strains isolated before and after its introduction. It would also be sensible to avoid treating chest infections with "Rifaprim" until any possible risk has been assessed in large numbers of patients under careful surveillance.

In this context it is interesting to draw a comparison with the use of streptomycin in the early 1950's. Due to the lack of antimicrobial agents at that time streptomycin had to be used for both tuberculous and non-tuberculous indications. Yet despite a widespread belief to the contrary the facts are that such usage did *not* cause the emergence of streptomycin-resistant strains of *M. tuberculosis* [9]. A further point is that the mutation rate to streptomycin resistance in the tubercle bacillus is ten to one hundred times higher than that for rifampicin [29].

Another interesting question which has to be answered is whether emergence of resistance of non-tuberculous bacteria will occur. Again, until the combination has been used the answer will not be known. A point which needs to be stressed is that careful laboratory control must be exercised to ensure that organisms treated are sensitive both to rifampicin and to trimethoprim.

It is always encouraging to be able to end on a positive note. The presence of certain R-factors in bacteria can actually increase their sensitivity to rifampicin [30]. Furthermore, rifampicin has been shown to be capable of "curing" certain bacterial plasmids [31]. Thus, it may materialise that the more widespread use of rifampicin will ultimately have a beneficial effect in reducing the incidence of R-factors in the community.

Acknowledgements

We are grateful to Dow-Lepetit for their help with parts of this research.

References

[1] Dowling, H. F. (1965), *Am. J. Med.*, **39**, 796-803.

[2] Wallick, H., Harris, D. A., Reagan, M. A., Ruger M. and Woodruff, H. B. (1955), *Antibiotics A*. 909.

[3] Percori, V., Attocci, P., Cinmo, R., Mazzacca, G. and Riccardi, C. (1958), *Boll. Soc. Ital. Biol.*, **34**, 531.

[4] Hirsch, H. A., Finland, M. (1960) *New Engl. J. Med.*, **262**, 209.

[5] Brumfitt, W., Pursell R. (1972) *Brit. med. J.*, **ii**, 673-676.

[6] Kasanen, A., Kaarsala, E., Hiltunen, R. and Soini, V. (1974) *Ann. Clin. Research.* **6**, 285-289.

[7] Hamilton-Miller, J. M. T., Brumfitt, W. (1977) in *The rational choice of antibacterial agents*, 89-101. (R.P. Mouton, W. Brumfitt, J. M. T. Hamilton-Miller) Bohn, Scheltema & Holkema, Utrecht, Holland

[8] Klastersky, J., Cappel R. and Daneau, D. (1972) *Antimicrob. Ag. Chemother.*, **2**, 470-475.

[9] Long, E. R. (1958) *The Chemistry and Chemotherapy of Tuberculosis*, Chapter 27, Bailliere, Tindall & Cox, London.

[10] Jawetz, E., Gunnison, J. B. and Speck R. S. (1952) *Antibiot. and Chemother.*, **2**, 243-248.

[11] Grey, D., Hamilton-Miller, J. M. T. and Brumfitt, W. (1979) *Chemotherapy*, in press.

[12] Cluzel, R., Sirot, J., Cluzel, M. and Joly, B. (1972) *Path. Biol.* **20**, 871-879.

[13] Kerry, D. W., Hamilton-Miller, J. M. T. and Brumfitt, W. (1975) *J. Antimicr. Chemo.*, **1**, 417-427.

[14] Kerry, D. W., *M. Phil. thesis*, University of London (1975).

[15] Hamilton-Miller, J. M. T., Kerry, D. W. and Brumfitt, W. (1977) *J. Antimicr. Chemo*, **3**, 193-194.

[16] Grüneberg, R. N., Emmerson, A. M. (1977) *J. Antimicr. Chemo.*, **3**, 453-457.

[17] Farrell, W., Wilks, M. and Drasar, F. A. (1977) *J. Antimicr. Chemo.*, **3**, 459-462.

[18] Study Group (1973) *J. Inf. Dis.*, **128**, (Suppl.), 502-507.

[19] Arioli, V., Berti, M., Carniti, G., Rossi, E. and Silvestri, L. G. (1977) *J. Antimicr. Chemo.*, **3**, 87-94.

[20] Bohni, E. (1969) *Postgrad. med. J.*, **45**, S18-21.

[21] Kaplan, S. A., Weinfeld, R. E., Abruzzo, C. W., McFaden, K., Jack, K. L. and Weissman, L. (1973) *J. inf. Dis.*, **128**, S547-55.

[22] Acocella, G., Scotti, R. (1976) *J. Antimicr. Chemo.*, **2**, 271-277.

[23] Hamilton-Miller, J. M. T., Brumfitt, W. (1976) *J. Antimicr. Chemo.*, **2**, 181-188.

[24] East African/British Medical Research Council Study, *Am. Rev. Resp. Dis.*, **116**, 3-8.

[25] Morrison-Smith, J. (1975) *J. Antimicr. Chemo.*, **1**, 353-354.

[26] Leading Article (1976) *Lancet*, **1**, 290-291.

[27] Gelzer, J., Loustalot, P. (1976) *Lancet*, **1**, 642.

[28] Acocella, G., Hamilton-Miller, J. M. T. and Brumfitt, M. (1977) *Lancet*, **1**, 740-742.

[29] Verbist, L., Gyselen, A. (1968) *Am. Rev. Resp. Dis.*, **98**, 923-932.

[30] Scotti R., Silvestri L. G. and Romero E. (1974) *Antimicr. Ag. Chemother.*, **6**, 121-123.

[31] Riva S. A., Fietta M., Berti M., Silvestri L. G. and Romero E. (1973) *Antimicr. Ag. Chemother.*, **3**, 456-462.

10

In vitro studies of antibacterial synergy

D. Greenwood

Department of Microbiology, University Hospital, Queen's Medical Centre, Nottingham, NG7 2UH, England

Summary

Synergic interactions between antibacterial agents may take many diverse forms: the two agents may interact at the biochemical level; one may facilitate entry of the other into the bacterial cell; one may protect the other from enzymic destruction; or the two components may simply act on different sections of the bacterial population.

Antibacterial drug interactions are commonly assessed *in vitro* by chessboard titrations, but such tests only detect certain forms of interaction and are of little help in evaluating the likely therapeutic relevance of the results. Alternative methods of testing antibiotic interactions may lead to different conclusions as to their relevance. For example, tests in an *in vitro* model of the treatment of bacterial cystitis suggest, paradoxically, that spectacular synergy of the trimethoprim-sulphamethoxazole type may have limited clinical relevance, whereas minimal 'synergy', such as that which may be demonstrated between rifampicin and nalidixic acid, may be of therapeutic value.

Introduction

Although synergy and antagonism between antimicrobial agents may be equally important, most of the literature on antibiotic interactions has concerned itself with synergic phenomena. The present communication will continue this optimistic tradition by concentrating on such combinations as may be therapeutically beneficial.

Before trying to evaluate the implications of any particular form of synergy it is necessary at the outset to recognize the wide variety of forms that mutually beneficial interactions can take. Some of these are listed in Table 10.1. For the most part these have been well characterized, so that the mechanism of the interaction is generally known. The extent and relevance of the synergy is, however, more problematical.

β-lactam antibiotics

In order to document more fully and underline the diverse nature of synergy, it may be useful to survey briefly the various types of interaction that can occur with β-lactam antibiotics. This group of antibiotics is particularly rich in synergic potential, showing several types of interaction between members of the group as well as with other antibacterial agents (Table 10.2).

Amongst the best known of those interactions occurring between β-lactam agents is the β-lactamase inhibition effect [1], [2] in which the antibiotic-inactivating β-lactamase enzymes of enterobacteria may be sequestered by one compound, allowing a second enzyme-labile β-lactam antibiotic to reach its target unscathed. Interest in this attractive principle has recently been revived by the discovery of some novel broad-spectrum enzyme inhibitors of which clavulanic acid is the best known example [3]. Studies of our own of the β-lactamase inhibition effect suggest that it may be more useful to combine the inhibitor with a cephalosporin rather than a penicillin,

Table 10.1 Types of synergy

Biochemical interaction	(i) sequential blockade (ii) different sites of action	: trimethoprim/sulphonamides : cephalexin/mecillinam
Permeabilization	(i) cell wall (ii) cell membrane	: β-lactam antibiotics : polymyxins
Protection	β-lactamase inhibition	: nafcillin; clavulanic acid
Differential population effects	suppression of resistance	: antituberculosis therapy; rifampicin + nalidixic acid

Table 10.2 Examples of synergy involving β-lactam antibiotics

Type of synergy	*Bacteria affected*	*Example of effective combination*
β-lactamase inhibition	Gram-neg. bacilli	clavulanic acid + broad-spectrum β-lactams
lysis enhancement	Gram-neg. bacilli	cephalexin + cephalothin
complementation	Gram-neg. bacilli	cephalexin + mecillinam
dual activity	various	Glaxo compound MCO
permeabilization	enterococci, etc.	penicillins + aminoglycosides
	Gram-neg. bacilli	ampicillin + erythromycin
mutual suppression of resistance	staphylococci	penicillin + fusidic acid

as cephalosporins alone often have a transient effect on enterobacteria which are totally unaffected by penicillins [4].

Another type of synergy which may be demonstrated between β-lactam antibiotics is lysis enhancement [5] in which an improvement of the lytic effect of some β-lactam antibiotics can be demonstrated if enterobacteria are pretreated with cephalosporins such as cephalexin or cephradine which transform the bacteria into long filaments. The effect depends on sequential exposure if maximum enhancement of lysis is to be achieved by the second agent, and is not likely to be of therapeutic value.

A third type of interaction, complementation, also depends on the unusual behaviour of cephalexin and that of another novel β-lactam antibiotic, the amidinopenicillin mecillinam. These antibiotics, in combination, produce morphological changes (and consequently rapid lysis) in gram-negative bacilli that neither agent can produce alone, apparently because the two agents act at different sites in the bacterial cell. The net effect is that together they produce a similar effect to a conventional penicillin or cephalosporin [6].

One of the most ingenious ways of combining antibacterial agents is exhibited by the Glaxo compound MCO [7], [8] a cephalosporin which carries a sub-group at the 3-position which itself has intrinsic antibacterial activity. When hydrolysed by β-lactamase, the cephalosporin disintegrates releasing the antibacterial sub-group and prolonging the antibacterial effect. This elegant solution to the problem of β-lactamase instability has unfortunately not yet yielded a compound which lacks toxicity.

Because β-lactam antibiotics act at the level of the bacterial cell wall a common form of synergy occurs between these compounds and others which are normally excluded from the intact bacterial cell. This appears to be the chief mechanism of synergy in the combination of penicillin and aminoglycosides which is successfully exploited therapeutically in the treatment of enterococcal endocarditis. Synergy between penicillins and aminoglycosides may also be readily demonstrated in gram-negative bacilli, but another type of interaction which depends on the facilitation of antibiotic entry into gram-negative cells has received less attention, presumably because it is not easily demonstrated by conventional *in vitro* tests: the inactivation of β-lactam antibiotic-induced spheroplasts by erythromycin. In conditions of suitably high osmolality, penicillins and cephalosporins can convert gram-negative bacilli into osmotically-stable spheroplasts, which can revert to bacillary form as the antibiotic concentration declines. Such spheroplasts may be penetrated by erythromycin (normally excluded from gram-negative bacilli) and reversion consequently prevented.

Finally, penicillins may interact synergically with inhibitors of protein synthesis, such as fusidic acid, against some penicillinase-producing staphylococci [9]. The basis of this interaction appears to be that fusidic acid inhibits the penicillinase-producing cells while the penicillin suppresses the minority population which is resistant to fusidic acid [9]. The two agents are thus acting on different sections of the bacterial population. Assessment of the significance of this effect is complicated by two factors: firstly, fusidic acid and penicillins are mutually antagonistic against many strains of staphylococci [10]; secondly, pharmacokinetic differences between these agents are such that conditions under which synergy and antagonism may be demonstrated *in vitro* may not exist in the treatment situation [10].

In vitro assessment of antibacterial synergy

Although antibacterial agents may interact in a wide variety of ways, only a handful of *in vitro* tests are generally used to detect such interactions. Most frequently used is the chessboard titration technique in which synergy is detected as a mutual lowering of minimum inhibitory concentration (MIC) values. The degree of synergy is often expressed in terms of the fractional inhibitory concentration (FIC) of each component or plotted as an isobologram. MIC end-points are notoriously variable and are determined by factors which vary widely between antimicrobial agents [11]. The interpretation of chessboard titrations is therefore complicated from the outset by its reliance on the MIC as the measure of activity. Furthermore, these tests only detect certain forms of antimicrobial interaction, and other approaches are needed to investigate the nature, extent and relevance of combined effects.

In order to avoid the complications presented by the MIC it is often helpful to investigate antibacterial interactions in terms of the minimum antibacterial concentration (MAC). The MAC is defined as the lowest concentration of an agent to cause an observable antibacterial effect; this may be as little as 1/32nd of the conventionally determined MIC [11]. The simplest way to measure these values is by continuous turbidimetric monitoring, in which the MAC is revealed as the smallest concentration to cause a deviation from the normal growth curve, and the MIC as the lowest concentration to suppress growth overnight.

Since the MAC is a measure of the intrinsic activity of the antibiotic, uncomplicated by subsequent events during the incubation period, it may be used as a measure of 'true' synergy between antimicrobial agents. If the trimethoprim-sulphamethoxazole combination is investigated in this way, synergy is still displayed, but the degree of synergy is reduced (Table 10.3). The reason for the reduction in the effect appears to be that the MIC end-point for both trimethoprim and sulphamethoxazole is influenced by the emergence of a bacterial population exhibiting a slightly increased resistance to the drugs; a significant component of the overall synergy detected by conventional chessboard titrations is the mutual suppression of such 'adaptive' resistance [12], [13].

The use of this type of approach can be further exemplified by the trimethoprim-rifampicin combination that has also been suggested [14]. MIC titrations of rifampicin do not reflect the intrinsic activity of the drug because the end-point is markedly influenced by the tendency of resistant mutants (in this case single-step, high-level mutants) to emerge during the incubation period. Trimethoprim suppresses this resistance, while rifampicin prevents the low-level adaptation typical of trimethoprim (Figure 10.1). When the two agents are tested together at levels just below the MAC, however, no interaction is detectable, suggesting that true biochemical synergy between these compounds does not exist.

Because the activity of rifampicin is influenced by the emergence of resistance, both in the laboratory and in the patient [15], [16], the possibility suggests itself that the drug might be usefully combined with another antibacterial agent which has similar limitations — nalidixic acid. This combination too shows no interaction in MAC-related terms and an interaction only becomes detectable in chessboard titrations as the inoculum is raised. However, the two agents do interact by mutually suppressing resistance (Figure 10.2), and this type of 'synergy', which has been exploited so successfully in the treatment of tuberculosis, may be of wider therapeutic significance.

Table 10.3 Synergy between trimethoprim and sulphamethoxazole expressed in terms of the minimum antibacterial concentration and minimum inhibitory concentration.

Smallest concentration producing					
deviation from normal growth			*suppression of growth overnight*		
TMP	+	SMX	TMP	+	SMX
MAC/2	+	MAC/8	MIC/8	+	MIC/32
MAC/4	+	MAC/2	MIC/16	+	MIC/8

TMP = trimethoprim; SMX = sulphamethoxazole.

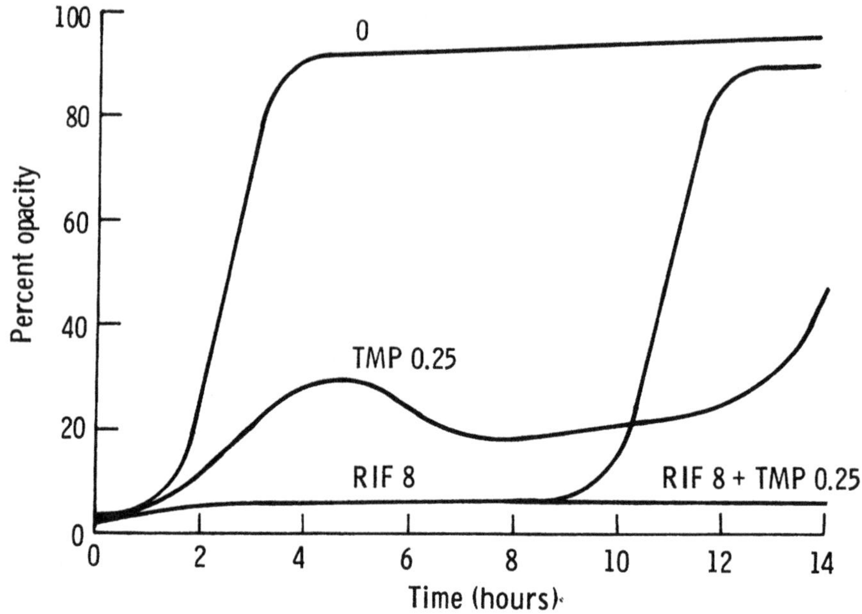

Fig. 10.1 - Continuous opacity records of a strain of *Escherichia coli* exposed to trimethoprim and rifampicin, alone and in combination.

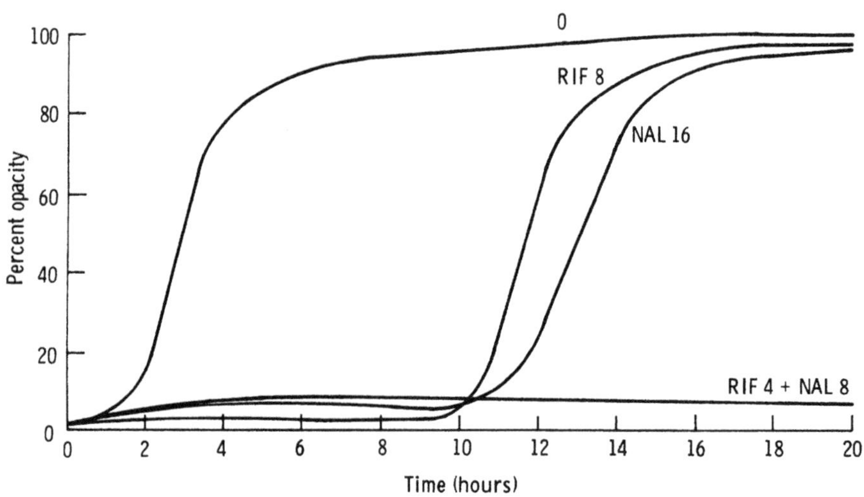

Fig. 10.2 - Continuous opacity records of a strain of *Proteus mirabilis* exposed to rifampicin and nalidixic acid, alone and in combination.

Relevance of *in vitro* studies to *in vivo* efficacy

When antibacterial combinations are used therapeutically, the conditions of exposure of bacteria to the drugs differs greatly from those of conventional *in vitro* tests. Differences in the pharmacokinetic properties of antimicrobial agents have already been alluded to: clearly conclusions drawn from static *in vitro* tests may not apply if the components of the mixture do not reach the same place in the right concentration at the same time. Even when pharmacokinetic properties are similar, striking *in vitro* synergy may not operate under *in vivo* conditions. A good example is the co-trimoxazole combination, widely used for the treatment of urinary tract infection. Synergy between the components of this mixture is usually detected as a spectacular mutual lowering of the MIC, but in the urine concentrations far above MIC values are easily maintained for long periods. Trimethoprim has a relatively rapid antibacterial effect because its action on dihydrofolate reductase quickly traps tetrahydrofolate in the unusable dihydrofolate form. Sulphonamides, on the other hand, cut off the folate supply at source, but do not affect the existing folate pool. Consequently they do not produce an observable effect for several generations because preformed folate is able to accommodate the needs of the organisms until continuing bacterial growth and division have diluted it to below a functional level. Thus, in the presence of excess trimethoprim (as in the urine during therapy), sulphonamides do not have the opportunity to exert any antibacterial effect and may, in this sense, be said to be antagonized by trimethoprim [12], [13]. Studies in an *in vitro* model of the urinary bladder [17] support this. If dense cultures of bacteria (such as are often found in infected urine) are exposed to sustained concentrations of trimethoprim and sulphamethoxazole in the dynamic conditions of dilution and discharge which obtain in the bladder, sub-MIC concentrations of the components act synergically, but supra-MIC concentrations do not. Using concentrations readily achievable in urine, the dominant effect of trimethoprim overrides that of the sulphonamide (Table 10.4).

In the case of rifampicin and nalidixic acid the situation is quite different. When these agents were tested in the bladder model, both showed high intrinsic activity in suppressing bacterial growth, but resistant bacteria readily emerged [18] [19]. When these agents were used in combination resistance was not evoked and advantage could be taken of the high intrinsic activity of the compounds to obtain long-term suppression of bacterial growth with much lower concentrations of the two components of the mixture (Table 10.5).

These results suggest that a good deal of caution is needed in interpreting synergy in therapeutic terms. The synergic interaction between trimethoprim and sulphamethoxazole is so marked *in vitro* that the possibility of using trimethoprim alone was at first neglected, although in urinary tract infection it is probably as effective alone as in combination [20]. Conversely, one should not overlook potentially useful combinations such as rifampicin and nalidixic acid merely because their interaction is not readily detected in conventional *in vitro* tests of synergy.

Table 10.4 Comparison of the ability of cultures of Escherichia coli to recover following exposure to sulphamethoxazole and/or trimethoprim for 8 hours in the bladder model.

Antibacterial agent	*Concentration (μg per ml) in infusion*	*Time (hours) to regrowth after switch to drug-free broth*
SMX	5	1.5
TMP	0.25	1.5
SMX + TMP	5 + 0.25	3
SMX	100	3
TMP	5	6
SMX + TMP	100 + 5	6

SMX = sulphamethoxazole; TMP = trimethoprim.

Table 10.5 Suppression of bacterial growth in the bladder model by nalidixic acid and rifampicin, alone and in combination.

Antibacterial agent	*Plateau level (μg per ml)*	*Time (hours) before growth detectable*
nalidixic acid	50	22
rifampicin	50	12
nalidixic acid	2	
plus	+	> 40
rifampicin	10	

References

[1] Hamilton-Miller, J. M. T., Smith, J. T. and Knox, R. (1964) *Nature*, **201**, 867-868.

[2] Sutherland, R. and Batchelor, F. R. (1964) *Nature*, **201**, 868-869.

[3] Reading, C. and Cole, M. (1977) *Antimicrob. Ag. Chemother.*, **11**, 852-857.

[4] Greenwood, D. and O'Grady, F. (1975) *Chemotherapy*, **21**, 330-341.

[5] Greenwood, D. and O'Grady, F. (1975) *J. Med. Microbiol.*, **8**, 205-208.

[6] Greenwood, D. and O'Grady, F. (1973) *J. Infect. Dis.*, **128**, 791-794.

[7] O'Callaghan, C. H., Sykes, R. B. and Staniforth, S. E. (1976) *Antimicrob. Ag. Chemother.*, **10**, 245-248.

[8] Greenwood, D. and O'Grady, F. (1976) *Antimicrob. Ag. Chemother.*, **10**, 249-252.

[9] Waterworth, P. M. (1963) *Clin. Med.*, **70**, 941-953.

[10] O'Grady, F. and Greenwood, D. (1973) *J. Med. Microbiol.*, **6**, 441-450.

[11] Greenwood, D. (1976) *J. Antimicrob. Chemother.*, **2**, 87-95.

[12] Greenwood, D. and O'Grady, F. (1976) *J. Clin. Pathol.*, **29**, 162-166.

[13] Greenwood, D. (1977) in: *Antibiotics and Antibiosis in Agriculture* (M. Woodbine, Ed.), Butterworths, London, pp. 82-88.

[14] Kerry, D. W., Hamilton-Miller, J. M. T. and Brumfitt, W. (1975) *J. antimicrob. Chemother.*, **1**, 417-427.

[15] Atlas, E. and Turck, M. (1968) *Amer. J. Med. Sci.*, **256**, 247-254.

[16] McCabe, W. R. and Lorian, V. (1968) *Amer. J. Med. Sci.*, **256**, 255-265.

[17] Greenwood, D. and O'Grady, F. (1978) *J. Antimicrob. Chemother.*, **4**, 113-120.

[18] Greenwood, D. and O'Grady, F. (1977) *Antimicrob. Ag. Chemother.*, **12**, 678-681.

[19] Greenwood, D. and Andrew, J. (1978), *J. Antimicrob. Chemother.*, **4**, 533-538.

[20] Brumfitt, W., Franklin, I., Hayek, L. J., Leigh, D. A., Pursell, R. and Tsekos, G. R. (1973) in: *Urinary Tract Infection* (Brumfitt, W. and Asscher, A. W. eds), Oxford University Press, London, pp. 229-239.

11

Drug combinations in hospital and domiciliary antibacterial therapy

D. S. Reeves, D. W. Bullock

Department of Medical Microbiology, Southmead Hospital, Bristol, England

Introduction

The increasing availability of a variety of antibacterial drugs in the 1950s lead to the introduction of numerous fixed combinations. While this gave convenience in dosing, many combinations were not based on sound pharmacological or microbiological principles which led to a reaction against them. A powerful anti-combination movement grew up in the 1960s culminating in the virtual removal of all combinations from the market in the U.S. under pressure from the F. and D. A. As with many shifts of opinion there was perhaps a degree of over-reaction at that time which lingers still in the minds of many academics and some regulatory authorities. This review will examine some aspects of therapy with antibiotic combinations and will attempt to provide a rational basis for their evaluation. Antituberculous therapy is not discussed as it is covered elsewhere in this symposium.

Ideally, antibiotic therapy should be aimed solely at the bacterial pathogen concerned by using a single agent with a narrow spectrum. This will cause the least possible disturbance to the commensal flora and will therefore minimise the risk of superinfection. Narrow spectrum therapy is possible for some infections with well defined causative bacteria, such as those caused by the haemolytic streptococcus Group A. Some infections require treatment with a single broad-spectrum agent. Sometimes neither approach is possible and combined therapy must be used.

There are two basic underlying reasons for combined antibiotic therapy. The first is when two or more antibiotics are given to increase the spectrum of antibacterial activity, and the agents in the combination are not expected to interact. The second is when two antibiotics are given in the expectation that they will interact to give an antibacterial effect which cannot be produced by either alone.

Non-interactive combinations

Serious sepsis of unknown cause. In severe infections the patient usually requires treatment before the identity or antibiotic susceptibility of the causative bacteria are known. Even wide spectrum antibiotics have deficiencies in their spectra which can assume importance in some clinical situations and therefore 2 or more antibiotics are often given for serious infections. Although there is no controlled evidence for the validity of this approach, experience has shown it is difficult to predict the causative bacteria from the clinical circumstances, and it would therefore be a foolhardy clinician who did not try to cover all reasonable possibilities by using a combination. A typical example would be the use of an aminoglycoside (to treat enterobacteria) with metronidazole or clindamycin (to treat anaerobes) as empirical primary therapy in abdominal sepsis, although the recent availability of cefoxitin has provided a satisfactory single alternative treatment]3]. Once the pathogens have been isolated and their sensitivities determined a change to a more directed antibiotic regime should be contemplated.

Infections with mixed bacteria. A second main indication for non-interactive combined antibiotic therapy is the existence of a mixed infection where the sensitivities of the individual bacteria are not covered by a single agent, or where there are other sites of infection with different bacteria. In these circumstances a combination of antibiotics chosen on the basis of sensitivity testing may be prescribed. As with the treatment of serious sepsis of unknown cause, no interaction between the agents is anticipated and they must therefore be given in full dosage as independent agents. Mixed infections are often found in lesions when there is dead tissue, or loculated pus, and the importance of surgical debridment and drainage should not be forgotten. In the absence of specific laboratory tests to exclude it, antagonistic activity between components of combinations cannot be excluded. Although much was made of this possibility in the 1960s during the reaction against combined therapy, except in a few well known instances, the clinical importance of antagonism has not been demonstrated.

Interactive combinations

Combined therapy using two antibiotics with the prospect of obtaining a favourable interaction is given for three main reasons: to delay the emergence of resistant bacteria; to give a bactericidal effect; to give enhanced activity by bacteristatic synergy.

Delayed emergence of resistance. Outside antituberculous therapy, combined therapy has been suggested as a method delaying the emergence of resistance. Table 11.1 shows that of the 6 suggested combinations only in one instance is there formal evidence for the suggestion that combined therapy delays the emergence of resistance outside the laboratory. Many of the recommendations are based on laboratory data using artificial conditions not found in the body. In treating urinary infections Acar and Brisset [1] showed that the use of rifampicin alone gave poor results with the bacterial pathogen becoming resistant to rifampicin in most of the failures of therapy. The addition of a second antibiotic improved the efficacy of the therapy and the incidence of rifampicin-resistant bacteria on failure was minimal. The lack of clinical evidence for the other combinations does not exclude their validity in preventing resistance emerging, but it does place the onus of proof on those who advocate them. Clinical trials of this nature are difficult to design and complete but nevertheless the combinations given in Table 11.1 are best considered as unproven.

Bactericidal therapy. There are few indications for mandatory bactericidal antibiotic therapy. They include conditions where the local

Table 11.1 Possible combinations to delay emergence of resistance (non-TB).

Staphylococcal infection	Fusidic acid and penicillin	? Clinical proof
	Erythromycin and a penicillin	
Pseudomonal infection	Aminoglycoside and carbenicillin or ticarcillin or azlocillin	no proof from trial
Meningococcal carriage	Rifampicin and minocycline	
Various infections	Trimethoprim and sulphonamide	no proof from trial
Urinary infections	Rifampicin and another agent	probably valid

body defences are impaired, such as in the depth of an endocarditic lesion or around an implanted prosthesis, or when the immune mechanisms are generally impaired, as in profound neutropenia or during immunosuppressive therapy. Bacterial endocarditis is one of the few conditions where the use of bacteristatic agents has been shown to give inferior results to bactericidal agents. In enterococcal endocarditis it is usually necessary to add an aminoglycoside to penicillin therapy in order to achieve a bactericidal effect using a feasible penicillin dosage. The aminoglycoside appears to increase the permeability of the bacterial cell to penicillin [15]. The synergistic effect is not necessarily present with all aminoglycosides with each strain and it is therefore important to check for the presence of bactericidal synergy by appropriate laboratory tests. Some clinicians combine penicillin with an aminoglycoside for treating endocarditis caused by viridans-type streptococci. Provided that satisfactory bactericidal activity can be demonstrated with penicillin alone this practice would seem to expose patients to unnecessarily toxic treatment, particularly elderly patients. In addition to demonstrating satisfactory bactericidal activity by the chosen therapy against the homologous isolate in the laboratory and to measuring satisfactory blood concentrations, a simple bactericidal titre of the patient's serum against the infecting bacterium of 1 in 4 or 8 is a valuable indicator of a successful outcome [8]. In staphylococcal endocarditis it is customary to use a combination of two agents, such as an isoxazoyl penicillin with an aminoglycoside or fusidic acid, but the necessity for this has yet to be demonstrated by clinical trial.

In neutropenic patients claims have been made that combined therapy gives superior clinical results to single treatment with an aminoglycoside. The contention is that, against *Ps. aeruginosa* in particular, aminoglycosides are not fully bactericidal in the absence of neutrophils [5]. Clearly enhanced bacteristatic activity cannot also be excluded from consideration. A further justification for using a combination is the fuller coverage of possible pathogens in a condition where early effective therapy may be critical to the outcome. In animals Andriole [2] demonstrated greater clinical efficacy by using combinations than by using single antibiotics alone. Early data by Schimpff et al. [24] and Rodriguez et al. [21] suggested an improved efficacy in an open, uncontrolled trial design. Table 11.2 shows data compiled from Klastersky et al [9]. There is a statistically significant difference between the results of combined carbenicillin/gentamicin over the results with either drug alone in a group of cancer patients infected with a variety of bacteria. Apart from the ethical problems of continuing therapy once the isolation of bacterium showing in vitro resistance to it, the study has been criticised because of the heterogeneity of the patients and their infections in the relatively

Table 11.2 Therapy of serious infections from Gram negative rods [9].

	Gentamicin + carbenicillin	*Gentamicin*	*Carbenicillin*
Patients treated	23	23	22
Successes	19	13	11
Deaths	3	2	5
Resistant isolates	2	2	5
Successes from patients with sens. isolates	19/21	12/21	(N.S.) 11/17
Peak Mean Bactericidal Dilution	1/8	1/4	1/4
Peak Mean Inhibitory Dilution	1/16	1/4	1/8

small groups treated. In spite of full doses of the antibiotics being used, presumably producing supra-inhibitory blood concentrations, there was an association of higher mean peak inhibitory and cidal titres of patients sera for the homologous bacterium with combined as opposed to single therapy. There is a need for clinical studies in this field so as to weigh possible benefits against the disadvantages of combined therapy. It should also be remembered that many deaths in neutropenic and immunosuppressed patients may occur from infections not treatable with antibacterial drugs.

Meningitis and osteomyelitis have been suggested as requiring treatment with combinations, presumably to increase spectrum of activity as well as to perhaps gain advantage from bactericidal interactions. There is no evidence that this approach is beneficial and it may even be harmful (see below).

Increased antibacterial activity. Rather than seeking a specific bactericidal effect, two antibiotics may be given together so as to produce increased antibacterial activity not achieved by giving either drug alone in feasible doses. It might thus be possible to reduce the doses of the component antibiotic, reducing dose-related toxicity. An outdated example of this is the triple combination of tri-sulphapyrimidines, with which the possibility of crystalluria is reduced without reducing antibacterial activity. More recently it has been suggested that combining flucytosine with amphotericin B might allow lower doses of amphotericin B to be effective.

Interactive combinations may also allow infection to be treated which would otherwise be inaccessible at tolerable doses. Many urinary pathogens are resistant to broad spectrum penicillins (ampicillin, carbenicillin) because of their ability to synthesise β-lactamase enzymes which destroy these penicillins. Isoxazoyl penicillins are virtually inactive against enterobacteria, but are resistant to destruction by many enterobacterial β-lactamases and at high concentrations can block the activity of these enzymes against ampicillin. Because of the high concentrations needed the urinary tract is the only site where the possible interaction could be exploited, and Sabath et al [22] reported some therapeutic success. This has been questioned by others [20], and in any case the combinations were superceded by more satisfactory single agents, such as the oral cephalosporins. The recent discovery of the β-lactamase inhibitor clavulanic acid which is effective at low concentrations has again opened up the therapeutic possibilities for combinations with β-lactam molecules [16], [7], [18].

The causative bacterium of gonorrhoea, *N. gonorrhoeae*, is inherently resistant to trimethoprim, and many strains display acquired resistance to sulphonamide. Treatment with either agent alone would therefore be expected to give low cure rates. Synergistic activity between trimethoprim and a sulphonamide for *N. gonorrhoeae* is regularly demonstrable *in vitro*. Csonka and Knight [6] treated patients with gonorrhoea with the agents alone and with 3 different combinations of trimethoprim and sulphatriad. There was a strong suggestion of *in vivo* synergy, although further treatment groups with higher doses of the individual drugs alone would have been needed to demonstrate synergy conclusively.

Finally, there is the argument that, in spite of achieving supra-inhibitory concentrations of the individual agents in the blood, combinations displaying additive or synergistic activity must have inherent superiority. Although there is little clinical proof that this is true for the majority of infections, there is logic in exploiting interactive combinations in situations where the concentrations of the individual drugs with regard to the required inhibitory concentrations are marginal and may even be below them. This might be true for the components of cotrimoxazole in the sputum or in the acini of the prostate, but not, for example, in the urine.

Possible disadvantages of combined therapy

Possible disadvantages of combined therapy are listed in Table 11.3, but not in order of importance.

1. Antagonism between agents might occur unknown to the

Table 11.3 Possible disadvantages of combined therapy

1	Undetected antagonism may occur.
2	Chemical interaction may inactivate one or both components.
3	Increased frequency of adverse reactions.
4	Costs are higher unless doses reduced.
5	Increased frequency of superinfection may occur.
6	False sense of security in user.
7	Fixed combinations not always in satisfactory ratios.

clinician, and therapy might therefore be inferior to the use of one of the agents alone. Although this is a theoretical possibility its importance was probably over rated at the time of the reaction against combinations. No doubt minor degrees of antagonism between various combinations can be sometimes demonstrated in the laboratory, but outside certain recognised conditions there is not evidence that it is important clinically. Nevertheless, in serious infection, it would be as well to check the bactericidal activity of the patients serum against the homologous isolate to obviate any problems arising from antagonism. In meningitis first Leper and Dowling [10] and then Mathies et al [13] showed that combination therapy may be inferior to treatment with a single agent. This is supported by the antagonism shown between gentamicin and chloramphenicol used to treat a Proteus meningitis in neutropenic mice [23]. There is some suggestion that in vitro antagonism may play a role in determining the outcome in pyelonephritis [14] (Table 11.4).

2. Chemical interactions inactivating antibiotics can occur in mixtures prepared before administration, but given the correct

Table 11.4 Combined therapy in pyelonephritis [14].

		In vitro results		
		Synergy	Additive	Antagonism
clinical	cure	7	16	1
	relapse	1	8	6

(Overall, p. < 0.01).

knowledge this should be avoidable. Gentamicin undergoes significant inactivation by carbenicillin in patients with renal failure [19], which means that blood concentrations of the former will need monitoring to ensure adequacy.

3. Giving more than a single agent will clearly increase the frequency of adverse reactions, unless perhaps the individual dosage of components is reduced. Interactions causing adverse reactions have been reviewed elsewhere in this symposium.
4. Unless lower doses can be used as a result of combined therapy, drug costs will be greater.
5. Combined therapy may well give a wider spectrum of activity which can in turn produce a more profound disturbance of the commensal body flora, encouraging colonisation with antibiotic-resistant bacteria and occasionally superinfection. There may be epidemiological consequences as a result of spreading such bacterial strains to the environment and to other patients.
6. Using broad-spectrum combinations can engender a false sense of security in clinicians. This is typified by surgeons who speak of operating under "a cover" or "an umbrella", little realising that without proper aseptic techniques these objects may prove not to be leak-proof against multiply-resistant bacteria. Combination may thus foster an uncritical approach to chemotherapy which should always be tailored to the needs of individual patients. The false sense of security is emphasised by the occurrence of meningitis in patients on cephalothin [12]. The use of apparently adequate antibiotic therapy can lead to diagnostic delay.
7. Finally there is the problem of fixed combinations. While these provide convenient dosage forms the combination available may not be in the most suitable ratio for a particular therapeutic situation. In the U K cotrimoxazole is the only interactive fixed combination used on a wide scale. Other non-interactive fixed combinations (amoxycillin/flucloxacillin; penicillin/streptomycin) are used by some practitioners as general antibacterial agents, but except in certain circumstances their use cannot be recommented.

Having discussed some of the general principles of combination therapy it would be useful to examine an interactive combination in order to ascertain how it conforms to those principles.

Cotrimoxazole

Cotrimoxazole is a fixed 1 : 5 combination of trimethoprim with sulphamethoxazole which has been available in the U K for some 9 years. In this time cotrimoxazole has established a reputation as a valuable combination for treating especially lower respiratory and urinary infections. Although it has not been provenby clinical trial, the regular occurrence of synergy between the components against respiratory bacterial pathogens makes it logical to use the combination for such infections to gain the advantage offered, especially as concentrations in bronchial secretions may be below those found in the blood. In the urine, however, where high concentrations of both agents are found the arguments for using the combination are not proven.

It has been questioned as to whether trimethoprim alone would not be as satisfactory a therapy as the combination for most urinary infections [4]. Trimethoprim alone has been used for some years in Finland without apparent disadvantage. Some adverse reactions to cotrimoxazole are due to the sulphonamide component and these would be avoided. Patients with sulphonamide allergy could be treated with trimethoprim which is not possible at present. Furthermore, on the basis of *in vitro* experiments using urine as culture medium, the advantage of adding sulphonamide to trimethoprim has been questioned [11]. Certainly bactericidal activity is absent or slow in the presence of both drugs in urine with many bacterial strains, and the rate seems to be independent of the supramaximal concentration achieved by various dosing regimes [17].

For most urinary infections the U K recommended dosage is probably too great and one tablet (80 mg TMP; 400 mg SMZ) 12-hourly would often suffice. This regime is used in Scandinavia. With a loading dose it is also the recommended dosage for use with the combinations of trimethoprim with sulphamoxole, apparently without therapeutic disadvantage in urinary infections. The lack of a loading dose is a disadvantage in tissue infections since, as whould be expected from the plasma half life of the components and the dosing interval, it takes some 5-6 doses before steady state plasma concentrations are achieved. These steady state concentrations are about twice as high as those after the first dose, which is too long to wait for in an ill patient.

The availability of ratios other than 1 : 5 would be an advantage in treating infections in some sites or with some pathogens. Because sulphamethoxazole diffuses into tissue and fluid less well than trimethoprim the optimal ratio of 1 : 20 found in the blood is disturbed. In treating lower respiratory infections and prostate infections more sulphonamide should be used, whereas infections with the gonococcus require a higher proportion of trimethoprim for optimal therapy. While a multiplicity of ratios might be confusing and impracticable for using in general practice, the single components should be available as supplements to the fixed combination for use in hospitals.

Finally, it might be questioned as to whether sulphamethoxazole is the ideal sulphonamide to combine with trimethoprim. The ostensible reason for chosing sulphamethoxazole was its similar plasma half-life to trimethoprim allowing the optimal plasma ratio of the components to be maintained without complex dosing arrangements. Sulphamethoxazole, although one of the most active sulphonamides, is highly protein-bound which restricts its ingress into the tissues. Other sulphonamides, for example sulphadiazine, are just as active, have lower protein binding, and a plasma half-life which is not too dissimilar to that of trimethoprim. This last point is probably not of great importance as the necessity of maintaining the plasma concentration ratio within narrow limits has not been demonstrated. On theoretical grounds sulphamethoxazole is not the ideal sulphonamide for urinary infections since it is more extensively metabolised than many others and thus a lower proportion of the administered dose reaches the urine in an active form.

Conclusions

Ever since clinicians have had access to more than a single antibacterial drug, combinations of them have been used for treating infections. Although there has been a strong reaction against combined therapy, and against fixed combinations in particular, no rational therapist would deny that combined therapy has a place and this is likely to remain so (until the unlikely discovery of the perfect single agent). Combined therapy with two or more antibiotics not intended to interact but just to provide a wider cover is logical and, provided it is not used unthinkingly, a valid treatment for serious infections. When combined therapy is given with the intention of the components interacting then strict criteria should be applied to ensure that the advantage sought is at least likely to be obtained and that any possible adverse effects should not outweigh it. In this review more questions about the validity of combinations have been raised than it is possible to answer, demonstrating our incomplete knowledge. It may not be practicable or ethical to answer all these questions by controlled clinical trials but blind acceptance of the applicability of laboratory investigation should be avoided. On the other hand the use of combinations should not be rejected for lack of clinical evidence if there are strong indicators for their use based on *in vitro* data. What is needed is an unemotional and rational examination of the interactive combinations presently in use and of those offered for appraisal. It is important to ensure that they will be of benefit to the patient and that they are not used to provide psychological security to the physician.

References

[1] Acar, J. F., and Brisset, J. M. (1975) *Combinations of Antibiotics in Urinary Tract Infections. Clinical Use of Combinations of Antibiotics*, ed. J. Klastersky, Hodder and Stoughton, London, pp. 126-134.

[2] Andriole, V. T. (1974) "Antibiotic synergy in experimental infection with *Pseudomonas*. II. The effect of carbenicillin, cephalothin or cephanone combined with tobramycin or gentamicin". *J. infect. Dis.*, **129**, 124.

[3] Bint, A. J., Holt, H. A., Reeves, D. S., and Stocks, P. J. (1978) "Clinical experience with cefoxitin", *Current Chemotherapy*, eds.) Siegenthaler and Lüthy, American Soc. for Microbiology, p. 783.

[4] Brumfitt, W., Franklin, I.S.N., Hayek, L. J., Leigh, D. A., Pursell, R., and Tsekos, G. R. (1973) "A comparative study of the treatment of urinary infection in domiciliary practice with ampicillin, cephalexin, co-trimoxazole and trimethoprim". *Urinary Tract Infection*, Brumfitt W., and Asscher A. W., eds. Oxford University Press, London, p. 229.

[5] Bodey, G. P., and Rodriguez, V. (1973) "Advances in the management of *Pseudomonas aeruginosa* infections in cancer patients", *Euro. J. Canc.*, **9**, 435-441.

[6] Csonka, G. W., and Knight, G. J. (1967) "Therapeutic trial of trimethoprim as a potentiator of sulphonamides in gonorrhoea", *Brit. J. vener. Dis.*, **43**, 161-165.

[7] Hunter, P., Reading, C., and Witting, D. A. (1977) "The in vitro and in vivo properties of BRL-14151, a novel β-lactam with β-lactamase inhibitory properties", Abstract 238, *16th Intersc. Conference Antimicrob. Ag. Chemother.*, Chicago.

[8] Jawetz, E. (1962) "Assay of antibacterial activity in serum", *Amer. J. Dis. Children*, **103**, 81-84.

[9] Klasterwky, J., Cappel, R. and Daneau, D. (1973) "Therapy with carbenicillin and gentamicin for patients with cancer and severe infections caused by gram-negative rods", *Cancer*, **31**, 331-336.

[10] Lepper, M. H., and Dowlig, H. F. (1951) "Treatment of pneumococci meningitis with penicillin compared with penicillin plus aureomycin", *Arch. int. Med.*, **88**, 489.

[11] Lewis, E. L., Anderson, J. D., and Lacey, R. W. (1974) "A reappraisal of the antibacterial action of cotrimoxazole *in vitro*", *J. clin. Pathol.*, **21**, 87-91.

[12] Mangi, R. J., Kundargi, R. S., Quintiliani, R., and Andriole, V. T. (1973) "Development of meningitis during cephalothin therapy", *Ann. int. Med.*, **78**, 347-351.

[13] Mathies, A. W., Leedom, J. M., Ivler, D., Wehrle, P. F. and Portnery, B. (1968) "Antibiotic antagonism in bacterial meningitis", *Antimicr. Ag. Chemother. - 1967*, 218-224.

[14] McCabe, W. R., and Jackson, G. G. (1965) "The treatment of pyelonephritis: Bacterial, drug and host factors in success or failure among 252 patients", *New Engl. J. Med.*, **272**, 1037.

[15] Moellering, R. C. and Weinberg A. N. (1971) "Studies on antibiotic synergism against enterococci. II. Effect of various antibiotics on the uptake of ^{14}C-labelled streptomycin by enterococci", *J. clin. Invest.*, **50**, 2580.

[16] Reading C., and Cole, M. (1977) "Clavulanic acid: a β-lactamase-inhibitory β-lactam from *Streptomyces clavuligerus*", *Antimicr. Ag. Chemother.*, **11**, 852-857.

[17] Reeves, D. S., Broughall, J. M., Bywater, M. J., and Holt, H. A. (1978) "Comparative antibacterial activity of two trimethoprim-sulphonamide combination in human urine", *Current Chemotherapy*, eds. Siegenthaler and Lüthy, Amer. Soc. for Microbiolgy, p. 205.

[18] Reeves, D. S., Bywater, M. J., and Holt, H. A. (1978) "Antibacterial synergism between β-lactam antibiotics: results using clavulanic acid (BRL, 14151) with amoxycillin, carbenicillin or cephaloridine", *Infection* (in press).

[19] Riff, L. J. and Jackson, G. G. (1972). "Laboratory and clinical conditions for gentamicin inactivation by carbenicillin", *Arch. int. Med.*, **130**, 887-891.

[20] Riff, L. J., Olexy, V. M., and Jackson, G. G. (1970) "Therapy with combinations of penicillins analogues in urinary-tract infections", *Antimicr. Ag. Chemother.*, 1969, 405-409.

[21] Rodriguez, V., Whitecar, J. P., and Bodey, G. P. (1970) "Therapy of infection with the combination of carbenicillin and gentamicin", *Antimicr. Ag. Chemother.*, 1969, 386-390.

[22] Sabath, L. D., McCall, C. E., Steigbigel, N. H., and Finland, M. (1967) "Synergistic penicillin combinations for treatment of human urinary-tract infections", *Antimicr. Ag. Chemother.*, 1966, 149-155.

[23] Sande, M. A., and Overton, J. W. (1973) "*In vivo* antagonism between gentamicin and chloramphenicol in neutropenic mice", *J. infect. Dis.*, **128**, 247-250.

[24] Schimpff, S., Satterlu, W., Young, V. M., and Serpick, A. (1971). "Empiric therapy with carbenicillin and gentamicin for febrile patients with cancer and granulocytopenia", *New Engl. J. Med.*, **284**, 1061-1065.

12

Factors influencing the use of antibiotic combinations

J. D. Williams, F. A. Drasar

Department of Medical Microbiology,
The London Hospital Medical College, London, England

Introduction

Laboratory tests may show that two or more antibiotics exert a synergistic, inhibitory or bactericidal effect against bacteria. However, it is sometimes difficult to decide whether or not the results of laboratory tests have any relevance to clinical practice. The problems which require clarification are, firstly, whether or not the laboratory test accurately reflects the activity which the antibiotics(s) have *in vivo*; secondly, if it is shown that the tests correspond to the *in vivo* effects, whether or not the increase in susceptibility demonstrable in resistant bacteria brings the susceptibility into a useful therapeutic range; thirdly, it is necessary to justify the use of multiple antimicrobial agents for treatment of infections which would respond satisfactorily to a single agent. In this paper we examine some of the suggested indications for the use of antibiotic combinations and consider whether they have proven or possible clinical relevance.

Indications for the use of combinations of antibiotics

Several arguments have been advanced to support the use of antibiotic combinations. Some of these are entirely justifiable, e.g. in the treatment of infections in which several organisms with different antibiotic susceptibilities are, or may be, involved and no single agent is suitable, or as 'broad-spectrum' antibiotic cover for critically ill patients in whom the aetiological agent is, as yet, unknown. Antibiotic combinations have also been used in order to allow reduction in dosages of potentially toxic antibiotics or in attempts to prevent the emergence of resistance which could theoretically occur if a single antibiotic were used. Synergy is not an essential component in these situations but is, however, of prime importance when combinations of antibiotics are used in order to achieve an increased killing effect against an infecting organism.

Laboratory methods for the demonstration of synergy

There are several methods which may demonstrate synergy *in vitro*. These include simple methods such as the use of antibiotic-containing paper strips and a cellophane tambour [14], [13], or the use of blotting paper discs carefully positioned on plates. More quantitative results can be obtained using the chequer board method by which bacteriostatic and bactericidal effects of two antibiotics singly and in combination in a given range of concentrations are determined in broth [13]. In this method only a minute volume of culture is tested for bactericidal effect. Solid media incorporating antibiotics may also be used to determine the bacteriostatic effect of combinations.

Quantitative methods which produce a numerical bactericidal or bacteriostatic value allow an isobologram to be drawn which indicates from the shape of the curve whether that combination gives a synergistic, additive or antagonistic effect [27], [13].

Fractional Inhibitory Concentrations (FICs) of the individual antibiotic combinations [22] may also be plotted as an isobologram.

There are verying opinions on what constitutes an acceptable degree of microbiological synergy. A threefold reduction in MIC is taken as synergy by some workers [22] while others require a fourfold reduction in the concentration of each antibiotic [38].

It is also necessary to distinguish between microbiological synergy and clinically obtainable synergy [11] especially when organisms being examined may be resistant to one or both of the antibiotics under test. For clinically useful synergy to occur antibacterial effects should be present at concentrations of antibiotics which can be readily obtained in the serum or other biological fluid of the patient undergoing treatment.

The use of laboratory techniques has led to the production of a voluminous literature on *in vitro* synergy between combinations of antibiotics. Table 12.1 lists some of the combinations of antibiotics which have been most widely studied. The mechanisms of the interaction of the antibiotics which may give rise to *in vitro* synergy are also very varied and have been widely studied (Table 12.2).

The significance of these laboratory results in the clinical situation is generally unclear. Most available information on the benificial effects of antibiotic combinations is anecdotal and not

Table 12.1 Some antibiotic concentrations showing synergistic activity in vitro.

Antibiotic combination	*Bacteria*	*Reference*
Aminoglycosides + carbenicillin	*Ps. aeruginosa* Gram-negative rods	[26], [23]
Netilmicin + carbenicillin	Gram-negative rods	[12]
Tobramycin/gentamicin + cephalothin	Gram-negative rods *Klebsiella* spp.	[12], [8]
Aminoglycosides + cephalosporins	*Proteus* and *Providencia* spp.	[17]
Amikacin + cefoxitin	*Serratia marcescens*	[28]
Streptomycin + penicillin	Viridans streptococci	[18], [29]
Gentamicin + penicillin	Enterococci	[39]
Aminoglycosides + amoxycillin	Group D streptococci and viridans streptococci	[3]
Amikacin + trimethoprim	Gram-negative rods *Ps. aeruginosa*	[33]
Trimethoprim + sulphamethoxazole	Most bacteria	[40]
Trimethoprim + rifampicin	Gram-negative rods and streptococci	[22], [11], [15]
Cotrimoxazole + polymixins	Gram-negative rods	[36]
Rifampicin + novobiocin	*S. typhi*	[37]
Rifampicin + naficillin/vancomycin	*S. aureus*	[41]
Erythromycin + fusidic acid	*S. aureus*	[27]
Mecillinam + β-lactams	Gram-negative rods	[30], [38]
Clavulanic acid + β-lactams	Gram-negative rods *Ps. aeruginosa* *Bacteroides spp.*	[43]
Clavulanic acid + ampicillin/cephalothin	*Bacteroides* spp.	[44]
Fosfomycin + ampicillin/chloramphenicol	*Salmonella* sp. *Shigella* sp.	[34]

Table 12.2 Methods by which synergy may occur.

Sequential blockade of a biochemical pathway	Trimethoprim - sulphamethoxazole - folate metabolism
Similar action at same site in bacterium	Mecillinam - ampicillin - cell wall
Alteration at one site allowing increased activity of another antibiotic	Penicillins or cephalosporins + aminoglycosides — increased permeability of cell wall to aminoglycoside by β-lactam antibiotics
Inhibition of β-lactamases produced by bacteria and some antibacterial effect	Cloxacillin or clavulanic acid + penicillins

related to results obtained from clinical trials. However, in the case of streptococcal endocarditis there appears to be a significant correlation between the combination of antibiotics used, the *in vitro* synergistic effect demonstrated by such a combination against the infecting organism and the cure rate in patients. An accepted antibiotic combination for *Strep. viridans* endocarditis is streptomycin and benzyl penicillin. The need to add streptomycin in viridans streptococcal endocarditis remains unconfirmed by comparative studies in man [42] but is widely used by experienced cardiologists. Because of the high resistance of enterococci to streptomycin, gentamicin is preferred by some cardiologists [32]. Certainly the combination of penicillin and gentamicin gave improved results in animals with endocarditis induced by a highly streptomycin-resistant enterococcus [6].

Toxicity of antibiotic combinations

The use of combined antibiotic therapy to give reduced toxicity has been mentioned earlier. However, the use of certain antibiotic combinations may lead to increased toxicity. Combination of cephalosporins and aminoglycosides may be given to achieve broad-spectrum antibacterial effect in patients with severe undiagnosed infection but in man this combination may be more nephrotoxic than either antibiotic given alone [24], [4]. In animals this effect is reversed as cephaloridine, cephalothin and cephazolin appear to protect rat kidneys against tubular damage by gentamicin [9], [25]. The reasons for the protective action of the combination in animals but increased toxicity in man have not been explained.

Pharmacological problems with antibiotic combinations

Despite the demonstration of a synergistic effect between two antibiotics at given concentrations of two drugs when they are administered together to a patient they may have differing pharmacokinetics creating a diversion from the *in vitro* ideal situation. This causes prescribing difficulties and leads to complex dose regimens to ensure that both drugs are present in the serum in the appropriate bactericidal concentration at the same time. There are also the possibilities that one drug may alter the excretion of the other, may adversely affect the other during administration or disease states especially of the liver or kidneys may affect one drug more than the other.

Two combinations which have been extensively studied are

sulphamethoxazole with trimethoprim and trimethoprim with rifampicin. Sulphamethoxazole was chosen to partner trimethoprim because its pharmacokinetic behaviour was similar to that of trimethoprim [5]. (It does not necessarily follow that sulphamethoxazole is an ideal partner for trimethoprim on microbiological grounds.) The pharmacokinetics of sulphamethoxazole and trimethoprim are similar when given in combination and when given separately [21]. The adsorption of sulphamethoxazole in combination is slightly lower than that when given alone but the extent of absorption is the same. However, when there is a disturbance of renal function the half-life of trimethoprim rises sharply at creatinine clearance less than 30 ml/min whereas the half-life of unconjugated sulphamethoxazole remains unchanged [35].

Several investigations have been carried out in volunteers receiving trimethoprim and rifampicin [16], [1], [10]. Using differing dose regimens all found the plasma half-life of rifampicin was significantly reduced when both drugs were given for a week, probably due to changes in biliary excretion. Emmerson, Grüneberg and Johnson [10] also found that the urinary clearance of trimethoprim from serum was significantly increased after 8 days. Clearance of rifampicin was unchanged.

Use of combined therapy to prevent emergence of resistance

There are several antibiotics including streptomycin, rifampicin and fusidic acid, to which bacteria develop resistance rapidly *in vitro* and *in vivo*. In order to minimise this occurrence it has been suggested that some antibiotics should not be used as a single agent but always in combination with another antibiotic [2].

Streptomycin is mainly now used for anti-tuberculous therapy when it is combined with other anti-tuberculous drugs. However, if streptomycin and isoniazid are being used to treat a patient who is able to acetylate isoniazid rapidly the effective treatment for a large proportion of therapy may be equivalent to using streptomycin alone.

The use of fusidic acid is exclusively directed against *Staph. aureus*. Fusidic acid resistant mutants of staphylococci are selected with great ease when this antibiotic is used alone and so it is therefore often combined with another anti-staphylococcal drug, e.g. cloxacillin, erythromycin or lincomycin, to try and prevent this occurrence. Most patients receiving fusidic acid, however, do not receive two antibiotics and the level of fusidic acid resistant staphylococci remains low except when extensive topical therapy has been carried out.

The combination of cloxacillin and fusidic acid has been shown to be effective in the suppression of fusidic acid resistant mutants [2] but *in vitro* the combination has been demonstrated to be antagonistic [31]. *In vivo* the antagonism seems to be insignificant as good recovery, even in staphylococcal pneumonia, has been reported in patients receiving this combination [20].

Fusidic acid which is concentrated in bone is widely used in treating staphylococcal osteomyelitis, sometimes in combination with other agents, e.g. erythromycin [27]. The isobolograms made using these isolates showed that in a high fusidic acid/erythromycin ratio synergy was obtained but in a high erythromycin/fusidic acid ratio the combination was antagonistic. In bone the former ratio would be present and presumably synergy would also be found but the complexity of the laboratory findings underlies the difficulty of translating *in vitro* results to the possible therapeutic effects in ill patients.

Fusidic acid and rifampicin have also been used in combination to treat staphylococcal infection [19] which was successful in 33 out of 38 patients. Rifampicin resistant staphylococci emerged in 2 patients.

Resistance is most likely to arise *in vivo* when a prolonged course of an antibiotic is necessary to eradicate the organism; staphylococcal osteomyelitis or endocarditis are two infections where fusidic acid resistance has been seen to emerge during treatment and we would not use fusidic acid as a single agent where more than 5-7 days treatment was envisaged.

Conclusions

Although there are various reasons why the use of antibiotic combinations has become widespread, many of these uses are based upon relatively insubstantial findings. Often the use of a mixture of agents is preferred to the discipline of making a precise clinical or laboratory diagnosis of the infection. We have indicated some of the possible problems which may be encountered in the use of combined therapy and which have to be taken into account when combination therapy is considered.

References

[1] Accocella, G. and Scotti, R. (1976) "Kinetic studies on the combination rifampicin - trimethoprim in man", *J. Antimicr. Chemo.*, **2**, 271.

[2] Barber, M. and Waterworth, P. M. (1962) "Antibacterial activity *in vitro* of fucidin", *Lancet*, **i**, 931.
[3] Basker, M. J. and Sutherland, R. (1977) "Activity of amoxycillin alone and in combination with aminoglycoside antibiotics against streptococci associated with bacterial endocarditis", *J. Antimicr. Chemo.*, **3**, 273.
[4] Bobrow, S. N., Jeffe, E. and Young, R. C. (1972) "Anuria and acute tubular necrosis associated with gentamicin and cephalothin", *J. Am. Med. Ass.*, **222**, 1456.
[5] Busby, S. R. M. (1969) "Combined antibacterial action *in vitro* of trimethoprim and sulphonamides. The *in vitro* nature of synergy", *Postgrad. Med. J.*, **45** (Suppl.), 10.
[6] Carrisoza, J. and Kaye, D. (1976) "Antibiotic synergism in enterococcal endocarditis", *J. Lab. Clin. Med.*, **88**, 132.
[7] Crawford, K. and Abraham, E. P. (1957) "Synergistic action of cephalosporin C and benzyl penicillin against a penicillinase-producing strain of *Staphylococcus aureus*", *J. Gen. Microbiology*, **16**, 604.
[8] D'Allesandri, R. M., McNeely, D. J. and Kluge, R. M. (1976) "Antibiotic synergy and antagonism against clinical isolates of *Klebsiella* species", *Antimicr. Ag. Chemother.*, **10**, 889.
[9] Dellinger, P., Murphy, T., Pinn, V., Barza, M. and Weinstein, L. (1976) "Protective effect of cephalothin against gentamicin induced nephrotoxicity in rats", *Antimicr. Ag. Chemother.*, **9**, 172.
[10] Emmerson, A. M., Grüneberg, R. N. and Johnson, E. S. (1978) "Pharmacokinetics in man of a combination of rifampicin and trimethoprim.", *J. Antimicr. Chemo.* (in press).
[11] Farrell, W., Wilks, M. and Drasar, F. A. (1977) "The action of trimethoprim and rifampicin in combination against Gram-negative rods resistant to gentamicin", *J. Antimicr. Chemo.*, **3**, 459.
[12] Fu, K. P. and Neu, H. C. (1976) "*In vitro* synergistic effect of netilmicin, a new aminoglycoside antibiotic", *Antimicr. Ag. Chemother.*, **10**, 511.
[13] Garrod, L. P., Lambert, H. P. and O'Grady, F. (1973) *Antibiotic and Chemotherapy*, 4th edition, Churchill-Livingstone.
[14] Garrod, L. P. and Waterworth, P. M. (1962) "Methods of testing combined antibiotic bactericidal action and the significance of the results", *J. Clin. Path.*, **15**, 328.
[15] Grüneberg, R. N. and Emmerson, A. M. (1977 "The interaction between rifampicin and trimethoprim: an *in vitro* study", *J. Antimicr. Chemo.*, **3**, 453.
[16] Hamilton-Miller, J. M. T. and Brumfitt, W. (1976) "Trimethoprim and rifampicin; pharmacokinetic studies in men", *J. Antimicr. Chemo.*, **2**, 181.
[17] Hyams, P. K., Simberkoff, M. S. and Rahal, J. J. (1974) "Synergy between cephalosporin and aminoglycoside antibiotics against Providencia and Proteus", *Antimicr. Ag. Chemother.*, **5**, 571.
[18] Jawetz, E. and Gunnison, J. B. (1950) "The determination of sensitivity to penicillin and streptomycin of enterococci and streptococci of the viridans group", *J. Lab. Clin. Med.*, **35**, 488.
[19] Jansen, K. (1968) "Methicillin resistant staphylococci", *Lancet*, **ii**, 1078.
[20] Jensen, K. and Lassen, H. C. A. (1969) "Combined treatment with antibacterial chemotherapeutic agents in staphylococcal infections", *Quarterly Journal of Med.*, **38**, 91.
[21] Kaplan, S. A., Weinfield, R. E., Abruzzo, C. W., McFaden, K., Jack, M. L. and Weissman, L. (1973) "Pharmacokinetic profile of trimethoprim-sulphamethoxazole in men", *J. Inf. Dis.*, **128**, 5547.
[22] Kerry, D. W., Hamilton-Miller, J. M. T. and Brumfitt, W. (1975) "Trimethoprim and rifampicin. *In vitro* activities separately and in combination", *J. Antimicr. Chemo.*, **1**, 417.

[23] KLASTERSKY, J., HENRI, A. and VANDENBORRE, L. (1973) "Antimicrobial activity of tobramycin and gentamicin used in combination with cephalothin and carbenicillin", *Amer. J. Med. Sc.*, **266**, 13.
[24] KLASTERSKY, J., HENRI, A., HENSGENS, C. and DANEAU, D. (1974) "Gram-negative infections in cancer; study of empiric therapy comparing carbenicillin-cephalothin with and without gentamicin", *J. Am. Med. Ass.*, **227**, 45.
[25] LUFT, R. C., PATEL, V., YUM, M. N. and KLEIT, S. A. (1976) "Nephrotoxicity of cephalosporin-gentamicin combinations in rats", *Antimicr. Ag. Chemother.*, **9**, 831.
[26] MARKS, M. I., HAMMERBERG, S., GREENSTONE, G. and SILVER, B. (1976) "Activity of newer aminoglycosides and carbenicillin alone and in combination against gentamicin-resistant *Pseudomonas aeruginosa*", *Antimicr. Ag. Chemother.*, **10**, 399.
[27] MCALLISTER, T. A. (1974) "Treatment of osteomyelitis", British Journal of Hospital Medicine, p. 535.
[28] MILLER, M. A., CRANE, J. K., GRIFFIN, P. S. and YOUSUF, M. (1978) "Antibiotic synergism and characterisation of R factors in *Serratia marcescens*", *Current Chemotherapy*, **1**, p. 437. Pub. American Society of Microbiology. Eds. W. Siegenthaler and R. Lüthy.
[29] MOELLERING, R. C. and WEINBERG, A. N. (1971) "Studies on antibiotic synergism against enterococci", *J. Clin. Invest.*, **50**, 2580.
[30] NEU, H. C. (1976) "Synergy of mecillinam a beta-amidino penicillanic acid derivative combined with beta-lactam antibiotics", *Antimicr. Ag. Chemother.*, **10**, 535.
[31] O'GRADY, F. and GREENWOOD, D. (1973) "Interactions between fusidic acid and penicillin", *J. Med. Microbiol.*, **6**, 441.
[32] OAKLEY, C. (1978) "Use of antibiotics - Endocarditis", *Brit. Med. J.*, **2**, 489.
[33] PARSLEY, T. L., PROVONCHEE, R. B., GLICKSMAN, C. and ZINNER, S. H. (1977) "Synergistic activity of trimethoprim and amikacin against Gram-negative bacilli", *Antimicr. Ag. Chemother.*, **12**, 349.
[34] PEREA, E. J., TORRES, M. A. and BOROBIO, M. V. (1978) "Synergism between fosfomycin-ampicillin and fosfomycin-chloramphenicol against salmonella and shigella", *Antimicr. Ag. Chemother.*, **13**, 705.
[35] REIDER, J., SCHWARTZ, D. E., FERNEX, M., BERGAN, T., BRODWALL, E. K., BLUMBERG, A., COTTIER, P. and SCHEITLIN, W. (1976) "Pharmacokinetic of the antibacterial combination sulphamethoxazole plus trimethoprim in patients with normal or impaired renal function", *Antibiot. and Chemother.*, **18**, 148.
[36] ROSENBLATT, J. F. and STEWARD, P. R. (1974) "Combined activity of sulphonamides, trimethoprim and polymixin B against Gram-negative bacilli", *Antimicr. Ag. Chemother.*, **6**, 84.
[37] SHANSON, D. C. and LEUNG, T. (1976) "Susceptibility of *S. typhi* to rifampicin and novobiocin", *J. Antimicr. Chemo.*, **2**, 81.
[38] SJÖBERG, B., BAMBERG, P., EKSTRÖM, B., FORSGREN, U., MAGNI, L. and PRING, B. G. (1978) "Synergistic action *in vitro* between amidinopenicillanic acids and penicillins and cephalosporins", *Current Chemotherapy*, p. 432. Pub. American Society for Microbiology. Eds. W. Siegenthaler, and R. Lüthy.
[39] STANDIFORD, H. D., DE MAINE, J. B. and KIRBY, W. M. M. (1970) "Antibiotic synergism of enterococci", *Arch. Int. Med.*, **125**, 258.
[40] *Supplement* (1973) "Trimethoprim-sulphamethoxazole", *J. Inf. Dis.*, **128**.
[41] TUAZON, C. U., LIN, M. Y. C. and SHEAGREN, J. N. (1978) "*In vitro* activity of rifampicin alone and in combination with nefcillin and vancomycin against pathogenic strains of *Staph. aureus*", *Antimicr. Ag. Chemother.*, **13**, 759.
[42] WATT, B. (1978) "Streptococcal endocarditis: a penicillin alone or a penicillin with an aminoglycoside", *J. Antimicr. Chemo.*, **4**, 107.
[43] WISE, R., ANDREWS, J. M. and BEDFORD, K. A. (1978) "*In vitro* study of clavulanic

acid in combination with penicillin, amoxycillin and carbenicillin", *Antimicr. Ag. Chemother.*, **13**, 389.
[44] WÜST, J. and WILKINS, J. D. (1978) "Effect of clavulanic acid on anaerobic bacteria resistant to beta-lactam antibiotics", *Antimicr. Ag. Chemother.*, **13**, 130.

13

Quantitative aspects of synergistic combinations of antibacterial agents

R. L. Then

Pharmaceutical Research Department, F. Hoffmann-La Roche & Co., Basel, Switzerland

Summary

The enhancement of *in vitro* activity resulting from synergistic combinations with underlying different modes of action has been investigated. Examples are given for synergism produced by sequential blockade in bacterial tetrahydrofolate – and cell wall biosynthesis, the β-lactamase principle and for synergism obtained by eliminating penetration barriers. Most combinations were studied by the tube titration method and led to a 5-to 10-fold increase in the activity of the individual compounds, the most powerful ones up to a 20-fold increase. Higher figures were seen with combinations of β-lactam antibiotics and β-lactamase inhibitors in penicillin-resistant strains. The *in vitro* findings are compared with results obtained in animal experiments and in human therapy, and the relevance of synergistic combinations is discussed.

Introduction

Antibacterial agents are administered simultaneously if they combine synergistically: *a*) to broaden the spectrum of antibacterial activity, an important consideration in the case of severe infections; *b*) to eradicate resistant strains that cannot be eradicated by single drug therapy; and *c*) to prevent or delay the development of bacterial resistance.

The *in vitro* study of antibacterial agent combinations provides a firm base for subsequent *in vivo* and clinical studies. In synergy studies of antibacterial agent combinations intended for clinical use, the following points should be borne in mind:

a) The degree of synergism of a combination should be as high as possible.
b) The combination should, in general, act synergistically against as

many strains of different species as possible. This involves the inclusion of broad-spectrum antibacterial agents in the combination. This may not however always be advisable because it may lead to imbalanced flora problems or secondary infections by yeasts and fungi [1]. The combination of narrow-spectrum antibacterial agents may be adequately effective against certain problem species or groups such as *Pseudomonas aeruginosa* or enterococci in bacterial endocarditis.

c) The synergism of a combination should be demonstrable *in vivo* as well as *in vitro*.

The pharmacokinetic properties of the drugs used in a combination have an important bearing on the question of whether it is advisable to use synergistic combinations as a single preparation or to combine the drugs ad hoc. This has to be carefully decided in every case.

In the present paper we compare the effects of a number of synergistic combinations under the heading of their mode of action. No attempt is made to review all the possible modes of synergism [2], [3], [4].

Materials and methods

By synergism we mean potentiation and refer to all instances where the concentration of a drug required to produce a given effect is reduced by over half when the drug is used in combination with another drug. If fractional inhibitory concentrations are used, as described by Elion et al. [5], the sum would be less than one. To account for experimental errors, an over 4-fold reduction of the minimal inhibitory concentration (MIC) is generally accepted as an indication of potentiation. In studying combinations, it is, however, advisable to use smaller steps than the usual geometric dilution [6].

We usually measure inhibition of growth. To avoid the problem of trailing end points, which is particularly important when studying sulfonamides and other antifolates, growth is monitored photometrically in liquid media, and concentrations producing 50% inhibition (IC_{50}), in some cases 75% (IC_{75}) or 90% (IC_{90}) inhibition, are determined. To check combinations, increasing concentrations of the first compound are titrated with the second, as described by Elion et al. [5]. To minimize labour, three concentrations of the second compound are usually used. The accuracy of these curves can, however, be increased by plotting more points. Plotting the concentrations which produce the IC_{50} (IC_{75}, MIC etc. as the case may be) either directly, or in the form of fractional inhibitory concentrations, results in the well known isobologram.

It should be borne in mind that if we only measure one parameter, namely growth, we can never be certain that a relevant answer is obtained. If we measure another parameter, such as, for example, the production of acid, the minimal bactericidal concentration (MBC), inhibition of respiration, spheroblast formation, etc., the results of the experiment as regards synergism, or the degree of synergism, may well be different [6]. What this means under *in vivo* conditions remains to be clarified. In our opinion, measuring the effect on growth is the simplest thing to do and will give useful answers.

The results of synergy studies are greatly affected by the methods used. Thus, for example, with cotrimoxazole 100-fold or even greater reductions of individual MIC have sometimes been reported. These reductions, which we have never observed, can most easily be explained by the trailing end points, especially of the sulfonamide, which makes the basis for the calculation uncertain.

To check the accuracy of the titration method we use, we tested some combinations where an additive effect was expected. Rifampicin, which affects RNA synthesis, and trimethoprim (TMP) which, by inhibition of dihydrofolate reductase, mainly affects DNA synthesis, showed an additive effect against *E. coli* B (Fig. 13.1). Pyrimethymine and TMP, which both inhibit the same enzyme, also showed an additive effect against *E. coli* B (Fig. 13.2). Finally, penicillin G and gentamicin, which are reported to act synergistically against enterococci, showed additive effects against *E. coli* B.

For rapid initial screening, simple qualitative methods, such as the filter paper strip method on agar plates [7], [8], are often useful.

Modes of action

Synergism by sequential blockade in bacterial tetrahydrofolate biosynthesis. We consider this mechanism of synergism first because cotrimoxazole type of combinations have been extensively studied in our laboratory. The synergism of a combination of sulfamonomethoxine and TMP against *E. coli* B in a semisynthetic medium is shown in the isobologram in Fig. 13.3. This combination produced an 8- to 10-fold reduction in the effective concentration of each component. Here, the composition of the medium used is important because a lower degree of synergism is observed in a mineral salts medium [9], and a higher degree of synergism (up to a 20-fold reduction in dose) in complex media containing thymidine. These differences are certainly due to the fact that the steps of metabolism affected and hence modes of action vary according to the type of medium used. Different sulfonamides are potentiated by

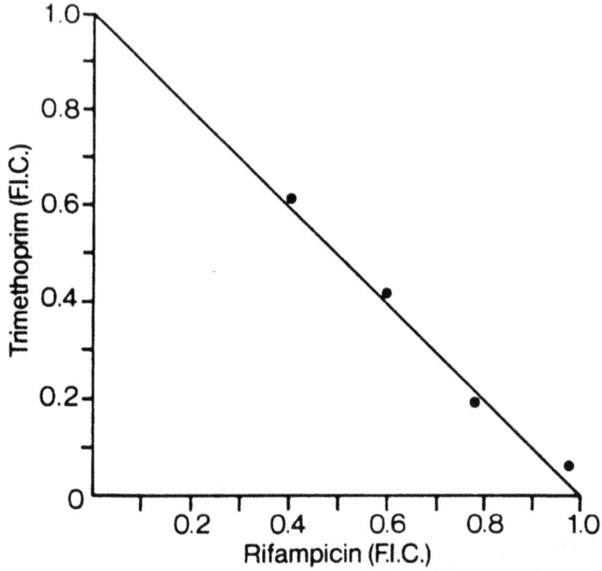

Fig. 13.1 - Additive effect of rifampicin/trimethoprim combinations against *E. coli* B. IC_{50} measured in a semisynthetic medium.

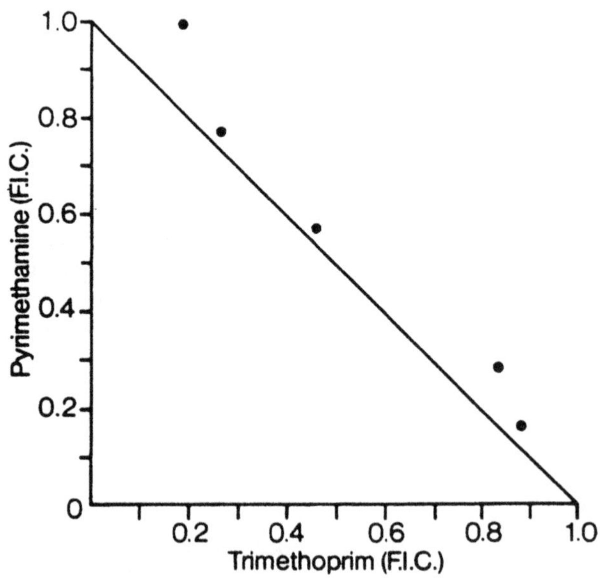

Fig. 13.2 - Additive effect of pyrimethamine/trimethoprim combinations against *E. coli* B. IC_{50} measured in a semisynthetic medium.

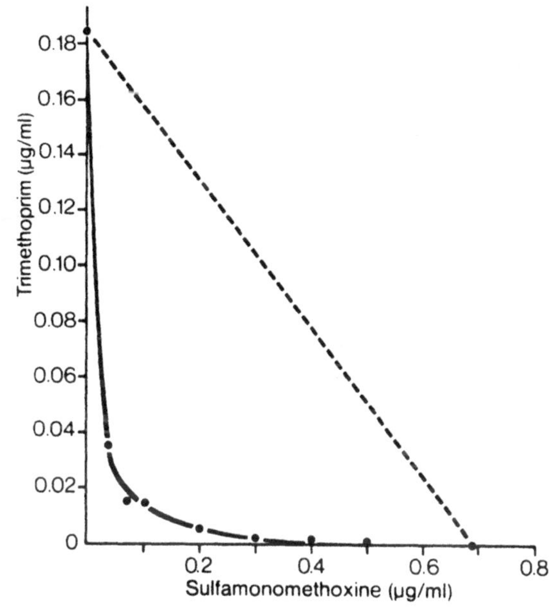

Fig. 13.3 - Synergism between sulfamonomethoxine (6-methoxy-4-sulfanilamidopyrimidine) and trimethoprim against *E. coli* B. IC_{50} (µg/ml) in a semisynthetic medium, containing amino acids, but no purines and thymidine.

TMP to the same extent (Table 13.1). The differences in the potentiation figures given in this table for the different sulfonamides were not statistically significant. We expected this because all these combinations are based on the same mechanism.

If the same sulfonamide is combined with different types of dihydrofolate reductase inhibitors, the same degree of synergism is observed as with TMP (Table 13.2). An 8- to 10-fold enhancement of activity was observed. The differences were not considered statistically significant. It should be mentioned that the somewhat higher degree of synergism obtained with pyrimethamine as compared to TMP coincides with the findings of Seydel and Wempe [10]. These investigators also found a higher degree of synergism with pyrimethamine. This cannot be readily explained and deserves further study.

In vivo findings with TMP - or pyrimethamine-sulfonamide combinations are similar to the *in vitro* findings. Reductions of the ED_{50} in experimental infections in mice of approximately 10-fold are common (the enhancement of activity of the single component depending on the ratio used and the scheme of treatment. A detailed study on the synergism of pyrimethamine + sulfadoxine in ex-

Table 13.1 Potentiation of different sulfonamides by trimethoprim. Mean of three to four experiments. E. coli *B, semisynthetic medium. Figures represent the optimal ratio, shown in isobolograms.*

Sulfonamide	*Sulfonamide* (IC_{50}, μg/ml)			*Trimethoprim* (IC_{50}, μg/ml)		
	Alone	*Combined*	*—Fold reduction*	*Alone*	*Combined*	*—fold reduction*
Sulfadiazine	2.8	0.297	9.4	0.204	0.064	3.2
Sulfanilamide	177	30	5.9	0.450	0.061	7.4
Sulfathiazole	3.02	0.267	11.3	0.330	0.027	12.2
Sulfisoxazole	3.75	0.565	6.6	0.340	0.053	6.4
Sulfadimethoxine	0.80	0.103	7.8	0.173	0.024	7.2
Sulfamethoxazole	2.05	0.285	7.2	0.330	0.042	7.9
Sulfamonomethoxine	0.76	0.090	8.4	0.337	0.044	7.7
Sulfadoxine	2.40	0.217	11.1	0.150	0.016	9.4
Isosulfisoxazole	14.0	1.20	11.7	0.158	0.015	10.5
Sulfacytine	2.07	0.190	10.9	0.150	0.017	8.8
Mean			9.0			8.0

Table 13.2 Potentiation of sulfamethoxazole by different dihydrofolate reductase inhibitors. Mean of four experiments. E. coli *B, semisynthetic medium.*

Potentiator	*Sulfamethoxazole* (IC_{50}, μg/ml)			*Potentiator* (IC_{50}, μg/ml)		
	Alone	*Combined*	*—Fold reduction*	*Alone*	*Combined*	*—Fold reduction*
Trimethoprim	2.05	0.285	7.2	0.33	0.042	7.9
Pyrimethamine	1.64	0.173	9.5	33.2	3.03	11.0
Triamterene	1.20	0.188	6.4	49.0	6.75	7.3
2,4-Diamino-6,7-diisopropylpteridine	1.50	0.118	12.7	3.28	0.355	9.2
Mean			9.0			8.8

perimental malaria in rhesus monkeys has been recently made [11]. Thirty to 100-fold enhancement in the activity of the individual compounds were observed against various organisms, including several types of resistant plasmodia. A study of the synergistic interactions between sulfalene and pyrimethamine in plasmodia-infected mice showed an approximately 10-fold enhancement of activity [12]. This well established principle is considered one of the most potent and interesting.

There are other examples of synergy within folic acid biosynthesis, which are currently only of basic research interest. We recently observed that the oxidized form of the natural precursor of dihydropteroic acid, 2-amino-4-hydroxy-6-hydroxymethylpteridine, which by itself does not inhibit growth of *E. coli* B, strongly potentiates sulfonamides and TMP (Fig. 13.4 and Fig. 13.5). The degree of synergism can only be extrapolated since no effect on growth was observed with the highest soluble concentration of the pteridine (this is indicated by the arrow). The mode of action has not yet been investigated; it is obviously different from that of TMP and sulfonamides, and may involve the inhibition of pteridine pyrophosphate-kinase, as Wood [13] has shown for similar compounds.

Synergism by sequential blockade in bacterial cell wall biosynthesis. Cell wall biosynthesis is the most attractive target site for specific antibacterial action. Over the last few years several new compounds interfering with cell wall biosynthesis have been developed and described. It is worth noting that some were deliberately developed for synergistic combination with other cell wall inhibitors.

3-fluoro-D-alanine and 2-deutero-3-fluoro-D-alanine are analogues, and were made as structurally related to D-alanine [14], a precursor of peptidoglycan. 3-Fluoro-D-alanine is a potent inhibitor of alanine racemase, and acts synergistically with D-cycloserine [15].

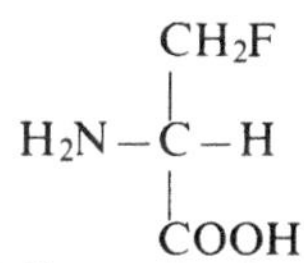

3-fluoro-D-alanine

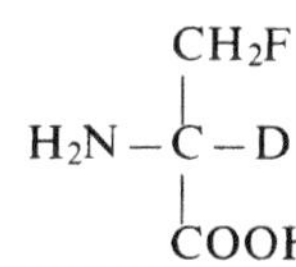

2-deutero-3-fluoro-D-alanine

The latter is an inhibitor of D-alanyl-alanine ligase, which catalyzes the step following the action of alanine racemase. Synergism between these two agents can therefore also be attributed to sequential blockade. The antibacterial action of fluoro-D-alanine is somewhat complicated by the effect described as self-reversal. This means that at higher concentrations this compound is not an inhibitor any more, but is used as a substrate. On agar plates, this results in a double ring phenomenon around discs or cylinders. D-cycloserine, in addition to its potentiating effects, prevents this phenomenon.

Tube titration showed strong synergism between 3-fluoro-D-alanine and D-cycloserine with a 10- to 20-fold reduction of individual effective concentrations against *E. coli* B (Fig. 13.6) and *Bacillus subtilis*, a representative gram-positive organism (Fig. 13.7). Though this combination has not been introduced into human chemotherapy, probably due to side effects of the alanine analogue,

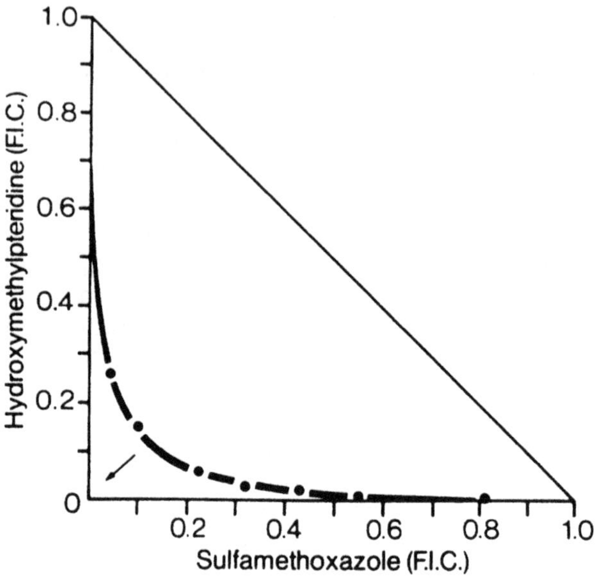

Fig. 13.4 - Synergism between 2-amino-4-hydroxy-6-hydroxymethyl-pteridine and sulfamethoxazole against *E. coli* B. IC_{50} measured in semisynthetic medium.

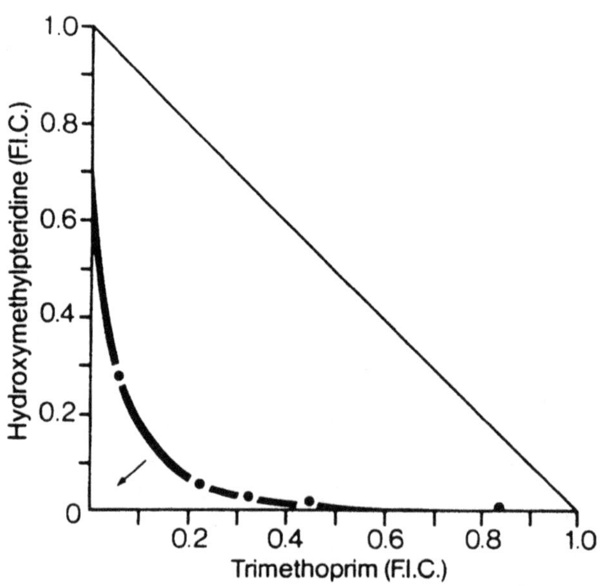

Fig. 13.5 - Synergism between 2-amino-4-hydroxy-6-hydroxymethyl-pteridine and trimethoprim against *E. coli* B. IC_{50} measured in semisynthetic medium.

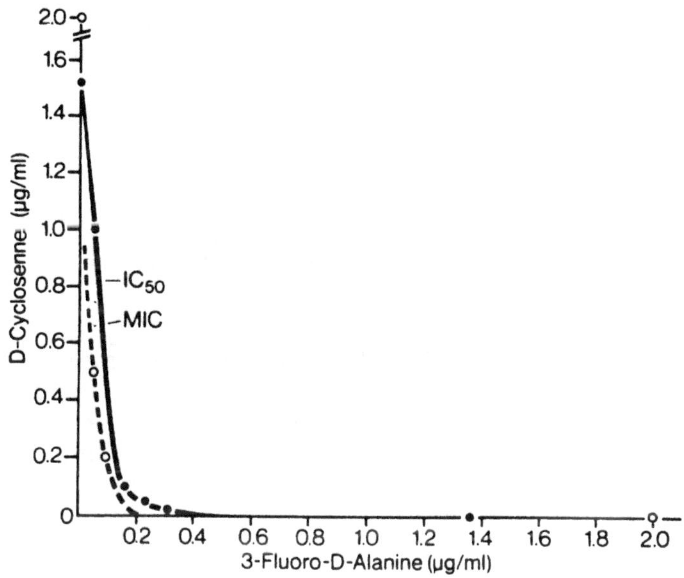

Fig. 13.6 - Synergism between 3-fluoro-D-alanine and D-cycloserine against *E. coli* B in mineral salts medium M9 (IC_{50} and MIC, $\mu g/ml$).

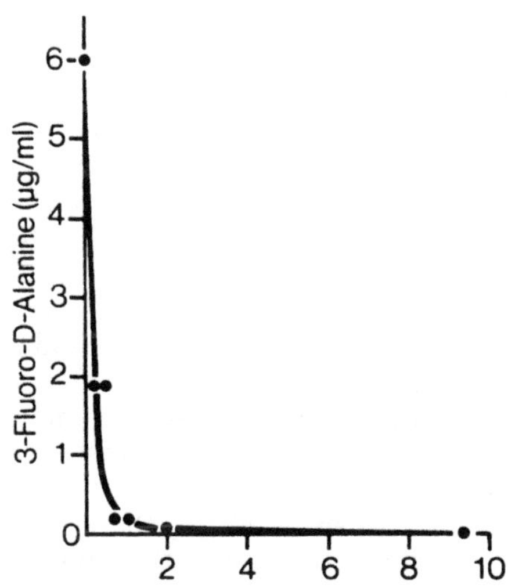

Fig. 13.7 - Synergism between 3-fluoro-D-alanine and D-cycloserine against *B. subtilis* in mineral salts medium M9 (IC_{50}, $\mu g/ml$).

it acts synergistically in experimental infections. Dose reductions of 10- to 20-fold were reported [16] and observed in our laboratory.

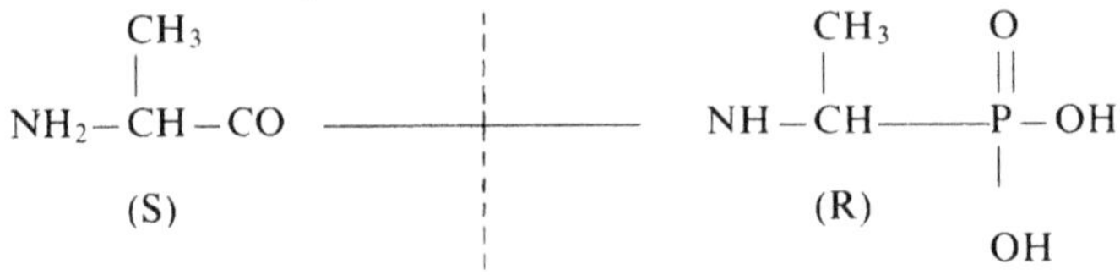

(S)-alanyl-(R)-1-aminoethylphosphonic acid
"alaphosphalin" (Ro 3-7008)

Another novel principle has been very recently described. A peptide mimetic, "alaphosphalin" was designed on a rational basis as a structural analogue of the terminal alanyl-alanine of peptidoglycan [17]. It is taken up by the bacterial cell by an active process, accumulated up to several thousand fold and cleaved intracellularly to release the active principle, 1-aminoethylphosphonate, which by itself does not penetrate the cell.

This compound is a potent inhibitor of alanine racemase and also affects some other steps in peptidoglycan synthesis. It acts synergistically with several other cell wall inhibitors, especially those which interfere with early steps in peptidoglycan synthesis. Combination with D-cycloserine results in a 10- to 12-fold enhancement of activity. Synergism was also found with mecillinam (Fig. 13.8). This combination led to a 7- to 10-fold enhancement of activity in *E. coli*. On the basis of experiments carried out so far, these combinations were also found to be synergistic in experimental infections in mice.

The mechanisms of other combinations of cell wall inhibitors which are synergistic are not understood as well. Though one can assume that sequential or concurrent blockades may strongly contribute to the synergism observed, other mechanisms such as autolysin activation may play a role and complicate the picture considerably. Thus, combinations of mecillinam with penicillins or cephalosporins which are often synergistic [18], [19], are more difficult to explain, as the mode of action of the individual compounds is still not fully understood.

In vitro the combination of ampicillin with mecillinam results in an 8- to 10-fold enhancement of activity against *E. coli* B (Fig. 13.9), but the effect is known to be strongly dependent on the medium composition, particularly its osmolality [20]. This combination and others are, however, synergistic in experimental animal infections [21].

Numerous examples of synergistic combinations of cell wall inhibitors could be given. In contrast to most of the examples mentioned so far, synergism is often restricted to certain strains or groups

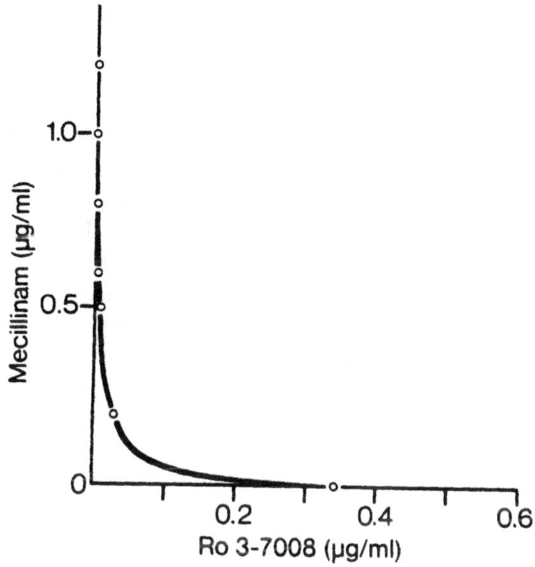

Fig. 13.8 - Synergism between mecillinam and alaphosphalin (Ro 3-7008) against *E. coli* B in mineral salts medium M9 (IC_{50}, µg/ml).

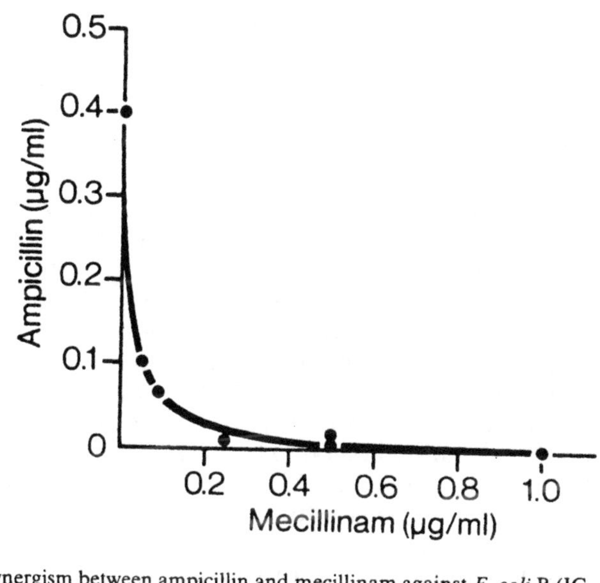

Fig. 13.9 - Synergism between ampicillin and mecillinam against *E. coli* B (IC_{50}, µg/ml).

of bacteria, thus necessitating confirmation of synergy for a specific germ, before using a particular combination in therapy.

As an example of weak synergism, the interaction between vancomycin and novobiocin against *E. coli* B is shown in Fig. 13.10. A 3- to 4-fold reduction in the effective concentrations is observed. This combination was also synergistic against some *Klebsiella* and *Proteus* strains, as judged by the filter paper strip method.

Novobiocin, combined with mecillinam, led to a 6- to 8-fold enhancement of activity against *E. coli* B (Fig. 13.11). Judging from results with the filter paper strip test, this combination was synergistic against 8 out of 18 strains, mostly gram-negative ones. The underlying mode of action is not easily explained, as novobiocin not only affects cell wall biosynthesis, but also DNA synthesis and other reactions. Recently, synergism between novobiocin and β-lactam antibiotics was observed in mice [22]. The underlying mode of action was, however, reported to be a probenecid effect of novobiocin, viz. blockade of renal tubular excretion, but not of glomerular filtration. The *in vivo* mechanism therefore appears to be completely different from the *in vitro* mechanism; both mechanisms may contribute to the synergism of such combinations.

Synergism produced by inhibition of β-lactamases. Inactivation of β-lactam antibiotics by β-lactamases is the reason for increasingly observed resistance to these antibiotics. Prevention of β-lactam degradation by inhibition of β-lactamase restores the original activity and combinations of β-lactam antibiotics with β-lactamase inhibitors are often highly synergistic against β-lactamase producers [23]. Marked reductions in the effective concentration of these combinations are reported – up to more than 100-fold. As this is a very well known principle, it will not be discussed in detail here. Ampicillin, combined with dicloxacillin against a resistant strain of *E. coli* (a β-lactamase producer, as verified with the chromogenic cephalosporin nitrocefin [24]), produced synergism with a 15- to 20-fold reduction of individual effective concentrations (Fig. 13.12). The degree of synergism may be influenced to some extent by the amount of inoculum. In our experiments low inocula of 5×10^4 CFU/ml were used. Synergism between penicillin G and dicloxacillin against the same strains and under the same assay conditions was much more pronounced, resulting in a 15- to 50-fold reduction (Fig. 13.13). Though this principle results in powerful synergism *in vitro*, its application in therapy is limited so far. The great variation of bacterial β-lactamases makes them inaccessible to only one inhibitor and the effective concentrations of currently known β-lactamase inhibitors mostly exceed therapeutically achievable levels, urine being the only site of sufficiently high concentrations to demonstrate a beneficial

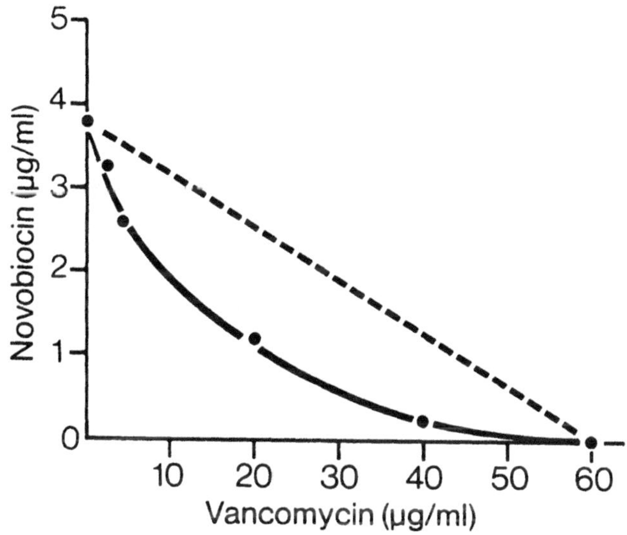

Fig. 13.10 - Synergism between vancomycin and novobiocin against *E. coli* B (IC_{50}, µg/ml).

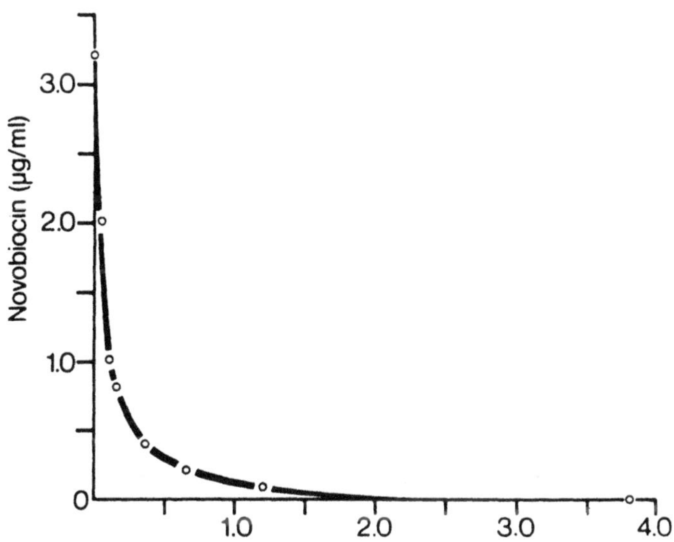

Fig. 13.11 - Synergism between mecillinam and novobiocin against *E. coli* B (IC_{50}, µg/ml).

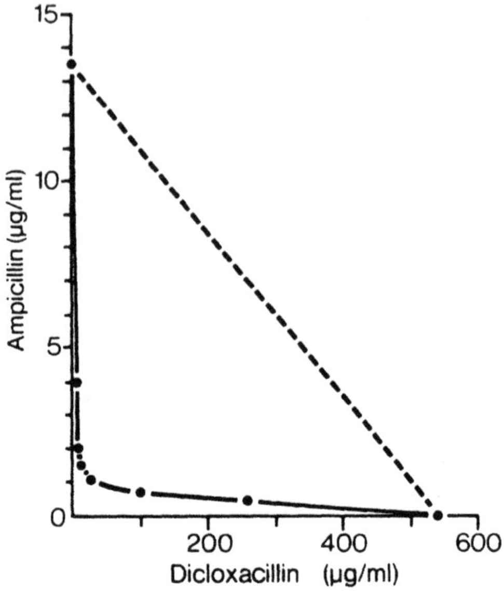

Fig. 13.12 - Potentiation of ampicillin by dicloxacillin against a β-lactamase producing *E. coli* strain (*E. coli* 1026, MIC, μg/ml).

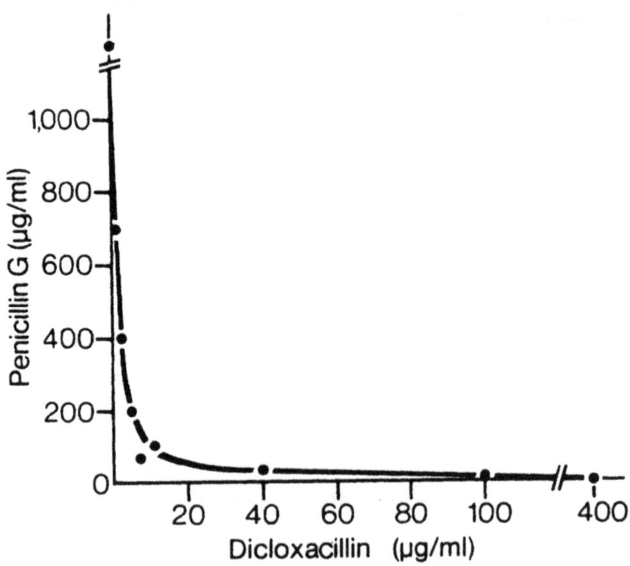

Fig. 13.13 - Potentiation of penicillin G by dicloxacillin against a β-lactamase producing *E. coli* strain (*E. coli* 1026, MIC, μg/ml).

effect [23]. This may change in the near future, particularly since the recent discovery of clavulanic acid and its derivatives, all potent β-lactamase inhibitors. Synergism between clavulanic acid and penicillins against β-lactam resistant organisms has been demonstrated not only *in vitro* but also in animal experiments [25-27].

Synergism produced by elimination of penetration barriers. Many antibacterial compounds are not or not readily taken up by certain species and are therefore scantly active or inactive against them.

Combining these compounds with an antibacterial compound that facilitates access to the target site often results in synergism. This mode of action seems to be the basis for the *in vitro* and *in vivo* synergistic effects of combinations of aminoglycosides and penicillins or cephalosporins against enterococci. Hence the use of such combinations in the therapy of bacterial endocarditis [28], [29], [30]. In this particular case the enhanced bactericidal effect of such combinations is important. In a similar way, gentamicin acts synergistically with carbenicillin against *Pseudomonas aeruginosa* [31]. The degree of synergism obtained seems to be in the order of what we have seen with other combinations, viz. a 5- to 10-fold enhancement of activity.

Most strains of *Proteus* are resistant to colistin [32]. Synergism was however observed between colistin and TMP against *P. vulgaris* (Fig. 13.14). The basis of this effect remains unclear, since colistin affects the membrane function. This particular combination is not synergistic against most other bacteria. Though synergism has been reported against *P. aeruginosa*, we were unable to confirm this against the particular strain of *P. aeruginosa* we tested [33].

Final remarks

The list of synergistic combinations is very long. Many combinations are of little importance in the therapy of infections, because their synergism is restricted to certain strains only and their therapeutic value is unknown. They are, however, useful for studying the mechanisms of action of antibacterial drugs and may also stimulate the development of new potent drug combinations.

In comparing the activity enlargements resulting from the combination of two synergistic drugs, it is obvious that there are limitations. Many of the combinations which we have studied show a 5- to 10-fold enhancement of activity. Only few, of the most powerful ones, showed a dose-reduction of up to 20-fold. This seems to be the limit for mechanisms based on sequential blockade or similar mechanisms.

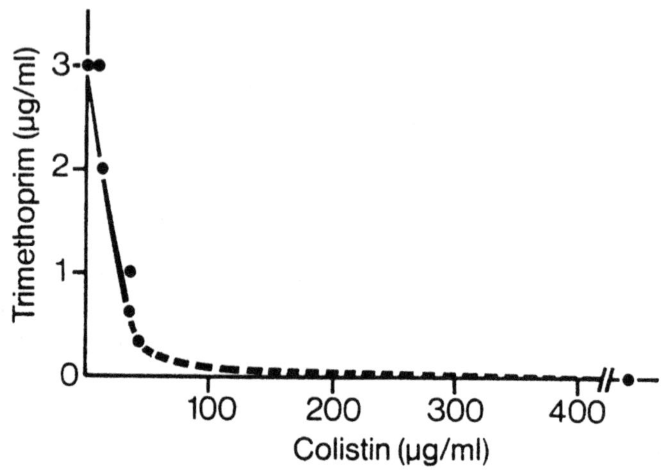

Fig. 13.14 - Synergism between colistin and trimethoprim against *Proteus vulgaris* (IC_{50}, µg/ml) [33].

For examples of dose-reductions resulting from synergism by sequential blockade in bacterial folate synthesis, see Tables 13.1 and 13.2, and by sequential blockade in bacterial cell wall biosynthesis, see Table 13.3.

Greater reductions seem to result from mechanisms whereby one drug, which need not even be active, prevents the destruction of the second (active) drug. This is the case with β-lactamase inhibitors against β-lactamase producing strains. Very marked reductions in the effective concentration of the labile drug result from this principle *in vitro*. Examples for this and some other mechanisms are given in Table 13.4. Though there are clear limitations on the clinical testing of combinations, there are well documented examples to show that impressive *in vitro* results obtained with synergistic combinations have clinical relevance. Beneficial effects of combinations can be shown by animal experiments, such as the one of Schmidt et al. [11] who treated malaria in rhesus monkeys with a combination of sulfadoxine and pyrimethamine.

Some clinical information on combinations, such as comparison of synergistic vs. nonsynergistic or even antagonistic combinations, is difficult or impossible to obtain for ethical reasons and will never be published. Despite these difficulties and restrictions reports on the beneficial effects of synergistic combinations are frequently published. Recently, Anderson et al. [35] found that in rapidly fatal bacteremias the clinical results obtained with synergistic combinations were significantly better than with nonsynergistic combinations. Similar results were obtained in a clinical study carried out by Klastersky et al. [36].

Table 13.3 Comparison of some synergistic combination of cell wall inhibitors. Mineral salts medium M9.

Synergistic combination	*Species*	*— Fold reduction in combination* A	B	*Parameter used*
3-Fluoro-D-alanine -	*E. coli*	9	15	IC_{50}
D-Cycloserine	*E. coli*	17	20	MIC
	B. subtilis	15	13	IC_{75}
Ampicillin - Mecillinam	*E. coli*	7	9	IC_{50}
Vancomycin - Novobiocin	*E. coli*	4	3	IC_{50}
Ro 3-7008 - D-Cycloserine	*E. coli*	10-15	10-15	IC_{50}
D-Cycloserine - Mecillinam	*B. subtilis*	8	7	IC_{50}
Novobiocin - Mecillinam	*E. coli*	9	8	IC_{50}

Table 13.4 Comparison of some synergistic combinations and underlying different modes of action.

Synergistic combination	*Species*	*— Fold reduction in combination* A	B	*Parameter used*
Ampicillin - Dicloxacillin	*E. coli* 1026 (β-lactamase producer)	13	22	IC_{50}
Penicillin G - Dicloxacillin	*E. coli* 1026 (β-lactamase	> 15	31	IC_{50}
	producer)	> 10	40-60	MIC
Carbenicillin - Phleomycin	*Ps. aeruginosa*	10	5	MIC
Carbenicillin - Gentamicin	*Ps. aeruginosa*	5-10	5-10	MIC
EDTA - Trimethoprim	*Ps. aeruginosa*	5-10	5-10	IC_{50}/MIC
Colistin - Trimethoprim	*Pr. vulgaris*	> 10	10	IC_{50}
5-Fluoro-2′-deoxyuridine-Trimethoprim	*E. coli*	3-5	3-5	IC_{50}*
5-Fluoro-2′-deoxyuridine-Sulfamethoxazole	*E. coli*	5-10	5-10	IC_{50}*

* Under certain conditions [34].

Only synergistic combinations have been considered in this short review. There are, of course, many examples, where combinations of two drugs produce reduced activity *in vitro*. The possibility of antagonism is always present, even though addition is much more common than antagonism in our experience.

Antagonism *in vitro* however need not necessarily mean antagonism *in vivo*. The pharmacokinetic properties of the two drugs and the immunological situation of the patient are of major importance here. This has to be considered before administering any combination in therapy.

References

[1] GRASSI, G. G. (1977) *J. Antimicr. Chemother.*, **3** (Suppl. C), 77-84.
[2] JAWETZ, E. and GUNNISON, J. B. (1953) *Pharmacol. Rev.*, **5**, 175-192.
[3] LOEWE, S. (1953) *Arzneim.-Forsch.*, **3**, 285-290.
[4] LACEY, B. W. (1958) *Symp. Soc. Gen. Microb.*, **8**, 247-288.
[5] ELION, G. B., SINGER, S. and HITCHINGS, G. H. (1954) *J. Biol. Chem.*, **208**, 477-488.
[6] ZIPF, H. F., and HALBEISEN, T. (1954) *Arzneim.-Forsch.*, **4**, 463-467.
[7] LORIAN, V. and FODOR, G. (1974) *Antimicr. Ag. Chemother.*, **5**, 630-633.
[8] OKUBO, T., INOUE, M. and MITSUHASHI, S. (1975) *J. Antibiot.*, **28**, 804-808.
[9] THEN, R. (1973) *Zbl. Bakt. Hyg. I. Abt. Orig.*, **A 225**, 34-41.
[10] SEYDEL, J. K. and WEMPE, E. (1975) *Chemotherapy*, **21**, 131-145.
[11] SCHMIDT, L. H., HARRISON, J., ROSSAN, R. N., VAUGHAN, D. and CROSBY, R. (1977) *Am. J. Trop. Med. Hyg.*, **26**, 837-849.
[12] PETERS, W. (1971) *Chemotherapy*, **16**, 389-398.
[13] WOOD, H. C. S. (1975) in: *Chemistry and Biology of Pteridines* (W. Pfleiderer ed.), de Gruyter, Berlin, pp. 27-44.
[14] KOLLONITSCH, J., BARASH, L., KAHAN, F. M. and KROPP, H. (1973) *Nature*, **243**, 346-347.
[15] KAHAN, F. M. and KROPP, H. (1975) *15th Interscience Conf. Antimicr. Ag. Chemother.*, Abstr. No. 100.
[16] KROPP, H., KAHAN, F. M. and WOODRUFF, H. B. (1975) *15th Interscience Conf. Antimicr. Ag. Chemother.*, Abstr. No. 101.
[17] ALLEN, J. G., ATHERTON, F. R., HASSALL, C. H., HALL, M. J., HOLMES, S. W., LAMBERT, R. W., NISBET, L. J. and RINGROSE, R. S. (1978) *Nature*, **727**, 56-58.
[18] BALTIMORE, R. S., KLEIN, J. O., WILCOX, C. and FINLAND, M. (1976) *Antimicr. Ag. Chemother.*, **9**, 701-705.
[19] NEU, H. (1976) *Antimicr. Ag. Chemother.*, **10**, 535-542.
[20] GREENWOOD, D. (1976) *Antimicr. Ag. Chemother.*, **10**, 824-826.
[21] GRUNBERG, E., CLEELAND, R., BESKID, G. andDELORENZO, W. F. (1976) *Antimicr. Ag. Chemother.*, **9**, 589-594.
[22] MILLER, A. K., CELOZZI, E. and PELAK, B. A. (1977) *J. Antibiot.*, **30**, 983-986.
[23] GREENWOOD, D. and O'GRADY, F. (1975) *Chemotherapy*, **21**, 330-341.
[24] O'CALLAGHAN, C. H., MORRIS, A., KIRBY, S. M. and SHINGLER, A. H. (1972) *Antimicr. Ag. Chemother.*, **1**, 283-288.
[25] HUNTER, P. A. and READING, C. (1976) *16th Interscience Conf. Antimicr. Ag. Chemother.*, Abstr. No. 101.
[26] HUNTER, P., READING, C. and WITTING, D. A. (1977) *10th Intern. Congr. Chemother.*, Abstr. No. 238.
[27] JACKSON, R. T. and ALFORD, R. H. (1977) *10th Intern. Congr. Chemother.*, Abstr. No. 239.
[28] SAPICO, F. L., KEYS, T. F. and HEWITT, W. L. (1972) *Am. J. Med. Sci.*, **262**, 128-135.
[29] LAMBELIN, G. (1972) *Chemotherapy*, **17**, 356-363.
[30] WEINSTEIN, O. (1976) *Antimicr. Ag. Chemother.*, **9**, 983-987.
[31] ANDERSON, E. L., GRAMLING, P. K., VESTAL, P. R. and FARRAR, W. E. (1975) *Antimicr. Ag. Chemother.*, **8**, 300-304.
[32] TEUBER, M. (1969) *J. Bacteriol.*, **98**, 347-350.
[33] SIMMONS, N. A. (1970) *J. Clin. Pathol.*, **23**, 757-764.
[34] THEN, R. (1974) *Arzneim.-Forsch.* **24**, 1949-1952.
[35] ANDERSON, E. T., YOUNG, L. S. and HEWITT, W. L. (1978) *Chemotherapy*, **24**, 45-54.
[36] KLASTERSKY, J., CAPPEL, R. and DANEAN, D. (1972) *Antimicr. Ag. Chemother.*, **2**, 470-475.

14

Evaluation of *in vitro* effects of subinhibitory concentrations of antibiotics as a function of their possible role in antibiotic combinations

G. Gialdroni Grassi, A. Ferrara, P. Sala

Cattedra di Chemioterapia, Facoltà di Medicina e Chirurgia, Università di Pavia - Italy

Summary

The effect that the association of some antibiotics and chemotherapeutic agents exerts on growth curves and the emergence of one-step resistant mutants of *E. coli* K 12 and *Staph. aureus* has been studied.

While the association of trimethoprim with rifampicin or fosfomycin results in a reduction of the mutation rate to fosfomycin and rifampicin and has a limited influence on the bacterial growth curves, the association of rifampicin and fosfomycin reduces both the mutation rate and the growth rate of the same bacteria.

Fosfomycin and ampicillin, which do not have any reciprocal influence on the emergence of one-step resistant mutants, in association induce a marked or complete inhibition of the bacterial growth.

The implications that these findings may have in the assessement of the "real" synergistic activity of the antibiotics in association are discussed.

Introduction

The effect that the subinhibitory concentrations of antibiotics or, better, that concentrations of antibiotics below the conventional MIC exert on bacteria both from a morphological [5], [7], [15], [16], [17], [18] and functional point of view, has drawn the attention of many researchers and some interesting observations have been made.

A more or less marked inhibition of bacterial growth [12], [22], prolongation of the log-phase, reduction in the rate of logarithmic growth and diminution in the size of the bacterial population have been observed.

In some experiences, by a method utilizing millipore filter membranes for growing and transferring bacteria in agar-plates containing subinhibitory concentrations of antibiotics (penicillin, ampicillin, cloxacillin, carbenicillin, gentamicin, chloramphenicol anf fosfomycin) we could demonstrate that they have no significant influence either on MICs or on the growth curves of bacteria or on the emergence of "one step" resistance [8], [9].

In order to confirm these results we repeated the experiments with the more conventional technique of growing bacteria in nutrient broth and of determining growth curves and emergence of resistance after different times of contact with subinhibitory concentrations of antibiotics. We obtained results that were practically superimposable on those obtained by means of the millipore filter membranes.

It is to be stressed that careful washing of bacterial cultures is of fundamental importance in these experiments.

The series of experiments that are referred to here concern some interactions existing among trimethoprim (TMP), rifampicin (RMP), fosfomycin (FM) and ampicillin (AMP), when associated in subinhibitory concentrations.

The reasons that induced us to study these agents will be briefly reviewed.

TMP is known to have a synergistic activity on sulfonamides, which has been attributed to the possibility to achieve a so-called "sequential blockade" [6] in the metabolic pathway involving folic acid in bacteria, as well as to its capacity of reducing the incidence of resistance. Successively TMP has also been shown to have a synergistic effect on polymyxins [21], [23], aminoglycosides [20], [25] and RMP [1], [14]. While for polymyxins and aminoglycosides no hypothesis on the mechanism of synergism has been put forward, for the association with RMP it has been postulated that the synergistic activity could be due to a sequential blockade, in that trimethoprim reduces the level of nucleic acids, and rifampicin, due to its action on the synthesis of RNA for competition with DNA-dependent RNA-polymerase, can be considered as acting on a successive step of the same metabolic pathway. In this case, however, the two steps are farther apart than in the case of association of trimethoprim with sulfonamides. In addition it has also been observed that TMP induces a reduction in the mutation rate of many bacterial species to RMP [1].

Fosfomycin is an antibiotic which demonstrates the capacity to act in a synergistic or ar least additive way with the most common antibiotics [2], [4], [9]. Therefore we considered that it would be interesting to investigate its behaviour in association with TMP and ampicillin.

The former could possibly reduce the incidence of fosfomycin-resistant mutants, which appears with a quite high frequency [24] in the same way as it does with rifampicin resistant mutants [1], [14]. The latter could achieve a sequential blockade with fosfomycin.

In fact while fosfomycin blocks the first steps of construction of the bacterial cell wall, by inhibiting the phosphoenolpyruvate synthetase [13], ampicillin acts on the last steps, by inhibiting the transpeptidation reaction [19].

The aim of the present study was not only to determine the existence or otherwise of a synergistic effect among the different combinations of the indicated antibacterial agents, but also the way in which this possible synergistic effect is achieved.

For this reason we investigated the effect that a previous or contemporaneous exposure to sub-MIC of these agents exerts on growth curves and the mutation rate of two common representatives of gram-positive and gram-negative organisms, *Staph. aureus* and *E. coli*.

Materials and methods

Bacterial strains. In all experiments *Staph. aureus* 209 P and *E. coli* K 12 were used.

Determination of one-step resistance. Overnight cultures of *Staph. aureus* 209 P and *E. coli* K 12 in Trypticase Soy Broth were exposed to subinhibitory concentrations of drugs (from 1/4 to 1/16 of MIC). The inoculum consisted of 10^6 colony forming units (CFU).

After incubation for 24 hours at 37 °C cultures were centrifuged, washed three times in saline, then diluted in order to obtain an inoculum of 10^6 CFU and spread on agar plate (Trypticase Soy agar) containing the selecting concentration of the same or of the other antibiotics (100-1000 μg/ml).

The mutation rate to resistance was calculated according to the method of Clowes and Hayes [3].

Determination of growth rate. Overnight cultures of *Staph. aureus* 209 P and *E. coli* K 12 in Trypticase Soy Broth were diluted in order to obtain a final concentration of 10^3 CFU, in 5 ml of Trypticase Soy Broth where TMP, RMP; FM and AMP were added at concentrations of 1/8-1/16 of MIC alone or in associations.

Counts of bacteria were performed by the streak agar method (Trypticase Soy Agar) at 1-2-3-4-5-8-24 h after inoculation.

Results

Effect on one step resistance. Previous exposure to TMP at 1/8 MIC does not influence the emergence of resistant mutants of *Staph. aureus* to the tested antibiotics, whilst it does when the concentration is 1/4 MIC (Table 14.1). TMP reduces the rate of mutation to RMP and FM resistance to *E. coli* (Table 14.2) in a more efficient manner, since it shows this effect even at 1/8 MIC.

Previous exposure to fosfomycin (1/8 MIC) reduces the mutation rate to RMP resistance both of *Staph. aureus* and *E. coli* (Table 14.3).

Previous exposure to RMP reduces the mutation rate to FM resistance of *Staph. aureus* and *E. coli* at 1/4 MIC, while at 1/8 of MIC there is only some reduction in the case of *Staph. aureus* (Table 14.4).

Previous exposure to fosfomycin did not modify the emergence of resistance to ampicillin and viceversa, as we already shown in preceding experiments.

Effect on growth rate. TMP and FM associated at 1/8 of MIC have only a slight influence on the growth rate of *Staph. aureus* and *E. coli*. Only a reduction in the rate of logarithmic growth and a decrease of a two log can be observed at the end of the period of growth (Fig. 14.1 and 14.2). The same pattern is observed when TMP is associated with rifampicin.

When FM and RMP are associated the growth rate is deeply affected. In fact the combination of 1/8 MIC both of FM and RMP reduces the final number of bacteria of 4-5 logs (Fig. 14.3) when the tested strain is *Staph. aureus*.

The association is still more active on *E. coli*, where the association of FM and RMP at 1/16 MIC completely abolishes the bacterial growth (Fig. 14.4).

Finally the effect of association of fosfomycin and ampicillin on the growth curve was determined. A remarkable effect on growth rate was observed, more evident on *E. coli* than on *Staph. aureus*. In fact the two antibiotics associated at 1/8 of MIC reduced the final population of *Staph. aureus* by 4 logs (Fig. 14.5) while *E. coli* was practically eliminated by the combined action of 1/16 of MIC of fosfomycin and ampicillin (Fig. 14.6).

Conclusion

It seems to us that some interesting conclusions can be drawn from the results obtained in these experiments with regard to the mechanism by which synergistic activity can be obtained through the use of antibiotic combinations.

For instance a reduction in the mutation rate to resistance to RMP

Table 14.1 Effect of previous exposure of Staph. aureus *209 P to subinhibitory concentrations of TMP on the rate of mutation to TMP, RMP and FM.*

Exposure to Subinh. Conc. of	*Selecting conc. of TMP, RMP, FM (μg/ml)*	*Mutation rate for* TMP		RMP		FM	
		Control	*Exposed culture*	*Control*	*Exposed culture*	*Control*	*Exposed culture*
TMP 1/8 MIC	100	$< 1 \times 10^{-11}$	$< 1 \times 10^{-11}$	3.4×10^{-6}	4.7×10^{-6}	4.1×10^{-6}	1.9×10^{-6}
	1000	$< 1 \times 10^{-11}$	$< 1 \times 10^{-11}$	2.5×10^{-6}	3.8×10^{-6}	7.6×10^{-7}	4.2×10^{-7}
TMP 1/4 MIC	100	$< 1 \times 10^{-11}$	$< 1 \times 10^{-11}$	2.6×10^{-6}	3.5×10^{-9}	4.6×10^{-6}	2.8×10^{-10}
	1000	$< 1 \times 10^{-11}$	$< 1 \times 10^{-11}$	3.5×10^{-7}	2.9×10^{-11}	3.1×10^{-7}	2.1×10^{-10}

Table 14.2 Effect of previous exposure of E. coli *K 12 to subinhibitory concentrations of TMP on the rate of mutation to TMP, RMP and FM.*

Exposure to Subinh. Conc. of	*Selecting conc. of TMP, RMP, FM (μg/ml)*	*Mutation rate for* TMP		RMP		FM	
		Control	*Exposed culture*	*Control*	*Exposed culture*	*Control*	*Exposed culture*
TMP 1/8 MIC	100	$< 1 \times 10^{-11}$	$< 1 \times 10^{-11}$	1.4×10^{-7}	3.4×10^{-10}	5.9×10^{-8}	6.8×10^{-10}
	1000	$< 1 \times 10^{-11}$	$< 1 \times 10^{-11}$	1.9×10^{-8}	$< 1 \times 10^{-11}$	6.2×10^{-8}	$< 1 \times 10^{-11}$

Table 14.3 Effect of previous exposure of Staph. aureus *209 P and* E. coli *K 12 to subinhibitory concentrations of FM on the rate of mutation to RMP.*

Exposure to Subinh. Conc. of	*Bacterial strain*	*Selecting conc. of RMP* (μg/ml)	*Mutation rate for RMP*	
			Control	*Exposed culture*
FM 1/8 MIC	*Staph. aureus* 209 P	100	6.5×10^{-7}	0.7×10^{-11}
		1000	2.4×10^{-7}	$< 1 \times 10^{-11}$
	E. coli K 12	100	9.6×10^{-7}	$< 1 \times 10^{-11}$
		1000	3.5×10^{-8}	1.7×10^{-10}

Table 14.4 Effect of previous exposure of Staph. aureus *209 P and* E. coli *K 12 to subinhibitory concentration of RMP on the rate of mutation to FM.*

Exposure to Subinh. Conc. of	*Bacterial strain*	*Selecting Conc. of FM* (μg/ml)	*Mutation rate for FM*	
			Control	*Exposed culture*
RMP 1/4 MIC	*Staph. au* 209 P	100	5.5×10^{-7}	4.9×10^{-9}
	Staph. au 209 P	1000	4.1×10^{-7}	$< 1 \times 10^{-11}$
	E. coli K 12	100	1.1×10^{-7}	$< 1 \times 10^{-11}$
	E. coli K 12	1000	1.2×10^{-9}	$< 1 \times 10^{-11}$
RMP 1/8 MIC	*Staph. au* 209 P	100	7.5×10^{-7}	5.2×10^{-8}
	Staph. au 209 P	1000	4.8×10^{-8}	$< 1 \times 10^{-11}$
	E. coli K 12	100	1.3×10^{-8}	1.7×10^{-8}
	E. coli K 12	1000	1.1×10^{-8}	2.4×10^{-8}

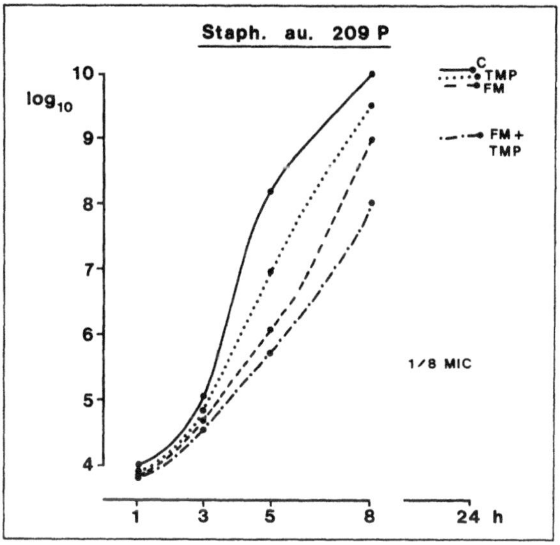

——— CONTROL — — — FOSFOMYCIN (FM)

Fig. 14.1 - Growth curves of *Staph. aureus* 209 P in presence of subinhibitory concentrations of trimethoprim and fosfomycin (1/8 MIC), alone and in association.

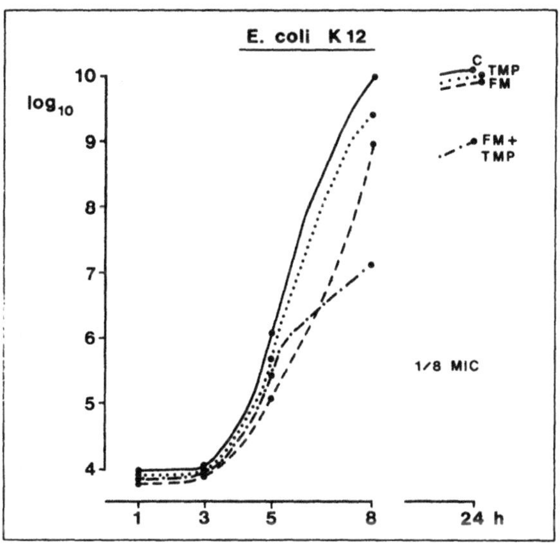

.... TMP —·—·— FM + TMP

Fig. 14.2 - Growth curves of *E. coli* K 12 in presence of subinhibitory concentrations of trimethoprim and fosfomycin (1/8 MIC), alone and in association.

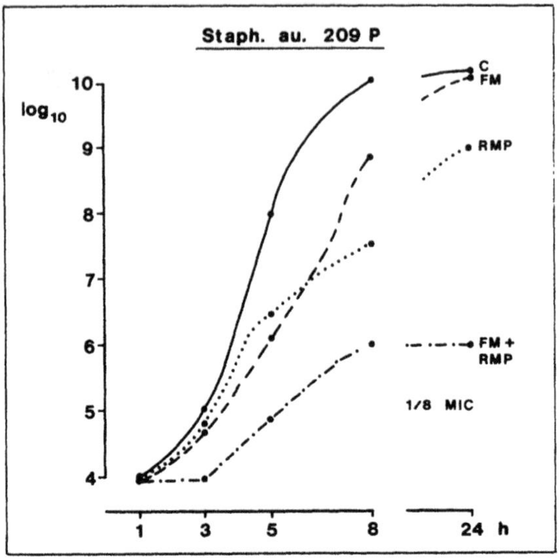

——— CONTROL ---- FOSFOMYCIN (FM)

Fig. 14.3 - Growth curves of *Staph. aureus* 209 P in presence of subinhibitory concentrations of fosfomycin and rifampicin (1/8 MIC), alone and in association.

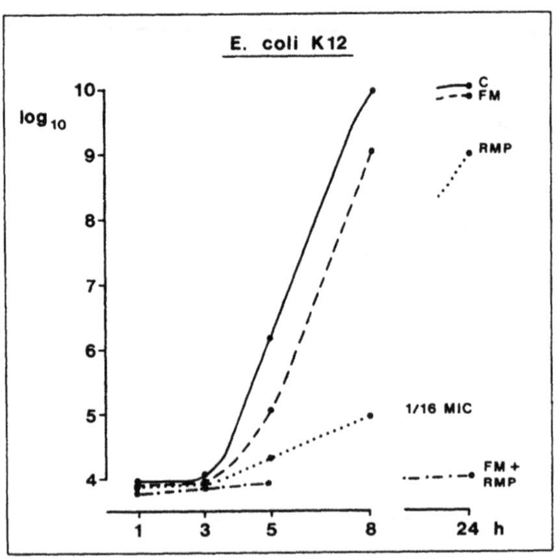

.......RIFAMPICIN (RMP) -·-·- FM + RMP

Fig. 14.4 - Growth curves of *E. coli* K 12 in presence of subinhibitory concentrations of fosfomycin and rifampicin (1/16 MIC), alone and in association.

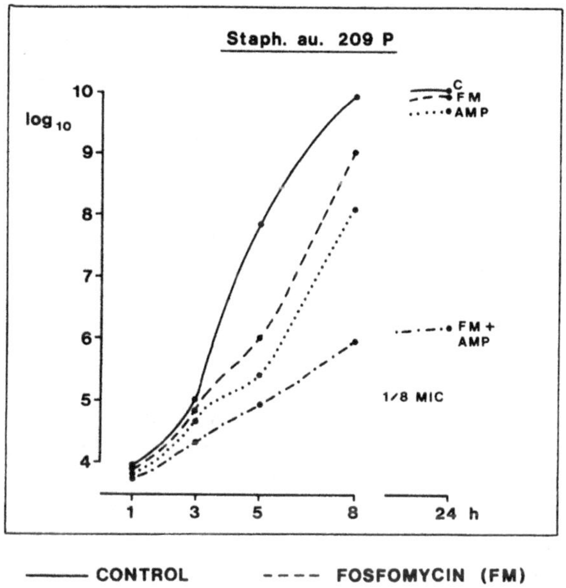

Fig. 14.5 - Growth curves of *Staph. aureus* 209 P in presence of subinhibitory concentrations of fosfomycin and ampicillin (1/8 MIC), alone or in association.

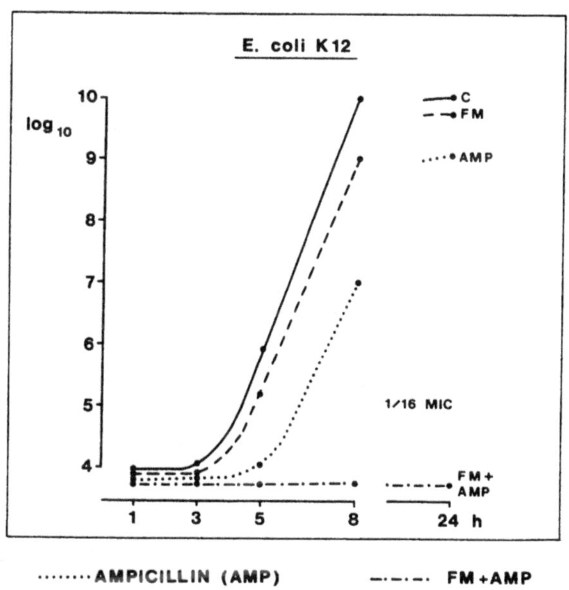

Fig. 14.6 - Growth curves of *E. coli* K 12 in presence of subinhibitory concentrations of fosfomycin and ampicillin (1/16 MIC), alone and in association.

and FM can be obtained in *Staph. aureus* and *E. coli* through a previous exposure to a subinhibitory concentration of TMP.

FM in subinhibitory concentrations causes a reduction in the mutation rate to RMP in *Staph. aureus* and *E. coli*. The reverse is also true, that is, RMP in subinhibitory concentrations reduces the mutation rate to FM-resistance.

On the contrary no effect on the mutation rate to ampicillin resistance was obtained through fosfomycin and conversely no influence on fosfomycin resistance was obtained through previous exposure to ampicillin.

If we now examine the effects that the same agents have on the growth rate some interesting features can be observed.

TMP which reduces the mutation rate to FM and RMP resistance on *Staph. aureus* and *E. coli*, has only a slight influence on the bacterial growth rate when associated with the same agents.

Fosfomycin and rifampicin reciprocally reduce the mutation rate to resistance and, in addition, they have a remarkable action on the growth rate when they are combined.

Fosfomycin and ampicillin, which do not affect the emergence of resistance to each other, have a remarkable influence on growth rate, as could be expected from the "sequential blockade" they exert on the construction of the bacterial cell wall.

It seems, therefore, that "synergistic" activity among antibiotics can be the result of different mechanisms.

The study of emergence of resistance and modifications of growth rate can give useful information in this regard, particularly when dosages far below the MIC can be used. In fact at these concentration levels these activities can be better differentiated.

A different trend can be observed according to antibiotics and bacterial species. By this it is possible to see a synergistic activity in the inhibition of bacterial growth but no effect in the inhibition of emergence of resistant mutants or viceversa. Each one of these effects can exert a different influence and have a different weight in determining the so-called synergism or antagonism among antibiotics.

It is possible that the knowledge of subtler mechanisms of interaction among antibiotics can elucidate the significance and the relevance of what is "macroscopically" seen as synergism or antagonism.

References

[1] ARIOLI, V., BERTI, M., CARNITI, G., ROSSI, E., SILVESTRI, L. G. (1977) "Interaction between rifampicin and trimethoprim in vitro and in experimental infections", *J. Antimicrob. Chemother.*, **3**, 87-94.

[2] BOROWSKI, J., LINDA, H. (1977) "Combined action of fosfomycin with β-lactam and aminoglycoside antibiotics", *Chemotherapy*, **23** (suppl. 1), 82-85.

[3] CLOWES, R. C., HAYES W. (1968) *Experiments in Microbial Genetics*, Blackwel Scientific Publ., Oxford, pag. 22-26.

[4] DAZA, R., MORENO LOPEZ, M., DAMASO, D. (1977) "Interaction of fosfomycin with other antibiotics", *Chemotherapy*, **23** (suppl. 1), 86-92.

[5] FLEMING, A., VOUREKA A., KRAMER, J. R. H., HUGHES, W. H. (1950) "The morphology and motility of *Proteus vulgaris* and other organisms cultured in presence of penicillin", *J. Gen. Microbiol.*, **4**, 257-269.

[6] GALE, E. F., CUNDLIFFEE., REYNOLDS, P. E., RICHMOND M. H., WARING, M. J. (1972) *The molecular basis of antibiotic action*, Wiley, London, p. 36 and 432-433.

[7] GARDNER A. D. (1940) "Morphological effects of penicillin on bacteria", *Nature*, **146**, 836-838.

[8] GIALDRONI GRASSI G., FERRARA, A., NAVONE, A., PEONA, V., SALA, P. (1977) "Attività antibatterica della fosfomicina. I. Determinazione della concentrazione minima inibente della fosfomicina in associazione con altri antibiotici", in *Aggiornamenti di Chemioterapia*, (E. Genazzani ed.), ESAM, Rome, p. 185-221.

[9] GIALDRONI GRASSI, G., FERRARA, A., NAVONE, A., PEONA, V., SALA, P. (1976) "Attività antibatterica della fosfomicina. II. Determinazione della concentrazione minima battericida in associazione con altri antibiotici", *Giorn. It. Chemiot.*, **23**, 87-94.

[10] GIALDRONI GRASSI, G. (1978) "Effects of subinhibitory concentrations of antibiotics on emergence of drug resistant bacteria in vitro", in: *Current Chemotherapy*. Proc. Xth Internat. Congr. of Chemotherapy, Zürich 1977, (W. Siegenthaler, R. Lüthy eds.), p. 77. American Society for Microbiology, Washington D.C.

[11] GREENFIELD, S., FEINGOLD, D. S. (1970) "The synergistic action of the sulfonamides and the polymyxins against *Serratia marcescens*", *J. Infect. Dis.*, **121**, 555-558.

[12] GREENWOOD, D. (1976) "Differentiation of mechanisms responsible for inoculum effects in the response of *Escherichia coli* to a variety of antibiotics", *J. Antimicrob. Chemother.*, **2**, 87-95.

[13] KAHAN, F. M., KAHAN J. S., CASSIDY, P. J., KROPP, H. (1974) "The mechanism of action of fosfomycin (phosphonomycin)", *Ann. N.Y. Acad. Sci.*, **235**, 364-386.

[14] KERRY, D. W., HAMILTON-MILLER, J. M. T., BRUMFITT, W. (1975) "Trimethoprim and rifampicin: in vitro activities separately and in combination", *J. Antimicrob. Chemother.*, **1**, 417-427.

[15] LORIAN, V. (1974) "Some effects of subinhibitory concentrations of penicillin on the structure and division of *Staphylococci*", *Antimicrob. Ag. Chemother.*, **7**, 864-870.

[16] LORIAN, V., ATKINSON, B. (1975) "Abnormal forms of bacteria produced by antibiotics", *Amer. J. Clin. Pathol.*, **64**, 678-688.

[17] LORIAN, V., ATKINSON, B. (1976) "Effects of subinhibitory concentrations of antibiotics on cross walls of cocci", *Antimicrob. Ag. Chemother.*, **9**, 1043-1055.

[18] LORIAN, V., ATKINSON, B. (1976) "Effects of subinhibitory concentrations of fosfomycin on bacteria", *Giorn. It. Chemiot.*, **23**, 65-74.

[19] PARK, J. T., STROMINGER, J. (1957) "Mode of action of penicillin", *Scienze*, **125**, 99.

[20] Parsley, T. L., Provonchee, R. B., Glicksman, C., Zinner, S. H. (1977) "Synergistic activity of trimethoprim and amikacin against Gram-negative bacilli", *Antimicrob. Ag. Chemother.*, **12**, 349-352.

[21] Rosenblatt, J. E., Stewart, P. R. (1974) "Combined activity of sulfamethoxazole, trimethoprim and polymyxin B against Gram-negative bacilli", *Antimicrob. Ag. Chemother.*, **6**, 84-91.

[22] Shah, P. M., Heetderks, G., Stille, W. (1976) "Activity of amikacin at sub-inhibitory levels", *J. Antimicrob. Chemother.*, **2**, 97-100.

[23] Simmons, N. A. (1970) "Colistin, sulphamethoxazole and trimethoprim in sinergy against Gram-negative bacteria", *J. Clin. Pathol.*, **23**, 757-764.

[24] Woodruff, H. B., Kahan, F. M., Wallick, H., Miller, A. K., Stapley, E. O. (1975) "Fosfomycin, a cell wall active antibiotic not inactivated by episome-mediated enzymes", in: *Microbial Drug Resistance* (S. Mitsuashi, H. Hashimoto eds.), University Park Press, Baltimore, 539-559.

[25] Zinner, S. H., Glicksman, C., Provonchee, R. B., Parsley, T. L. (1978) "Synergistic effects of trimethoprim and aminoglycosides against *Serratia marcescens*", in: *Current Chemotherapy.* Proc. Xth Internat. Congr. of Chemotherapy, Zürich 1977 (W. Siegenthaler, R. Lüthy eds.), 438-439. American Society for Microbiology, Washington D.C.

15

Synergistic activity *in vitro* and *in vivo* of a combination of ampicillin and cloxacillin

R. Sutherland, K. R. Comber

Beecham Pharmaceuticals Research Division,
Brockham Park, Betchworth, Surrey, RH3 7AJ, England

Summary

A combination of equal parts of ampicillin and cloxacillin was tested for antibacterial activity *in vitro* and *in vivo* against a range of ampicillin-sensitive and ampicillin-resistant bacteria. In general, the combination produced additive effects against bacteria sensitive to the individual penicillins but synergy was demonstrated between ampicillin and cloxacillin against a number of β-lactamase-producing ampicillin-resistant strains of gram-negative bacilli. Combined therapy was effective in the treatment of experimental infections and the combination of ampicillin and cloxacillin was significantly more active than ampicillin against a variety of infections. The results suggest that combined therapy with ampicillin and cloxacillin may be efficacious in the treatment of selected infections caused by certain ampicillin-resistant bacteria.

Introduction

Ampicillin is active against a broad-spectrum of gram-positive and gram-negative bacteria, whereas cloxacillin is effective only against gram-positive bacteria and shows little activity of consequence against gram-negative bacteria. However, ampicillin is inactivated by the β-lactamases produced by staphylococci and gram-negative bacilli but cloxacillin is usually stable to these enzymes. The resistance of gram-negative bacilli to ampicillin is often associated with the β-lactamase activity of these organisms, but, in the case of cloxacillin, resistance appears to be due to an intrinsic insensitivity of the cell to the compound.

A combination of ampicillin and cloxacillin would be expected to be active against many bacteria including β-lactamase-producing strains of *Staphylococcus aureus*. In addition, synergistic effects may be produced by ampicillin + cloxacillin against certain β-lactamase-producing ampicillin-resistant strains of gram-negative bacilli as a result of the competitive inhibition of the β-lactamase by the penicillinase-stable penicillin, cloxacillin [1], [2], [3], [4].

In this study, the antibacterial effects produced by a combination of equal parts of ampicillin and cloxacillin were compared *in vitro* and against experimental infections with the antibacterial activities demonstrated by the individual compounds.

Materials and methods

Minimum Inhibitory Concentrations. Two-fold serial dilutions of ampicillin, cloxacillin and of a combination of equal parts of ampicillin and cloxacillin were prepared in 18 ml volumes of molten nutrient agar and poured in petri dishes. When set, the plates were inoculated with 0.001 ml of an undiluted overnight broth culture of the test organism, an inoculum of 10^5-10^6 cells. Minimum Inhibitory Concentrations (MIC) were determined after overnight incubation at 37° C.

Intraperitoneal mouse infections. Mice were injected by the intraperitoneal route with 0.5 ml of a suspension in mucin of a dilution of an overnight broth culture of the test organism standardised to give an infective inoculum of 100-1000 Median Lethal Doses. The antibiotics were administered immediately after infection to groups of 10 mice by gastric intubation or by subcutaneous injection. The numbers of animals surviving 4 days after infection were recorded and the dose required to produce 50% protection of infected animals was calculated.

Mouse pyelonephritis. Mice were infected intravenously with 0,2 ml of a broth dilution of an ampicillin-resistant strain of *Escherichia coli* (JT 415), an inoculum of 10^8 cells. Iron sorbitol citrate (Jectofer) 30 mg/kg, was administered by the intramuscular route, 18, 42, 66 and 90 hours after injection to enhance the virulence of the organism for the kidney [5]. Treatment with subcutaneous injection of the penicillins was started 19h after infection and continued 4 times a day at 2 hourly intervals for 4 days. Groups of 5 mice were killed 7 days and 14 days after infection, the kidneys were examined for abscesses and bacterial counts were performed on homogenates of the infected kidneys.

Results

Activity in vitro. The antibacterial activities of ampicillin and of cloxacillin against a range of gram-negative and gram-positive bacteria are illustrated in Table 15.1. Ampicillin was active at low concentrations against gram-positive cocci, except penicillin-resistant staphylococci which were resistant as a result of β-lactamase activity. Gram-negative bacteria sensitive to ampicillin included gonococci, *Haemophilus influenzae*, *Escherichia coli*, *Proteus mirabilis* and *Salmonella* and *Shigella* species. Strains of *Pseudomonas aeruginosa*, Klebsiella and Enterobacter species, indole-positive *Proteus* species, *Serratia marcescens* and *Bacteroides fragilis* were relatively insensitive. Cloxacillin was active against pyogenic streptococci, pneumococci and staphylococci, including penicillinase-producing strains of *Staphylococcus aureus*, but was relatively inactive against enterococci. In general, gram-negative bacteria were relatively insensitive to cloxacillin with the exception of gonococci which were quite sensitive to the compound.

In general, ampicillin + cloxacillin produced additive effects against ampicillin-sensitive bacteria, so that the combination showed the activity to be expected from the activities of its constituent penicillins. In a number of cases, however, it was observed that the combination appeared to be as active as ampicillin although cloxacillin was usually relatively inactive against the test organisms (Table 15.2). It can be seen that about half of the strains of *E. coli* and *Shig. sonnei* tested appeared to be as sensitive to the combination as to ampicillin although cloxacillin was relatively inactive against these

Table 15.1 Antibacterial Spectrum of ampicillin and cloxacillin MIC (μg/ml)

Organism	*Ampicillin*	*Cloxacillin*
Escherichia coli	5.0	500
Salmonella typhi	1.25	500
Shigella sonnei	2.5	250
Klebsiella aerogenes	250	500
Pseudomonas aeruginosa	> 500	500
Proteus mirabilis	2.5	250
Proteus morganii	250	250
Haemophilus influenzae	0.25	12.5
Neisseria gonorrhoeae	0.02	0.1
Bacteroides fragilis	25	250
Staphylococcus aureus	0.1	0.1
*Staphylococcus aureus**	250	0.5
Streptococcus pyogenes	0.02	0.1
Streptococcus faecalis	1.25	25

* Penicillinase-producing strain.

Table 15.2 Activity of ampicillin + cloxacillin against ampicillin-sensitive bacteria.

Organism	*No. of strains*	*Increase in activity with ampicillin/cloxacillin**					
		–2	0	2	4	8	16
E. coli	43	3	16	22	2		
Shigella sp.	24		12	12			
Salmonella sp.	20	1	18	1			
P. mirabilis	10		8	2			
S. faecalis	20		2	18			
Staph. aureus	6		4	2			

$* \dfrac{\text{MIC of ampicillin}}{\text{Conc}^{n}\text{ of ampicillin in MIC of Combination}}$

Table 15.3 Examples of synergism between ampicillin and cloxacillin against ampicillin-resistant strains of gram-negative bacilli MIC (μg/ml).

Organism	*Ampicillin*	*Cloxacillin*	*Ampicillin/Cloxacillin*
Escherichia coli JT477	250	500	25 + 25
Shigella sonnei 5239	250	500	25 + 25
Enterobacter cloacae 7627	250	500	12.5 + 12.5
Citrobacter freundii W18	100	500	12.5 + 12.5
Proteus morganii 718	125	1000	12.5 + 12.5
Proteus vulgaris K	50	> 500	12.5 + 12.5
Serratia marcescens US4	250	1000	12.5 + 12.5
Acinetobacter anitratus 1025	50	1000	5 + 5
Acinetobacter lwoffii T834	100	> 1000	12.5 + 12.5
Pseudomonas cepacia 1211469	50	1000	5 + 5

bacteria (MIC 250 μg/ml or greater). Similarly, strains of enterococci (*Strep. faecalis*) were as sensitive to ampicillin/cloxacillin as to ampicillin although again the enterococci are generally 20 times less sensitive to cloxacillin than to ampicillin. Gram-negative bacilli that failed to show enhanced sensitivity to ampicillin/cloxacillin included *P. mirabilis* and *Salmonella species*, and staphylococci were generally no more sensitive to the combination than would be expected from the activities of the individual penicillins.

Examples of synergism demonstrated *in vitro* by the ampicillin + cloxacillin against ampicillin-resistant bacteria are illustrated in Table 15.3, and it can be seen that a wide range of gram-negative bacteria are represented in the Table. Cloxacillin was relatively inactive against these organisms (MIC 500->1000 μg/cloxacillin ml) and ampicillin MIC values varied from 50-250 μg ampicillin/ml. The combination showed a distinct degree of synergism against the test organisms and onc-tenth or less of the inhibitory concentration of ampicillin combined with one-tenth or less of the inhibitory concentration of cloxacillin produced inhibition. The frequency and extent of synergism demonstrated by ampicillin/cloxacillin against ampicillin-resistant bacteria are illustrated in Table 15.4. The most pronounced effects were seen with certain ampicillin-resistant strains of *E. coli*, *Shigella sonnei* and indole-positive Proteus species, and with many of these organisms the combination showed a 4-16 fold increase in activity compared with ampicillin. With Klebsiella/Enterobacter species the extent of synergism was generally less but was evident with a number of strains. Results are also given in Table 15.4 for penicillinase-producing strains of *Staphylococcus aureus* and it can be seen that the combination of ampicillin and cloxacillin was no more active than would be expected, in this case, no more active than the activity of its cloxacillin content. Similarly, synergism was not demonstrated with strains of R-factor enterobacteria with transferable ampicillin resistance, and the combination was inactive against these bacteria. Examples of this are seen in Table 15.4 where 15 strains of *E. coli* with ampicillin R-factors were no more sensitive to ampicillin/cloxacillin than to ampicillin alone. Synergism between ampicillin and cloxacillin was demonstrated solely against certain gram-negative bacilli possessing non-transferable chromosomally-mediated β-lactamases.

Activity in vivo. Data in Table 15.5 show the activities of ampicillin and of the combination of equal parts of ampicillin and cloxacillin against infections due to ampicillin-sensitive bacteria. Cloxacillin was relatively inactive compared with ampicillin against the gram-positive cocci and was inactive against the gram-negative bacilli.

Table 15.4 Antibacterial activity of ampicillin + cloxacillin against ampicillin-resistant bacteria

Organism	*No. of strains*	*Increase in activity with ampicillin/cloxacillin**					
		—2	0	2	4	8	16
E. coli	29		15		2	11	1
Shig. sonnei	18				6	6	6
Proteus sp. (indole +)	30		3	11	8	4	4
Klebsiella/ Enterobacter	81	1	11	45	19	5	
Staph. aureus	19		18	1			

$$* \; \frac{\text{MIC of ampicillin}}{\text{Conc}^{n} \text{ of ampicillin in MIC of Combination}}$$

Table 15.5 Activity of ampicillin + cloxacillin against experimental infections caused by ampicillin-sensitive bacteria, CD_{50} (mg/kg)

Organism	*Route of dosing*	*Ampicillin*	*Ampicillin/cloxacillin*
Staphylococcus aureus 1	oral subcutaneous	0.25 0.44	0.25 + 0.25 0.5 + 0.5
Streptococcus pyogenes 6	oral subcutaneous	0.36 0.12	0.28 + 0.28 0.18 + 0.18
Escherichia coli 8	oral subcutaneous	63 140	85 + 85 58 + 58
Salmonella typhimurium 10	oral subcutaneous	17 7	15 + 15 5 + 5
Proteus mirabilis 13	oral subcutaneous	46 55	60 + 60 35 + 35

Cloxacillin was relatively inactive (Cd_{50} ca. 20 mg/kg) against the Gram-positive cocci compared with ampicillin, and was inactive ($CD_{50} > 500$ mg/kg) against the Gram-negative bacilli.

There was no significant difference ($p < 0.05$) between ampicillin and ampicillin/cloxacillin in any test.

Table 15.6 Activity of ampicillin + cloxacillin against an experimental mouse infection produced by a penicillinase-producing strain of Staph. aureus 1555 CD_{50} (mg/kg).

Route of dosing	*Ampicillin*	*Cloxacillin*	*Ampicillin/Cloxacillin*
Oral	3250 (1410-7450)*	71 (49-104)	80 + 80 (100-256)
Subcutaneous	175 (116-262)	9 (—)	5.5 + 5.5 (6-20

* 95% Confidence Limits.

Ampicillin was effective against the test bacteria and, in general, ampicillin + cloxacillin showed the activity to be expected from the ampicillin content of the combination.

Ampicillin was inactive against an infection due to a penicillinase-producing strain of *Staph. aureus* whereas cloxacillin showed good activity, as would be expected from the *in vitro* sensitivity of the organism to the compounds. Ampicillin + cloxacillin showed activity corresponding to the cloxacillin moiety of the combination (Table 15.6).

The combination of ampicillin and cloxacillin was compared with ampicillin for activity against infections caused by ampicillin-resistant strains of gram-negative bacilli. The results in Table 15.7 show the relative activities against three strains of *E. coli*, two strains of *Shig. sonnei* and two strains of *P. morganii*, all of which were more sensitive to ampicillin/cloxacillin *in vitro* than to ampicillin alone. Ampicillin failed to protect the mice at very high dose levels (5000 mg/kg) by the oral route against the strains of *E. coli* and *Sh. sonnei* whereas the ampicillin/cloxacillin combination was effective at dose levels containing one-tenth to one-twentieth these levels of ampicillin. Ampicillin showed greater activity against these infections when given by the subcutaneous route rather than the oral route but ampicillin/cloxacillin showed a statistically significant ($p < 0.05$) greater activity than ampicillin alone. Similarly, the combination was significantly more active than ampicillin by both routes against the infections caused by the ampicillin-resistant strains of *P. morganii*.

The effect of treatment of a pyelonephritis infection produced by *E. coli* (JT 415) is illustrated in Fig. 15.1. This strain was relatively resistant to ampicillin *in vitro* (MIC 250 μg/ml) and to cloxacillin

Table 15.7 Synergism between ampicillin and cloxacillin against experimental intraperitoneal mouse infections caused by ampicillin-resistant Gram-negative bacilli CD_{50} (mg/kg).

Organism	*Strain*	*Route*	*Ampicillin*		*Ampicillin/Cloxacillin* *
Escherichia coli	JT92	oral	>5000	1	320 + 320
		subcutaneous	500	S	195 + 195
	131	oral	>5000	1	280 + 280
		subcutaneous	740	S	220 + 220
	JT487	oral	>5000	1	235 + 235
		subcutaneous	500	S	89 + 89
Shigella sonnei	3464	oral	>5000	1	210 + 210
		subcutaneous	150	S	22.5 + 22.5
	3606	oral	>5000	1	90 + 90
		subcutaneous	700	1	90 + 90
Proteus morganii	BS55	subcutaneous	320	S	56.5 + 56.5
	DU718	oral	760	S	175 + 175
		subcutaneous	>500	1	80 + 80

* Cloxacillin was inactive, $CD_{50} > 1000$ mg/kg.
1 Unable to calculate confidence limits.
S Statistically significant, $p < 0.05$.

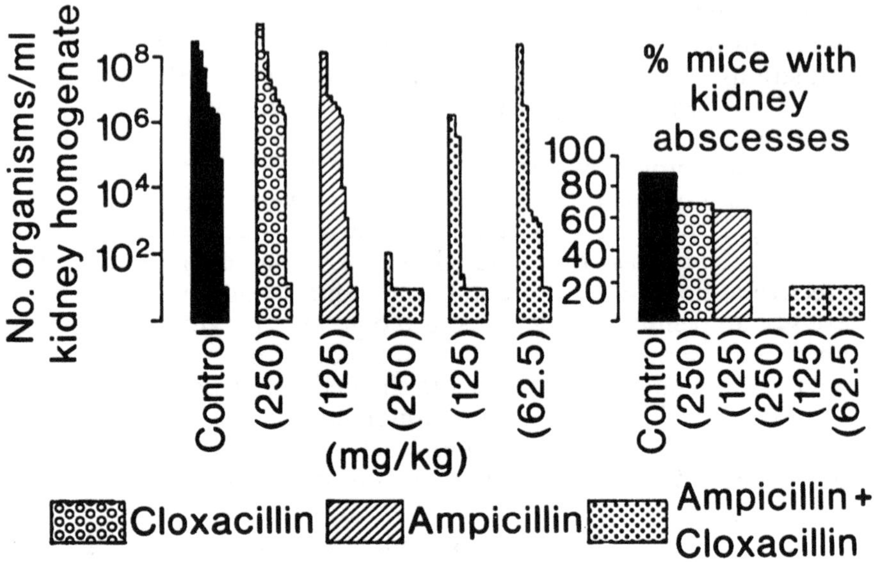

Fig. 15.1 - Comparative effects of ampicillin, cloxacillin and ampicillin + cloxacillin (equal parts) against an experimental pyelonephritis in mice infected with *Escherichia coli* JT 415.

(MIC $> 500\ \mu g/ml$) but was sensitive to a combination of ampicillin and cloxacillin containing 12.5 μg ampicillin/ml and 12.5 μg cloxacillin/ml. In untreated infected mice, 90% of the kidneys showed macroscopic lesions and 60-70% of animals treated with ampicillin (125 mg/kg) and 70-100% of those given cloxacillin (250 mg/kg) were similarly infected. Treatment with the combination of equal parts of ampicillin and cloxacillin resulted in significant reductions in the numbers of animals with kidney abscesses (0-20%). The bacterial counts of kidney homogenates were in agreement with these findings and high counts (10^5 to $> 10^8$ organism/ml of homogenate) were found in untreated animals or animals treated with ampicillin or cloxacillin, whereas the bacterial counts were significantly lower in the animals treated with combinations of equal parts of ampicillin and cloxacillin.

Discussion

The results reported here show that a combination of equal parts of ampicillin and cloxacillin is active against a broad spectrum of gram-positive and gram-negative bacteria including penicillinase-producing strains of staphylococci that are resistant to ampicillin.

There is no evidence of antagonism between the two penicillins and the combination displays, additive effects, and in a number of cases, synergistic effects. The combination of ampicillin + cloxacillin therefore extends the spectrum of ampicillin to incorporate penicillin-resistant staphylococci, and a number of ampicillin-resistant gram-negative bacilli.

Ampicillin + cloxacillin shows greatest synergistic effects against certain strains of *E. coli*, *Shig. sonnei* and *P. morganii* which have MIC values of 125-500 μg/ml, and the combination may be from five to ten times more active than ampicillin against these organisms. Other bacteria that appear to produce a β-lactamase relatively susceptible to cloxacillin include *Serratia marcescens* and *Acinetobacter* species, both of which may be associated with infections but are not generally considered among the more common pathogens. It is of interest that this group of organisms is generally more resistant to cephalosporins than to ampicillin and is relatively sensitive to carbenicillin as well as to ampicillin/cloxacillin. The combination demonstrates limited enhancement of activity against *Klebsiella aerogenes* and to a greater extent against Enterobacter and Citrobacter species, but in general the combination is only as active or twice as active as ampicillin against these organisms (MIC 50-250 μg/ml), although a number of strains show a rather greater degree of sensitivity. Strains of *E. coli* and other bacteria with plasmid-mediated β-lactamase show a high level of resistance to ampicillin (MIC > 1000 μg/ml) and the combination is rarely active at concentrations less than 500 μg/ml or greater. R. factor strains constitute a high proportion of ampicillin-resistant strains of *E. coli*, and ampicillin + cloxacillin would not be expected to be effective against these, or other bacteria, with transferable ampicillin-resistance.

The good correlation between the *in vitro* sensitivy of ampicillin-resistant bacteria to ampicillin/cloxacillin and the results of treatment of experimental mouse infections is of interest as it appears to show that bacteria with an increased sensitivity to ampicillin + cloxacillin *in vitro* also display an increased response to the combination *in vivo*. In some clinical studies, therapy of infections due to resistant organisms with combinations of penicillins has been considered to be successful, but in other studies no enhancement of activity was observed [6], [7], [8], [9]. However, as it appears that the organism was not always sensitive to the combination of penicillins *in vitro* it is unlikely in these cases that a clinical response would be expected. The *in vitro* and *in vivo* data reported here suggest that combined therapy with ampicillin and cloxacillin would be successful against certain clinical infections.

References

[1] Bornside, G. H. (1968) *Appl. Microbiol.*, **16**, 1507-1511.
[2] Nishida, M., Mine, Y. and Kuwahara, S. (1969) *J. Antibiot.*, **22**, 144-150.
[3] Regamey, C., Barrelet, L. and Waldvogel, F. A. (1976) *Proc. 9th Int. Congr. Chemother.*, London 1975, vol. 5, 7-12.
[4] Sutherland, R. and Batchelor, F. R. (1964) *Nature*, Lond. 201, 868-869.
[5] Comber, K. R. (1976) *Proc. 9th Int. Congr. Chemother.*, London 1975, vol. 2, 311-316.
[6] Klastersky, J., Debusscher, L. and Ruhl, D. (1972) *Chemotherapy*, **17**, 382-391.
[7] Riff, L., Olexy, V. M. and Jackson, G. G. (1970) *Antimicrob. Ag. Chemother.*, **1969**, 405-409.
[8] Sabath, L. D., Elder, H. A., McCall, C. E. and Finland, M. L. (1967) *New Engl. J. Med.*, **277**, 232-238.
[9] Sabath, L. D., McCall, C. E., Steigbigel, N. H. and Finland, M. (1967) *Antimicrob. Ag. Chemother.*, **1966**, 149-155.

16

Mecillinam/pivmecillinam and amoxycillin as combined treatment in purulent exacerbations of chronic bronchitis

A. Pines

East Herts Hospital, Hertford, U.K.

Summary

Mecillinam may potentiate the action of other β-lactam antibiotics against *Haemophilus influenzae* both *in vitro* [13], [12] and *in vivo*. In preliminary experience with patients suffering from purulent exacerbations of chronic bronchitis and who had failed to benefit from previous antibiotic treatment, 40 were given the oral form — pivmecillinam — plus amoxycillin. Another 63 patients were given injections of mecillinam together with oral amoxycillin or ampicillin. Most patients improved greatly with the treatment.

One hundred and thirty-two patients with purulent exacerbations of chronic bronchitis were randomly allotted to treatment in three groups. They received a) amoxycillin 250 mg and pivmecillinam 200 mg; or b) amoxycillin 500 mg; or c) amoxycillin 500 mg and pivmecillinam 400 mg: three times daily for 10 days. By the seventh day of treatment there was significant improvement over amoxycillin alone for both groups given combined chemotherapy in conversion of sputum to mucoid and in general improvement; at the end of treatment results in patients given the higher doses of both antibiotics were still superior to amoxycillin alone.

Patients were observed 2-4 weeks later, when those given amoxycillin alone relapsed much more frequently. The three treatments were well tolerated and succeeded equally in clearing potential pathogens from the sputum. Combined treatment may be superior due to synergy against *Haemophilus influenzae* or to the elimination of β-lactamase producing organisms and should be investigated further.

Introduction

Conventional laboratory investigations have shown that mecillinam is inactive against *Haemophilus influenzae*, the most frequently encountered pathogen in bronchitis. Recently however, it has been shown both *in vitro* [13], [12] and *in vivo* [1] that mecillinam together with ampicillin or amoxycillin may act synergistically against this organism. Mecillinam has to be given by injection. Pivmecillinam is a derivative which is well absorbed by mouth.

It is a common problem that patients with purulent exacerbations of chronic bronchitis may not improve with currently used antibiotics. We gave pivmecillinam by mouth with either ampicillin or amoxycillin to 40 such patients who remained ill despite courses of ampicillin, amoxycillin, tetracycline or co-trimoxazole. Twenty-three improved greatly, pus disappearing from the sputum. This was especially striking in three patients with severe purulent bronchiectasis who remained ill despite treatment with the previously mentioned antibiotics, but improved greatly when given pivmecillinam and ampicillin. They relapsed each time treatment was stopped and so have continued with it for 12 months now without incident.

Following this success, we treated 63 patients with far more acute illnesses. Because of the severity of the illness, treatments with intramuscular injections of mecillinam were advisable. This was given as 400 mg eight-hourly for a mean of 8.2 days - total dosage 9.8 g. Amoxycillin by injection gives twice the blood levels of the oral preparation but is painful in our experience: patients received 500 mg eight-hourly by mouth, but a few who were desperately ill were given the intramuscular preparation.

The characteristics of the 63 patients are described in Table 16.1. Most were elderly bronchitics with severe purulent exacerbations. Twenty-three had received previous antibiotic treatment without response.

In 55 patients (89%) there was good or excellent clinical improvement (see later criteria of assessments). Treatment failed in only 11%. Details are recorded in Table 16.2. Bacteriological findings are noted in Table 16.3. Response of *E. coli* was especially striking, even though in a few these were resistant to ampicillin. There was bacteriological clearance in 19 out of 30 instances (63%).

One patient had diarrhoea. No patient complained of discomfort with injection of mecillinam, which was exceptionally well tolerated. Full blood counts and liver function tests (SGOT, SGPT, alkaline phosphatase and bilirubin) showed no evidence of toxicity.

Following these successes, a formal 'blind' study was undertaken. This compared amoxycillin alone with two different dosages of amoxycillin combined with pivmecillinam.

Materials and Methods

Patients. Patients with a long standing history of chronic bronchitis and who had a recent purulent exacerbation were admitted to the study. They were moderately ill on clinical assessment; those who

Table 16.1 Diagnosis, age and sex of patients treated with mecillinam and amoxycillin.

Diagnosis	*Males*		*Females*	
	Number	*Mean Age*	*Number*	*Mean Age*
Acute Bronchitis	1	33	—	—
Pneumonia	4	71	4	51
Bronchiectasis	6	72	2	55
Chronic Bronchitis	38	69	8	75
Total	49		14	

Table 16.2 Diagnosis and results in patients treated with mecillinam and amoxycillin.

Diagnosis	*Success*	*Partial success*	*Failure*
Chronic Bronchitis			
previous therapy	11	5	1
no prior therapy	14	9	5
Bronchiectasis			
previous therapy	4	2	1
no prior therapy	2	—	—
Pneumonia			
previous therapy	1	—	—
no prior therapy *	2	4	—
Acute Bronchitis			
no prior therapy	1	—	—
Total	35	20	7

* In a further patient, efficacy could not be evaluated.

Table 16.3 Bacteriological isolates in patients treated with mecillinam and amoxycillin.

Organism	*Eradicated*	*Persisted*
E. coli	8	2
H. influenzae	5	4
E. coli/H. influenzae	—	1/1
Proteus spp.	2	—
Pseudomonas	2	2
Citrobacter	—	1
S. aureus	1	—
S. pneumoniae	1	—
Total	19	11

were only slightly ill were excluded as were those severely ill and who required parenteral therapy. Patients were also excluded who had known penicillin or cephalosporin sensitivity, tuberculosis or hepatic disease. One hundred and thirty-two patients were treated.

Treatment. Three groups of treatment were devised. In group "B", amoxycillin 500 mg was given three times daily, a standard and successful treatment [10]. Two groups received combined therapy. In group "A", patients were given amoxycillin 250 mg and pivmecillinam 200 mg three times daily and group "C" received amoxycillin 500 mg and pivmecillinam 400 mg three times daily. All drugs were administered with food for a total period of 10 days. Patients were randomly allocated to treatment, which was allotted according to sealed envelopes drawn in strict numerical order.

Assessments. The general clinical condition of the patient was assessed daily, in two states of improvement or deterioration, by an independent physician who did not know which treatment each patient was receiving. A 24 hour specimen of sputum was examined each day by the naked eye for purulence, assessment being in three grades: purulent (50% or more of pus), mucopurulent (less than 50% pus) and mucoid or at the most with a trace of pus. The quantity of the specimen was measured daily and several samples were sent for bacteriological examination before and after treatment. The peak flow rate was measured before and after treatment, the best of three efforts being taken. The haemoglobin, white count, erythrocyte sedimentation rate and p Co_2 were determined before and after treatment, but as these assessments did not help greatly in evaluating the results of treatment, they are not described further. After the end of the course of therapy, successfully treated patients were observed in hospital for 14 to 28 days.

Comparison of Patients Before Starting Treatment. Comparison of the three groups of age, sex, chemotherapy during the previous week, degree of sputum purulence and quantity showed almost no significant differences between the three groups. (Tables 16.4, 16.5. 16.6). However, there was a significant difference between groups B and C in relation to sputum purulence ($p < 0.05$).

Table 16.4 The patients.

Group	*A*	*B*	*C*
Male (No. of patients)	36	28	31
Female (No. of patients)	9	16	12
Mean age (Years)	.69	69	62
Total No. of patients	45	44	43

Table 16.5 Comparison of groups.

Group	*A*	*B*	*C*
No. Patients	45	44	43
Chemotherapy previous week	12	19	16
Sputum purulent	15	20	10
Sputum mucopurulent	30	24	33

Table 16.6

Sputum Volume before treatment (24 hour specimen)	*Group (No. of patients)* *A*	*B*	*C*
0-10 mls	20	19	13
10-20 mls	9	12	14
20-30 mls	4	7	2
30-40 mls	2	3	5
> 40 mls	1	0	3
No data	9	3	3
Mean Sputum/Volum	16.6 mls	16.7 mls	18.5 mls
Total	45	44	43

Results

Figure 16.1 shows improvements in general condition, assessed on the 7th and 11th day after starting treatment. The group receiving the higher dose of combined treatment showed significant improvement over that containing amoxycillin alone on the 7th ($p < 0.05$) and 11th days ($p < 0.05$).

Table 16.7 shows the number of patients whose sputum became mucoid. By the 7th day the sputum had become mucoid in a significantly high proportion of patients given pivmecillinam

Table 16.7 Sputum conversion.

	Group A		Group B		Group C	
	7 days	*11 days*	*7 days*	*11 days*	*7 days*	*11 days*
Mucoid	26	32	13	0	23	34
Purulent	14	7	27	10	16	5
Dead/ Ceased Therapy	5/0	5/1	3/1	3/1	1/3	1/3
Success rate %	58	71	30	68	54	79
Mean time for sputum conversion	4.7 days		6.7 days		5.8 days	

($p < 0.01$ and $p < 0.02$ for the low and high doses respectively), but by the 11th day there were no significant differences. Two thirds of the patients in each group achieved mucoid sputum.

Conversation of sputum to mucoid was quicker in those receiving combination therapy. In patients with purulent sputum before treatment the mean times for sputum conversion were 5, 5.1 and 6.8 days for the low dose and high dose combination and amoxycillin alone respectively. For patients with mucopurulent sputum before treatment the corresponding values were 4.1, 5.8 and 6.9 days.

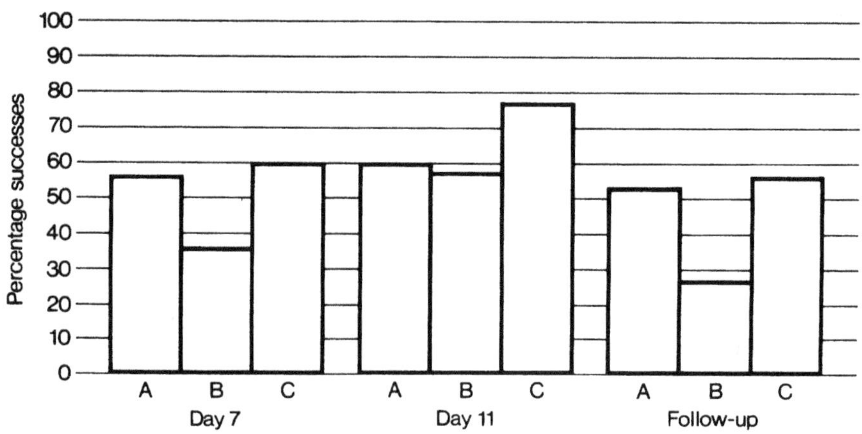

Fig. 16.1 - General assessment.

When patients were grouped according to the degree of sputum purulence before treatment (an indication of the degree of infection) there was less success in those patients given the small doses of amoxycillin plus pivmecillinam and who had sputum containing more than 50% pus: otherwise there were no differences within each group (Table 16.8). However, in patients with less than 50% pus in their sputum before treatment those given the higher dose of pivmecillinam improved more than with amoxycillin alone ($p = 0.05$).

Sputum volume decreased substantially in nearly two thirds or more in each group (Table 16.9).

Improvement in peak flow rate was greatest in patients receiving the higher doses of pivmecillinam and amoxycillin ($p < 0.05$) (Table 16.10).

All patients who were successfully treated were observed in hospital for a further two to four weeks. Patients who relapsed and were given further chemotherapy are shown in Table 16.11. Fewer relapses were recorded in both groups receiving combined therapy. This was most pronounced in patients with less than 50% pus in their sputum before treatment. At the end of the follow up period, both combined therapy groups had significantly greater success rates than

Table 16.8 Response in relationto pre-treatment sputum purulence.

	No. Patients	*Successes*	*% Success*
Group A			
Purulent	15	5	33%
Mucopurulent	30	22	73%
Group B			
Purulent	20	12	60%
Mucopurulent	24	13	54%
Group C			
Purulent	10	7	70%
Mucopurulent	33	26	79%

Table 16.9 Sputum volume after treatment.

Group	*A*	*B*	*C*
Vol. reduced by 50%	64%	71%	82%
Vol. same	25%	18%	8%
Vol. increased	11%	11%	10%

Table 16.10 Peak flow rate - Percentage improvement.

Group	A	B	C
Mean initial PFR (L/min)	180	167	173
Over 50% improvement	6%	6%	24%
Up to 50% improvement	37%	38%	46%
No improvement	57%	56%	30%

Table 16.11 Relapses in patients successfully treated.

	Group A			Group B			Group C		
	Purulent	Mucopurulent	Total	Purulent	Mucopurulent	Total	Purulent	Mucopurulent	Total
Relapsed	3	0	3	6	7	13	3	5	8
No relapse	2	22	24	6	6	12	4	20	24
% relapse	60	NIL	11	50	54	52	43	20	25

Table 16.12 General assessment after follow up in relation to pre-treatment sputun purulence

	Group A			Group B			Group C		
	Purulent	Mucopurulent	Total	Purulent	Mucopurulent	Total	Purulent	Mucopurulent	Total
Successes	2	22	24	6	6	12	4	20	24
Failures	13	8	21	14	18	32	6	13	19

the group receiving amoxycillin alone, not only when all patients were considered ($p < 0.02$ and $p < 0.01$) (Fig. 16.1) but especially when those with mucopurulent sputum before treatment were considered ($p < 0.01$ and $p < 0.01$). (Table 16.12)

Side Effects

Side effects were reported by 10 patients. In group "A", 3 had gastrointestinal upsets and one a rash. In group "B", 2 had a gastric upset and one a rash. In group "C" 2 had rashes (both later volunteered a history of previous rashes on ampicillin) and three had gastric upsets. A few patients (one, two and three respectively), had to stop treatment in each group.

Routine liver function tests performed on each patient before and after treatment revealed no evidence of toxic effect.

Bacteriology

There were no differences in the number of potential pathogens isolated in each group nor, with the possible exception of *E. coli*, in their clearance by treatment (Table 16.13).

Discussion

Our patients were severely disabled with long standing chronic bronchitis and with definite purulent exacerbations. Significant differences appeared in favour of the two groups treated with both pivmecillinam and amoxycillin, including those where the dose of amoxycillin was halved. Both their general state and the sputum improved more frequently by the seventh day of treatment. However, the patients given amoxycillin alone often had worse infections, as judged by the greater proportion of those who had over 50% of pus in their sputum before starting treatment. By the 11th day only the patients receiving the higher dose of combination therapy did significantly better, including improvement in respiratory function as measured by peak flow rate.

There were interesting differences in response relating to the degree of sputum purulence before treatment — a rough measure of the degree of infection present. In patients with purulent sputum, response was correlated with the higher dose of amoxycillin, whilst in

Table 16.13 Bacterial isolates.

Group	*Before Treatment*			*After Treatment*		
	A	*B*	*C*	*A*	*B*	*C*
H. influenzae	4* (9)	3 (7)	3 (9)	0	0	0
H. influenzae/E. coli	1	0	0	0	0	0
S. pneumoniae	0 (3)	0 (5)	0 (3)	0	0	0
E. coli	2 (7)	5 (5)	4 (8)	0	3	1
Proteus spp.	1	3	0	2	4	2
S. aureus	1	0	1	0	2	1
Pseudomonas	1	2	1	0	2	2

* First figure = no. of isolates in sputum specimen immediately before treatment. Figure in parenthesis = isolates found at any time during the four weeks preceding treatment.

those with mucopurulent sputum, the addition of pivmecillinam was important.

During the follow up period, there were fewer relapses in patients receiving combination therapy, especially in those patients with mucopurulent sputum before treatment. Amoxycillin alone gave a higher relapse rate than in previous experience [10], but the patients had more severe illnesses in the present study.

Increasing the dose of amoxycillin might give higher sputum levels and more efficacy [6], [11], [3]. However, the results in our patients who had mucopurulent sputum and were given a reduced amount of amoxycillin plus pivmecillinam were equal in most respects and superior in some to those when twice the dose of amoxycillin alone was given. Further comparison of a still higher amount of amoxycillin with our higher combined dosage would be worthwhile.

Thus we have found evidence that the combination of amoxycillin and pivmecillinam confers benefit over amoxycillin alone. This may be due not only to synergy against *Haemophilus influenzae*, but to a suppressive effect on β-lactamase producing Enterobacteriaceae, which were regarded by Maddocks and May [5] as indirect pathogens. Certainly in those patients where *E. coli* was isolated from sputum, combination therapy with mecillinam/pivmecillinam and amoxycillin, appeared very effective in eliminating the organism.

Other antibiotic combinations in which synergy has been claimed in the treatment of chronic chest infections though with varying substantiation, include sulphamethoxazole plus trimethoprim [2], [4], [7]; penicillin plus streptomycin [8] and novobiocin plus tetracycline [9]. The antibiotics in these combinations are each separately effective *in vitro* against *H. influenzae*, whereas mecillinam is not.

Conclusion

Further experience with such combined chemotherapy is required, with more detailed bacteriological assessment, before our promising results can be fully accepted.

Meanwhile in patients with purulent chest infections and who have failed to benefit from previous chemotherapy, the combination of amoxycillin with pivmecillinam, both in high dosage, may often succeed. In severely ill patients the intramuscular mecillinam may be preferred, together with intramuscular or oral amoxycillin.

References

[1] Grunberg, E., Cleeland, R., Beskid, G., and De Lorenzo, W. F. (1976) "In vivo synergy between 6 β-amidino penicillanic acid derivatives and other antibiotics", *Antimicr. Ag. Chemother.*, **9**, 589.

[2] Hughes, D. T. D. (1969) "Single-blind comparative trial of Trimethoprim - sulphamethoxazole and Ampicillin in the treatment of exacerbations of chronic Bronchitis", *Brit. Med. Journal*, **4**, 470.

[3] Ingold, A. (1975) "Sputum and serum levels of amoxycillin in chronic bronchial infections", *Brit. J. Dis Chest*, **69**, 211.

[4] Lal, S. and Bhalla, K. K. (1969) "Comparison of tetracycline and trimethoprim — sulphamethoxazole in acute episodes in chronic chest infections", *Postgrad. Med. Journal*, **45** (suppl.), 91.

[5] Maddocks, J. L. and May J. R. (1969) "Indirect Pathogenicity of penicillinase producing Enterobacteria in chronic bronchial infections", *Lancet*, **i**.

[6] May, R. and Ingold, A. (1972) "Amoxycillin in the treatment of chronic non-tuberculosis bronchial infections", *Brit. J. Dis. Chest*, **66**, 185.

[7] Pines, A. (1973) "Trimethoprim - sulphamethoxazole in the treatment and prevention of purulent exacerbations of chronic bronchitis", *J. Inf. Dis.*, **28** (suppl.), 706.

[8] Pines, A., Raafat, H., Plucinski, K., Greenfield, J. S. B. and Solari, M. (1968) "Antibiotic regimes in severe and acute exacerbations of chronic bronchitis", *Brit. Med. J.*, **ii**, 735.

[9] Pines, A., Raafat, H., Plucinski, K. and Solari, M. (1969) "A comparison of erythromycin, novobiocin, tetracycline and a novobiocin - tetracycline combination in purulent exacerbations of chronic bronchitis", *Brit. J. Dis. Chest.*, **63**, 206.

[10] Pines, A., Nandi, A. R., Raafat, H. and Rahman, M. (1977) "Amoxycillin and co-trimoxazole in acute purulent exacerbations of chronic bronchitis", *Chemotherapy* (Basel), **23**, 58.

[11] Stewart, S. M., Anderson, I. M. E., Jones, G. R. and Calder, M. A. (1974) "Amoxycillin levels in sputum, serum and saliva", *Thorax*, **29**, 110.

[12] Tybring, L. (1977) "Special aspects of laboratory investigations with mecillinam", *J. Antimicr. Chemo.*, **3** (Suppl. B), **23**.

[13] Wise, R., Ayliffe, G. A. J., Andrews, J. M. and Bedford, K. A. (1976) "The activity of mecillinam (FL 1060), in combination with other antibiotics, against *H. influenzae* and *Streptococcus faecalis*", in *Chemotherapy*, vol. 5, Edited by Williams J. D., and Geddes, A. M., Plenum Press, New York, p. 191.

17

Clavulanic acid in combination with penicillins: an *in vitro* study

R. Wise

Department of Medical Microbiology, Dudley Road Hospital, Birmingham, England

Summary

Clavulanic acid is a potent inhibitor of β-lactamases. The activity of this compound alone and when combined with different penicillins was studied. Marked synergy was noted against a wide range of different β-lactamase producing clinical isolates. Of particular note were the decreased penicillin MICs of *Bacteroides fragilis* and of β-lactamase producing strains of *Neisseria gonorrhoeae* when combined with clavulanic acid. β-lactamase producing strains of different Enterobacteriaceae also showed an increased susceptibility to amoxycillin when in combination with low amounts of clavulanic acid.

Introduction

Clavulanic acid was isolated from *Streptomyces clavuligarius* and has a structure which closely resembles the β-lactam nucleus [1]. Clavulanic acid itself is only a weak antibacterial agent but it was found that it is a potent inhibitor of many of the clinically important β-lactamases. It would appear that clavulanic acid is a progressive inhibitor of these enzymes.

Materials and Methods

Sodium clavulanate was obtained from Glaxo Laboratories (Greenford, Middlesex, U.K.). Benzylpenicillin, amoxycillin and carbenicillin were obtained from Beecham Research Laboratories. The activities of these compounds alone and in combination were studied by an agar plate dilution technique. The organisms were grown overnight in a nutrient broth (supplemented where necessary) and the organisms inoculated onto the antibiotic containing agar using a replicating device that transferred 1 μl. Two inocula were used either the undiluted overnight culture or a 1 : 1000 dilution. This gave a final number of colony forming units (cfu) per inoculum of approximately 10^6 and 10^3 respectively. The agar used throughout was Oxoid 'Isosensitest' pH 7.4. This was enriched with 10% lysed blood for the growth of *Bacteroides fragilis* and was chocolated in order to grow the *Haemophilus influenzae* and *Neisseria gonorrhoeae*. A wide range of penicillin concentrations were used and the concentrations of clavulanic acid incorporated were either 10, 5 or 1 μg/ml. Clavulanic acid and penicillin only control plates were included.

The plates were incubated at 37 °C for 24 hours. Ten per cent CO_2 was added in order to culture the *H. influenzae* and *N. gonorrhoeae* and a BBL 'GasPak' system used for anaerobes. The minimum inhibitory concentration (MIC) of the penicillin alone and when in combination with clavulanic acid was defined as the lowest amount of antimicrobial which gave a decrease in colony count of 99.9% compared with the original inoculum. Any inhibitory effect of the clavulanic acid alone was also noted.

Results

Clavulanic acid alone was found to be a weak antibacterial agent, having no activity against any of the Enterobacteriaceae tested. Two strains of *B. fragilis* were inhibited by 10 μg/ml of clavulanic acid. Most strains of the *N. gonorrhoeae* we tested were inhibited by 1 μg/ml or more of clavulanic acid. Several strains of *Staphylococcus aureus* were inhibited by 5 or 10 μg/ml especially when the low inoculum was used.

Haemophilus influenzae. As can be seen in Table 17.1 those strains which were β-lactamase producers showed a marked and progressive fall in the amoxycillin MICs when in combination with clavulanic acid. The non-β-lactamase producing strains showed a slight fall in MIC in combination with clavulanic acid.

Neisseria gonorrhoeae. Two β-lactamase producing strains were tested and there was a marked fall in MIC in the presence of clavulanic acid.

Staphylococcus aureus. Little effect of clavulanic acid in combination with benzylpenicillin was noted for the penicillin-sensitive strains that were tested. The known β-lactamase producers showed a mark-

Table 17.1 Effect of clavulanic acid on susceptibility of bacteria to amoxycillin.

Lab. No.	Amoxycillin		Amoxycillin + Clav 10 μg/ml		Amoxycillin + Clav 5μg/ml		Amoxycillin + Clav 1 μg/ml	
	10^3*	10^6	10^3	10^6	10^3	10^6	10^3	10^6
Esch. coli								
I3	4	4	1	4	2	4	4	4
I5	512	512	8	16	32	32	128	256
I11	4	4	1	2	4	4	4	4
I17	512	1024	8	32	32	64	128	256
124	>1024	>1024	1	8	16	16	128	256
I63	2	4	1	1	2	2	2	2
I64	2	4	1	1	2	2	2	2
I56	>1024	>1024	1	16	4	32	32	64
Klebsiella spp								
H1	64	256	0.5	1	1	8	2	16
H11	64	512	1	128	1	64	2	64
H18	64	512	0.5	1	1	64	2	64
H21	64	256	0.5	1	1	1	2	64
H24	64	512	0.5	1	1	2	2	4
Proteus mirabilis								
J11	0.5	0.5	0.25	0.25	0.25	0.5	0.5	0.5
J12	256	>512	0.25	0.5	0.5	1	4	16
J13	0.5	1	0.25	0.5	0.25	0.5	0.25	1
J15	0.5	1	0.25	0.5	0.25	0.5	0.25	1
J18	256	>512	0.5	0.5	1	1	1	2
J19	64	256	64	256	128	256	128	256
J40	128	>512	0.25	64	0.5	128	4	64
H. influenzae								
B127	0.25	0.25	0.06	0.12	0.25	0.25	0.25	0.25
S335	0.25	0.25	0.12	0.25	0.25	0.25	0.25	0.25
767	8	>16	0.25	0.25	0.5	0.5	0.5	>32
744	8	>16	0.25	0.25	0.5	0.5	0.5	1
12214	2	>16	0.06	0.12	0.12	0.25	0.25	0.5
1549	1	4	0.5	4	1	4	1	4
A106	0.25	0.5	0.25	0.25	0.25	0.25	0.25	0.5

* = Inoculum 10^3 or 10^6 cfu/μl.

ed MIC fall in the presence of clavulanic acid as can be seen in Table 17.2.

Bacteroides fragilis (Table 17.3). A wide range of sensitivity to benzylpenicillin was seen amongst these organisms. It was interesting to note that even those strains which were highly resistant to penicillin showed a greater than 1000 fold fall in penicillin susceptibility when combined with 10 μg/ml of clavulanic acid. At an inoculum of 10^3 cfu/μl one half of the strains tested had a penicillin MIC of $\leqslant 0.12$ μg/ml when combined with 5 μg/ml of clavulanic acid.

Enterobacteriaceae (Table 17.1). There was the usual bimodal dis-

Table 17.2 Effect of clavulanic acid on susceptibility of six strains of Staph. aureus *to penicillin.*

Strain	*MIC Penicillin*		*MIC Penicillin + Clav 10 μg/ml*		*MIC Penicillin + Clav 5 μg/ml*		*MIC Penicillin + Clav 1 μg/ml*	
	10^3X*	10^6	10^3	10^6	10^3	10^6	10^3	10^6
'Oxford' Staph	0.008	0.03	I	I	I	0.008	0.008	0.008
F111**	8	1024	0.06	8	0.12	8	1	32
F112**	4	1024	0.03	4	0.06	8	0.5	32
F107**	1	1024	I	2	0.06	8	0.12	8
F121	0.06	256	I	I	I	0.015	0.008	0.12
F15	0.015	0.03	I	0.008	0.008	0.015	0.015	0.03
F69*	0.25	1024	I	0.008	0.015	0.06	0.06	4

* = Inoculum 10^3 or 10^6 cfu/μl.
** = Penicillin and methicillin resistant strains.
I = Inhibited by clavulanic acid alone.

Table 17.3 Effect of clavulanic acid on susceptibility of Bacteroides fragilis *to penicillin.*

Lab. No.	*Penicillin*		*Penicillin + Clav. 10 μg/ml*		*Penicillin + Clav. 5 μg/ml*		*Penicillin + Clav. 1 μg/ml*	
	10^3*	10^6	10^3	10^6	10^3	10^6	10^3	10^6
Bacteroides fragilis								
B2	8	32	<0.01	1	2	1	1	1
B12	8	32	<0.01	4	4	16	8	16
Whale	16	32	<0.01	0.25	0.12	4	0.5	8
B16	16	32	<0.01	0.5	0.12	0.5	0.5	0.5
B14	16	32	<0.01	0.25	0.5	0.5	2	1
B4	4	8	<0.01	0.25	0.25	1	1	2
B1	8	16	<0.01	1	0.12	0.5	0.5	1
B22	16	32	<0.01	2	0.12	0.5	0.5	2
Whitlock	8	16	<0.01	0.06	0.12	0.5	0.5	0.5
B13	8	16	<0.01	0.25	0.06	0.5	0.5	1
B26	256	1024	<0.01	0.5	0.12	0.5	1	4
B18	8	16	<0.01	0.5	4	4	8	8
B9	8	16	<0.01	0.5	4	1	8	4
B24	8	16	<0.01	0.5	0.12	0.5	0.5	0.5
B25	512	1024	<0.01	1	0.25	2	2	4
B41	8	16	<0.01	<0.01	0.25	0.5	1	1
B3	8	16	<0.01	<0.01	0.25	0.25	0.5	0.5

* = Inoculum 10^3 or 10^6 cfu/μl.

tribution of amoxycillin sensitivities in the *E. coli* strains tested. This reflected the β-lactamase production. There was a progressive fall in the amoxycillin MIC with increasing amounts of clavulanic acid in those strains which were initially resistant to amoxycillin. This fall was more marked at the lower inoculum. The *Klebsiellae* strains tested were all resistant to amoxycillin and a marked fall in MIC was noted when clavulanic acid was present. A similar effect was also seen when the strains of *Proteus mirabilis* were tested. Clavulanic acid had no effect on the amoxycillin MIC of four strains of *Serratia marcescens*, three strains of *Enterobacter aerogenes* or one strain of *Enterobacter cloacae*. Two β-lactamase producing strains of *Pseudomonas aeruginosa* were also tested. In this case carbenicillin and clavulanic acid were used in combination. The carbenicillin MIC of the two β-lactamase producing strains were > 1024 μg/ml and there was no reduction in the MIC in the presence of 10 μg/ml of clavulanic acid.

Discussion

It has been known for many years that a number of penicillins and cephalosporins can inhibit β-lactamase production [2]. This inhibition is both reversible and competitive. It is interesting to note that the inhibition seen with clavulanic acid is progressive and also occurs at low concentrations.

Although clavulanic acid alone has low antibacterial activity it was interesting to note that against some non-β-lactamase producing strains of *N. gonorrhoeae* and the Enterobacteriaceae there did appear to be some synergistic activity between the penicillin and the clavulanic acid as increasing amounts of clavulanic acid was accompanied by a fall in the MIC to penicillin. It was also interesting to note that the β-lactamase producing strains of *Ps. aeruginosa* were not inhibited by clavulanic acid in combination with carbenicillin. This could be explained by the lack of permeability of clavulanic acid in this bacterial species.

Clavulanic acid is, therefore, an extremely interesting compound meriting further investigation.

References

[1] Howarth, T. T., Brown, A. G. and King, T. J. (1976), *J. Chem. Soc. Chem. Commun.*, 266-267.

[2] Sykes, R. B., Matthews, M. (1976), *J. Antimicrob. Chemother.*, **2**, 115-157.

18

Disappearance of resistance to aminoglycosides

J. H. Driessen

Regional Public Health Laboratory, Gemeenteziekenhuis
Bankastraat 57 - Dordrecht, The Netherlands

Summary

Resistance to aminoglycoside antibiotics disappeared in about 16/32 strains of 8 species of gram-negative bacteria after 15 to 30 subcultures at weekly intervals. Since almost no strains were resistant to amikacin, and at least 6/32 strains continued to remain resistant to gentamicin after 30 subcultures, amikacin, or one of the other newer aminoglycosides such as tobramycin or sisomicin, should possibly replace gentamicin as the drug of choice in the treatment of severe gram-negative infections. Resistance to aminoglycosides tends to develop rather quickly. Hence the need for combining these antibiotics with beta-lactam antibiotics. A simple, rapid and reliable method for performing synergy studies is urgently required.

Introduction

Kanamycin and gentamicin, two aminoglycosides, are quite often used in the treatment of life-threatening infections before the causative organism has been identified. As a result of this practice, the resistance rate of bacteria to these drugs has tended to increase. We therefore thought it important to determine whether two newer aminoglycosides, tobramycin and amikacin, should replace kanamycin and gentamicin as drugs of choice in the treatment of certain severe infections [1]. At the same time we were interested in the suggestion of Waterworth [2] that combinations of carbenicillin and aminoglycosides sterilize cultures because resistance to both of these types of antibiotics never exists in the same cell. This suggestion agrees with the findings of Lowbury [3] that intrinsic resistance to some antibiotics, such as kanamycin and chloramphenicol, is unusual in most species of gram-negative bacilli unless they have been exposed to these antibiotics. In the light of these considerations, we decided:

a) to compare the *in vitro* resistance pattern of unselected gram-negative strains and of gentamicin- or kanamycin-resistant strains to kanamycin, gentamicin, tobramycin, amikacin, ampicillin, cephaloridine and carbenicillin;
b) to study the extent of synergism between beta-lactam antibiotics and aminoglycosides.

Materials and Methods

Test organisms
a) 1098 gram-negative Dordrecht strains, i.e. strains isolated at random from patient material in the Dordrecht region, Netherlands, between February and August 1975.
b) 12 gram-negative, gentamicin-resistant Leuven strains, i.e. strains provided by Prof. Dr. J. Vandepitte of Leuven University Hospital, Belgium.
c) 20 gram-negative, gentamicin-resistant Bristol-Meyers strains, i.e. strains from the world-wide collection of Bristol-Meyers (Dr. K. E. Price).
d) 20 gram-negative, gentamicin-resistant Utrecht strains, i.e. strains provided by Dr. A. Kayser of St. Antonius Hospital, Utrecht, Netherlands.

Antibiotic disks. The following disks, supplied by Oxoid, Mast Laboratories or Bristol-Meyers Laboratories, were used: kanamycin (10 mcg), gentamicin (10 mcg), tobramycin (10 mcg), amikacin (10 mcg), ampicillin (25 mcg), cephaloridine (25 mcg), and carbenicillin (100 mcg).

Sensitivity testing
a) Sensitivity testing was performed on Oxoid DST agar (method of Kirby Bauer [4], modified by Garrod and Waterworth [5]).
b) Plates were inoculated by the method of Felmingham and Stokes [6].
c) Inhibition zones were measured down to 1 mm. With aminoglycosides, diameters greater than 15 mm were recorded as sensitive (positive) [7].

Synergy studies. Three different methods were used:
a) Rowsof disks are placed at right angles to each other [8]. To obtain straight borders of growth, disks are cut into halves, and, by using a special configuration, about 5 different antibiotics can be tested on the same plate (Fig. 18.1).
b) Disks are placed in a special configuration, about 15 mm apart, to

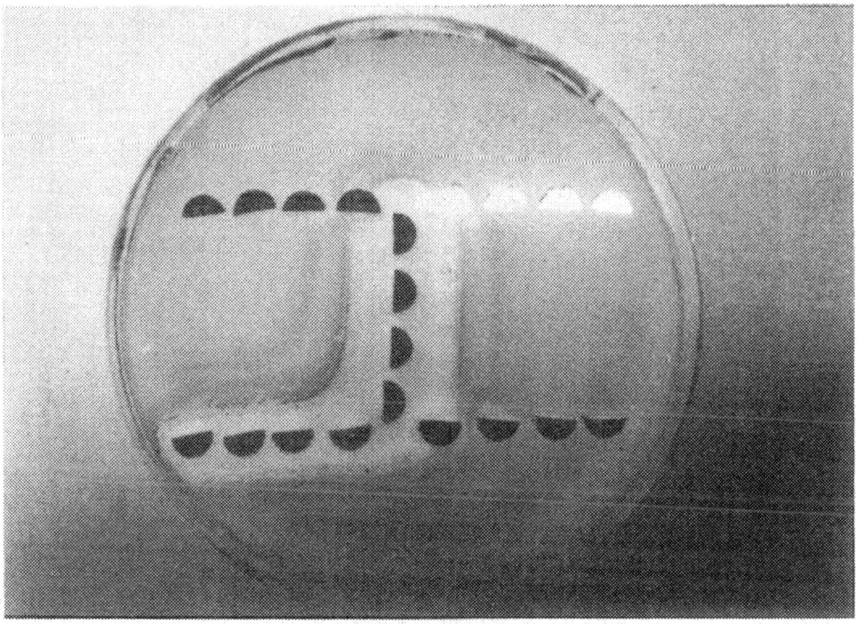

Fig. 18.1 - Synergism experiments, method (a).

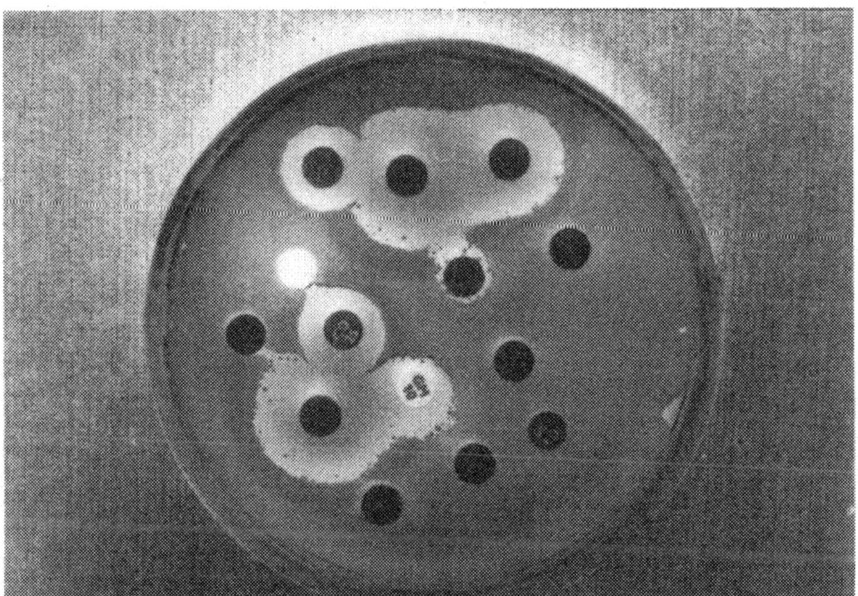

Fig. 18.2 - Synergism experiments, method (b).

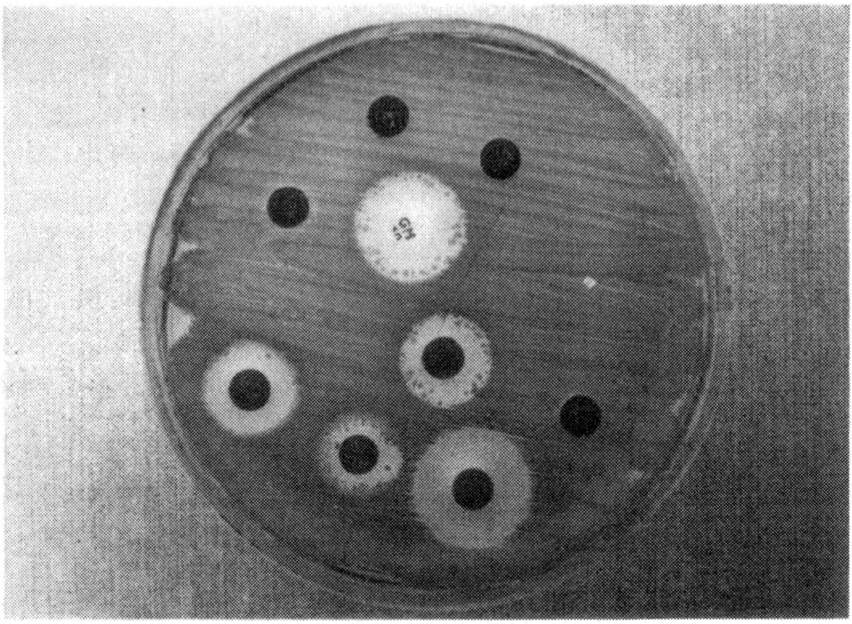

Fig. 18.3 - Loss of resistance: resistant colonies inside inhibition zone disappear after repeated subculturing.

study significant synergism or antagonism patterns (Fig. 18.2).

c) Disks of a non-aminoglycoside antibiotic are overlaid by 1/4 of a disk of an aminoglycoside, and the zones of inhibition of these overlaid disks are compared with those of non-overlaid disks.

Results

Of the 1098 Dordrecht strains, 282 proved resistant to kanamycin and none to gentamicin.

Of the 32 Leuven and Bristol-Meyers gentamicin-resistant strains that were weekly subcultured on antibiotic-free medium, about half became sensitive to various antibiotics after 15 to 30 subcultures (Table 18.1). The remarkable step-wise disappearance of resistance deserves a few words of explanation. After a varying number of subcultures, we observed the appearance of inhibition zones around disks that had not previously produced such zones. At the same time we also observed that these zones of inhibition contained resistant colonies (Fig. 18.3). After a few more subcultures these resistant colonies completely disappeared. At sensitivity testing, these colonies proved resistant to the antibiotic under consideration and proved to

Table 18.1 *Stepwise disappearance of resistance in 32 gram-negative strains to ampicillin* (Pn), *carbenicillin* (Py), *cephaloridine* (Cr), *kanamycin* (K), *gentamicin* (G), *tobramycin* (T), *and amikacin* (A) *after 15 to 30 subcultures.*

Strains	*Resistant to following antibiotics when isolated*	*Resistant to following antibiotics after 15 to 30 subcultures*	*Stepwise disappearance of resistance ** after 15 to 30 subcultures*
Leuven strains			
1 *Prov. stuartii*	1 Pn - Pr - Cr - K - G - T	1 Pn - Py - K	1 Cr - G - T
5 *Serratia marcescens*	1 Cr - G	1 Cr	1 G
	4 Pn - Py - Cr - K - G	1 sensitive to all	1 Pn - Py - Cr - K - G
		1 Cr - K	1 Pn - Py - G
		2 Cr	2 Pn - Py - K - G
1 *Proteus morganii*	1 Pn - Py - Cr - K	1 Cr	1 Pn - Py - K
2 *Klensiella*	1 Pn - Py - K	1 Pn - Py - K	
	1 Pn - Py - Cr - K - G	1 Pn - Py - Cr - G	1 K
1 *Citrobacter*	1 Pn - Py - Cr - K - G	1 Pn - Cr - K	1 Py - G
2 *Ps. aeruginosa*	1 Pn - Cr - K - G - T - A	1 Pn - Cr - K - T *	1 G - A
	1 Pn - Cr - K	1 sensitive to all	1 Pn - Cr - K
Bristol-Meyers strains			
8 *Klebsiella*	3 Pn - Py - Cr - G	2 Pn - Py - Cr	1 G
		1 Pn - Py - Cr - G	
	3 Pn - Py - Cr - K - G	3 Pn - Py - Cr - K	3 G
	2 Pn - Py - Cr - K - G - T	1 Pn - Py - Cr - K	1 G - T
		1 Pn - Py - Cr - K - G	1 T
4 *Enterobacter*	4 Pn - Py - Cr - K - G	2 Pn - Py - Cr - K	2 G
		2 Pn - Py - Cr - K - G	
3 *E. coli*	2 Pn - Py - Cr - K - G - T	1 Pn - Py - Cr	1 K - G - T
		1 Pn - Py - Cr - K	1 G - T
	1 K - G - T	1 K - G - T	
5 *Ps. aeruginosa*	3 Pn - Py - Cr - K	3 Pn - Py - Cr - K	
	1 Pn - Cr - K	1 Pn - Cr - K	
	1 Pn - Cr - K - G - T	1 Pn - Cr - K - G - T	

* This strain proved sensitive to G and A, without subculturing, after storage at 4 °C.
** Colonies inside inhibition zone gradually disappear during repeated subculturing (see text).

belong to the original species, as shown by re-determination. Further, some of these colonies proved sensitive to a beta-lactam antibiotic. In a number of strains, the stepwise disappearance of resistance was noted with gentamicin, tobramycin, cephaloridine, alone or in various combinations, so that no logical rule could be deduced to predict what would happen (Table 18.1). Some strains remained resistant even after 30 subcultures.

Of the 20 Utrecht gentamicin-resistant strains, 9/11 of the *Pseudomonas aeruginosa* strains proved resistant to kanamycin, gentamicin and tobramycin. Six of these strains were isolated from patients who had received gentamicin therapy. None of the Utrecht strains were subcultured to determine whether resistance disappeared.

The preliminary results of the synergy studies showed that synergism is often produced by combinations of beta-lactam and aminoglycoside antibiotics.

Comment

Experience throughout the world clearly shows that the increasing use and misuse of antibiotics renders more and more strains resistant. To prevent the misuse of antibiotics, prescription of antibiotics in the Dordrecht region is strictly monitored. The effectiveness of this measure in reducing bacterial resistance to antibiotics is clearly reflected in the reduced rates of the Dordrecht strains.

Most of the 20 Utrecht gentamicin-resistant strains were isolated from patients treated with gentamicin. The remaining strains appeared to be isolated from cross-infected patients. We were struck by the fact that 15 strains proved resistant to tobramycin and 3 to amikacin, two antibiotics that had never been used in the Utrecht region.

Some investigators have reported the disappearance of resistance to a specific antibiotic in a hospital ward after the antibiotic in question ceased to be used for some time [3], [9]. This agrees with the results of our *in vitro* studies on the stepwise disappearance of resistance to certain antibiotics.

We found the stepwise disappearance of resistance in the 32 Leuven and Bristol-Meyers strains very interesting. Nearly all the strains that remained resistant to gentamicin and kanamycin were sensitive to amikacin. This agrees with the findings of other investigators [10], [11], [12]. From this we conclude that kanamycin can be replaced by amikacin and that, since the introduction of the newer aminoglycosides, the position of gentamicin as the drug of choice in the treatment of severe gram-negative infections has become questionable.

We were not able to draw any definite conclusions from the results of our synergy studies on combinations of beta-lactam and aminoglycoside antibiotics. We feel the time has come for the Development of quicker, simpler and more reliable methods of performing these studies.

We feel it is very important that antibiotic prescription monitoring should be guided by the following conclusions of Lowbury [3]:

1. Short antibiotic course at fully effective dosage. Eradicate bacteria before resistance develops.
2. Combined antibiotic therapy can prevent or delay resistance.
3. Resistant strains can be eradicated fairly quickly by rotating the use of antibiotics and antibiotic combinations.

Acknowledgements

We are indebted to Prof. Dr. J. Pijck and Mr. H. H. Soep for discussing and correcting the manuscript, to Prof. Dr. J. Vandepitte of Leuven University Hospital and Dr. A. Kayser of St. Antonius Hospital, Utrecht, for providing gentamicin-resistant strains, and to Bristol-Meyers for supplying materials and strains.

References

[1] Reynolds, A. V., Hamilton-Miller, J. M. T. and Brumfitt, W. (1974) "Newer Aminoglycosides - Amikacin and Tobramycin: An in-vitro comparison with Kanamycin and Gentamicin", *Brit. Med. J.*, **3**, 778-780.

[2] Waterworth, P. M. (1972) "The in-vitro activity of Tobramycin compared with that of other aminoglycosides", *J. Clin. Path.*, **25**, 979-983.

[3] Lowbury, E. J. L. and Ayliffe, G. A. J. (1974) "Drug resistance in antimicrobial therapy", American Lecture series, Charles Thomas publ., Springfield, Illinois.

[4] Bauer, A. W., Kirby, W. M.M., Sherris, J. C. and Turck, M. (1966) "Antibiotic susceptibility testing by a standardized single disk method", *Amer. J. Clin. Path.*, **45**, 493-496.

[5] Garrod, L. P. and Waterworth, P. M. (1971) "A study of antibiotic sensitivity testing with proposals for simple uniform methods", *J. Clin. Path.*, **24**, 779-789.

[6] Felmingham, D. and Stokes, J. (1972) "Sterile Swabs as essential equipment for rapid reliable antibiotic sensitivity Tests", *Med. Lab. Technol.*, **29**, 198-200.

[7] Price, K. E. and Acar, J. F. (1976) *J. Infect. Dis.*, Amikacin Suppl.

[8] Ericsson, H., Tunevall, G. and Wickman, K. (1960) "A paper disc method for determination of bacterial sensitivity to antibiotics", *Scand. J. Clin. and Labo Invest.*, **12**, 414.

9] Geddes, A. M. and Williams, J. D. (1973) *Current Antibiotic Therapy* (Churchill Livingstone), p. 135.

[10] Price, K. E., Chisholm, D. R., Misiek, M., Leitner, F. and Tsai, Y. H. (1972)

"Microbiological Evaluation of BB-K8, A New Semisynthetic Aminoglycoside", *The Journal of Antibiotics*, **25**, 709-730.

[11] PRICE, K. E., PURSIANO, T. A., DE FURIA, M. D. and WRIGHT, G. E. (1974) "Activity of BB-K8 (Amikacin) Against Clinical Isolates Resistant to One or More Aminoglycoside Antibiotics", *Antimicrob. Ag. Chemother.*, **5**, 143-152.

[12] YOUNG, L. S. and HEWITZ, W. L. (1973) "Activity of five aminoglycoside antibiotics in-vitro against Gram negative Bacilli and Staphylococcus aureus", *Antimicrob. Ag. Chemother.*, **4**, 617-625.

19

Comparative *in vitro* activity of cefoxitin, cefuroxime and cephalothin and of their interactions with sisomicin or gentamicin

H. H. Schassan, A. Sukardi, G. Bentfeld

Institute for Medical Microbiology and Immunology, Hamburg University, Germany

Summary

The *in vitro* effectiveness of cefoxitin, cefuroxime and cephalothin, alone or in combination with sisomicin or gentamicin were compared against 150 strains of Enterobacteriaceae commonly encountered in severe gram-negative infections. Of the cephalosporins, cefoxitin proved the most effective against indole-positive *Proteus* spp and *Serratia*, cefuroxime the most effective against *E. coli* and *Klebsiella*. Of the combinations, cefuroxime and sisomicin proved the most effective, acting synergistically against most strains of the 8 species tested, and cefoxitin and sisomicin showed the best effectiveness against resistant strains. The combination of cefuroxime or cefoxitin and sisomicin is recommended for initial treatment of severe infections.

Introduction

Combinations of beta-lactam and aminoglycoside antibiotics are often used in the initial treatment of severe infections due to gram-negative or resistant organisms [4], [5]. The causative organisms most commonly encountered in these infections are *E. coli*, *Klebsiella*, *Enterobacter*, *Proteus* spp, *Serratia* and *Ps. aeruginosa*. Some of these pathogens have high resistance rates to beta-lactam antibiotics such as ampicillin and cephalothin, so that use of these antibiotics alone before cultural and sensitivity studies is no longer justified.

After cultural and sensitivity studies, a penicillin or cephalosporin is often selected for the treatment of severe gram-negative infections. In cases of severe gram-negative septicemia or bacterial endocarditis, a synergistic combination of a cephalosporin and an aminoglycoside is usually preferred.

In the present study, we compared the *in vitro* effectiveness of cefoxitin, cefuroxime and cephalothin, alone and in combination with gentamicin or sisomicin, against 8 species of gram-negative bacteria, in order to determine which agent or combination of agents was likely to be most effective against severe gram-negative infections.

Materials and Methods

Organisms. 150 clinical isolates of the following gram-negative bacteria were tested: *E. coli* (20), *Klebsiella* (20), *Enterobacter* (20), *Serratia* (20), *Proteus mirabilis* (20), *Proteus morganii* (20), *Proteus vulgaris* (20) and *Proteus rettgeri* (10).

Antibiotics. The following antibiotics were used: cefoxitin, cefuroxime, cephalothin, alone and in combination with gentamicin or sisomicin.

Minimum inhibitory concentrations (MIC). These were determined by the tube dilution method in geometric dilutions of the agents in Mueller-Hinton broth (Difco) adjusted to pH 7.4. The inoculum was standardized to a final concentration of 10^5 to 10^6 CFU/ml. The tubes were incubated at 37° for 18 h.

Minimum bactericidal concentrations (MBC). These were determined by subculture of clear tubes on blood agar after further incubation for 18 h at 37°. The first tube in the dilution range that yielded sterile subcultures was recorded as the total MBC.

Synergy studies. To each tube in the dilution range of the cephalosporins was added a constant subinhibitory concentration of 0.2 mcg of gentamicin or sisomicin. The MIC and MBC of the combinations were determined as above. The effects of antibiotic combinations were considered: *Indifferent*, if there was no change in MIC values, *Additive*, if MIC values fell 1- to 3-fold, *Synergistic*, if MIC values fell 4-fold or more, and *Antagonistic*, if MIC values rose 2-fold or more.

Results

Activity against *E. coli* (Fig. 19.1)

Single drugs: At a concentration of 8 mcg/ml, cefuroxime inhibited the growth of 100% of the *E. coli* strains, cefoxitin 95%, and cephalothin 40%.

Combinations: The combinations of a cephalosporin and sisomicin proved more active than the combinations of a cephalo-

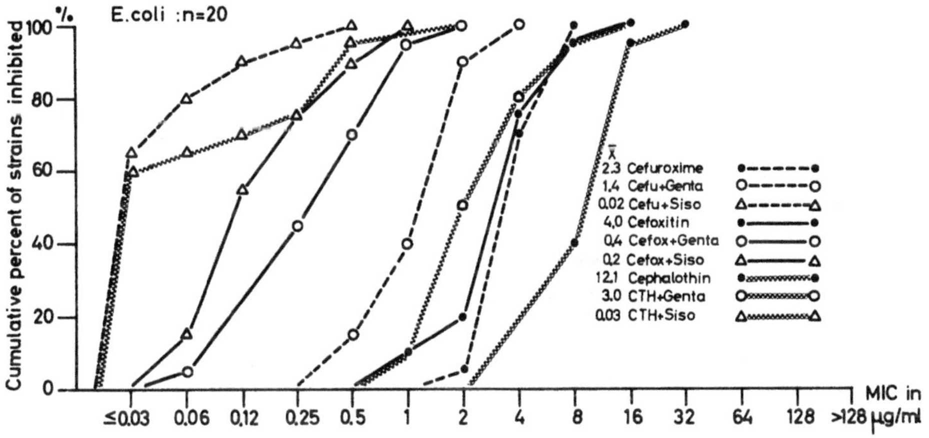

Fig. 19.1 - Combination of cephalosporins with aminoglycosides. In vitro activity - *E. coli*.

sporin and gentamicin. Of the former, the combination of cefuroxime with sisomicin proved the most effective, inhibiting the growth of all *E. coli* strains at 0.5 mcg/ml. The combinations of cefoxitin or cephalothin and sisomicin required a concentration of 1 to 2 mcg/ml to produce the same result.

Activity against *Klebsiella* (Fig. 19.2)

Single drugs: At a concentration of 16 mcg/ml, cefuroxime and cefoxitin inhibited the growth of 90% of sensitive and moderately sensitive *Klebsiella* strains, cephalothin only 50%.

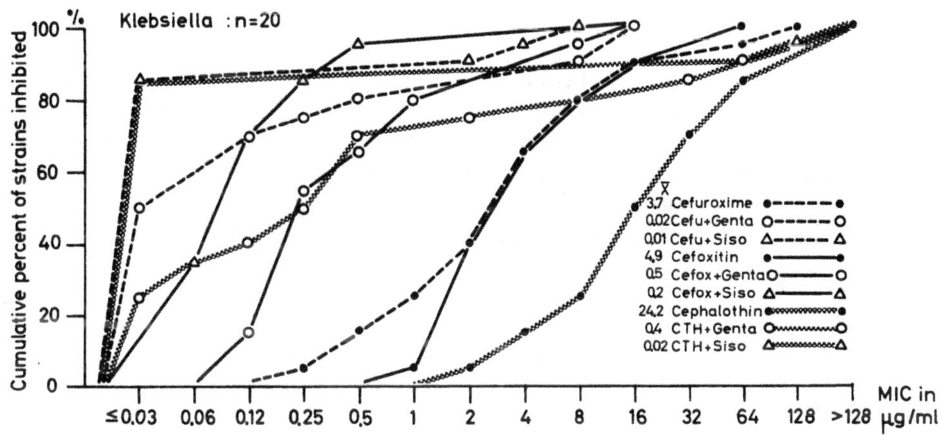

Fig. 19.2 - Combination of cephalosporins with aminoglycosides. In vitro activity - *Klebsiella*.

Combinations: The combination of cefuroxime and sisomicin proved the most effective, inhibiting the growth of all *Klebsiella* strains at 8 mcg/ml. Cefoxitin combined with sisomicin or gentamicin required a concentration of 16 mcg/ml to produce the same result. At low MIC levels, cephalothin and sisomicin proved as effective as cefuroxime and sisomicin, except that 15% of the *Klebsiella* strains proved resistant to the former.

Activity against *Enterobacter* (Table 19.1)

Single drugs: At a concentration of 16 mcg/ml, cefoxitin inhibited the growth of 50% of the strains of *Enterobacter*, cefuroxime and cephalothin only 10%. Judging by geometric mean MIC values, cefoxitin (27.8 mcg/ml) was far more effective than cefuroxime and cephalothin.

Combinations: At a concentration of 16 mcg/ml, cefoxitin and sisomicin proved the most effective, inhibiting the growth of 90% of the strains of *Enterobacter*, followed by cefuroxime combined with sisomicin or gentamicin (85%). Judging by geometric mean MIC values, cefuroxime and sisomicin proved strikingly effective (0.1 mcg/ml). Regarding the geometric mean MBC values, cefuroxime and sisomicin proved the most effective (0.4 mcg/ml), followed by cephalothin and sisomicin (1.8 mcg/ml).

Activity against *Serratia* (Table 19.2)

Single drugs: At a concentration of 16 mcg/ml, cefoxitin inhibited 70% of the strains of *Serratia*, cefuroxime and cephalothin failing to inhibit any. Judging by mean MIC values (13.0 mcg/ml) and MBC values (34.3 mcg/ml), cefoxitin proved far more effective than cefuroxime and cephalothin.

Combinations: At 8 mcg/ml the combination of cefoxitin and sisomicin or gentamicin proved the most effective, inhibiting the growth of 95% of all strains of *Serratia*, cefuroxime and sisomicin inhibiting 75% and cefuroxime and gentamicin 70%. Judging by geometric mean MIC values, the combination of cefuroxime and sisomicin proved the most effective (0.5 mcg/ml), followed by cefuroxime and gentamicin (2.5 mcg/ml). Judging by geometric mean MBC values, the combination of cefuroxime and gentamicin proved the most effective (27.8 mcg/ml).

Activity against *Proteus mirabilis* (Fig. 19.3)

Single drugs: At a concentration of 8 mcg/ml, cefoxitin inhibited the growth of 100% of this indole-negative species of *Proteus*.

Combinations: At 4 mcg/ml cefoxitin and sisomicin or gentamicin proved the most effective combinations, inhibiting the growth of 100% of the strains. The combination of cefuroxime and sisomicin or

Table 19.1 Antimicrobial activity of cefoxitin, cefuroxime and cephalothin and their combinations with sisomicin and gentamicin against Enterobacter.

MIC μg/ml	*Cumulative % of strains inhibited*							*Geom. Mean* (μg/ml)	
	0.03-0.5	*1-2*	*4-8*	*16*	*32*	*64*	*≥ 128*	*MIC*	*MBC*
Cefoxitin (CFX)			40	50	55	70	100	27.8	64
Cefuroxime (CFU)			10	10	30	45	100	84.4	> 128
Cephalothin (CTH)		5	10	10	15	20	100	> 128	> 128
CFX/Sisomicin	60	75	80	90	95	95	100	0.5	4.0
CFX/Gentamicin	10	25	55	85	90	90	100	6.0	19.7
CFU/Sisomicin	60	65	85	85	85	95	100	0.1	0.4
CFU/Gentamicin	35	55	85	85	85	95	100	1.1	8.0
CTH/Sisomicin	60	60	60	60	60	65	100	0.3	1.8
CTH/Gentamicin	20	25	35	45	55	55	100	12.1	36.7

Table 19.2 Antimicrobial activity of cefoxitin, cefuroxime and cephalothin and their combinations with sisomicin and gentamicin against Serratia.

MIC µg/ml	*Cumulative % of strains inhibited*							*Geom. Mean* (µg/ml)	
	0.03-0.5	*1-2*	*4-8*	*16*	*32*	*64*	*128*	*MIC*	*MBC*
Cefoxitin (CFX)			50	70	90	95	100	13.0	34.3
Cefuroxime (CFU)					10	50	100	84.5	128
Cephalothin (CTH)							100	> 128	> 128
CFX/Sisomicin		35	95	95	100			3.7	55.7
CFX/Gentamicin		25	95	95	100			4.3	55.7
CFU/Sisomicin	40	45	75	80	95	100		0.5	78.8
CFU/Gentamicin	20	30	70	85	90	95	100	2.5	27.8
CTH/Sisomicin	40	40	40	45	55	55	100	3.0	> 128
CTH/Gentamicin	15	15	15	15	15	15	100	52.0	> 128

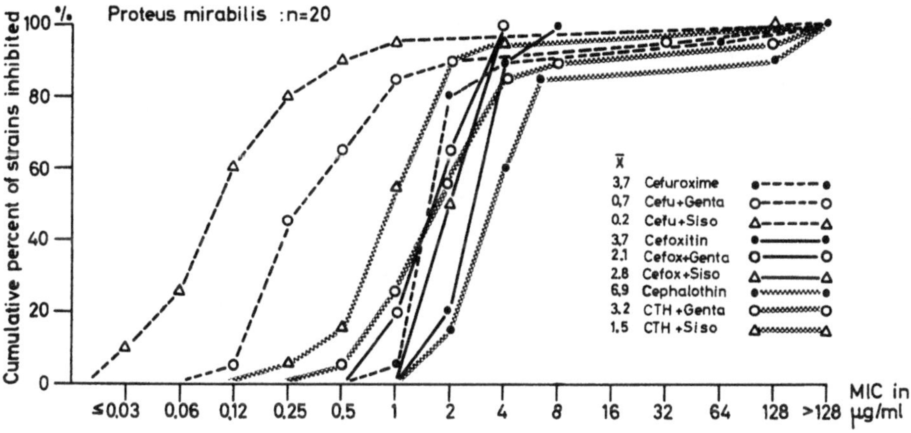

Fig. 19.3 - Combination of cephalosporins with aminoglycosides. In vitro activity - *P. mirabilis*.

gentamicin, and the combination of cephalothin and sisomicin or gentamicin, inhibited 85 to 95% of the strains. Judging by geometric mean MIC values (0.2 mcg/ml) and MBC values (0.3 mcg/ml), the combination of cefuroxime and sisomicin proved the most effective.

Activity against *Proteus vulgaris* (Fig. 19.4)

Single drugs: At 4 mcg/ml, cefoxitin inhibited 100% of all strains, cefuroxime and cephalosporin inhibiting none.

Combinations: The activity of cefoxitin was not enhanced by combination with sisomicin or gentamicin. In contrast, cefuroxime,

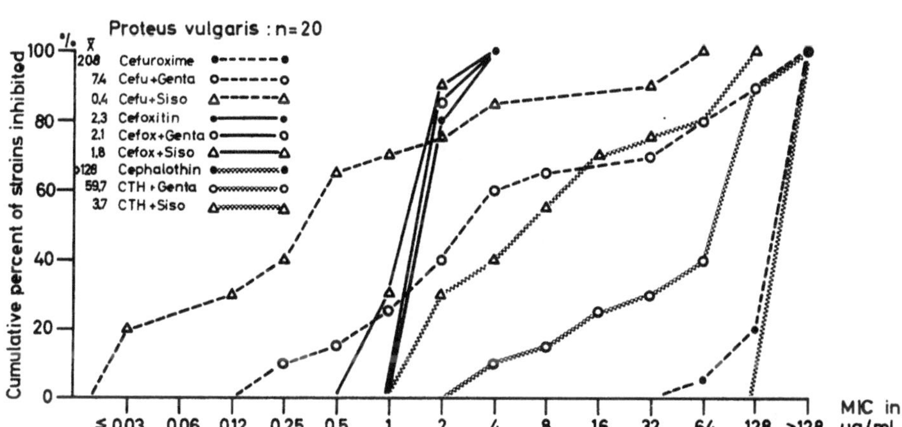

Fig. 19.4 - Combination of cephalosporins with aminoglycosides. In vitro activity - *P. vulgaris*.

which proved ineffective alone, proved effective when combined with sisomicin, inhibiting 85% of the strains at 4 mcg/ml and inhibiting about 70% of all strains at a lower MIC level than that of cephalothin alone or in combination with sisomicin or gentamicin.

Activity against *Proteus morganii* (Fig. 19.5)

Single drugs: As against *P. vulgaris*, cefoxitin proved far more effective than the other cephalosporins against *P. morganii*. At 16 mcg/ml cefoxitin inhibited 100% of all strains, cefuroxime only 20%.

Combinations: As against *P. vulgaris*, cefuroxime combined most effectively with sisomicin, synergistically inhibiting 100% of all strains at 0.5 mcg/ml. At higher MIC levels, cefoxitin and cephalothin also combined effectively with sisomicin.

Activity against *Proteus rettgeri*

Cefoxitin and cefuroxime proved effective against *P. rettgeri*. Cefuroxime and sisomicin inhibited 80 to 90% of the strains.

Results of synergy studies The results of our synergy studies are summarized in Table 19.3. It was rather surprising to find that one combination, cefuroxime and sisomicin, was so much more effective than the other combinations. Cefuroxime and sisomicin in fact combined synergistically against most of the strains of the 8 species tested. It is also worth noting that: *a*) cephalothin and sisomicin

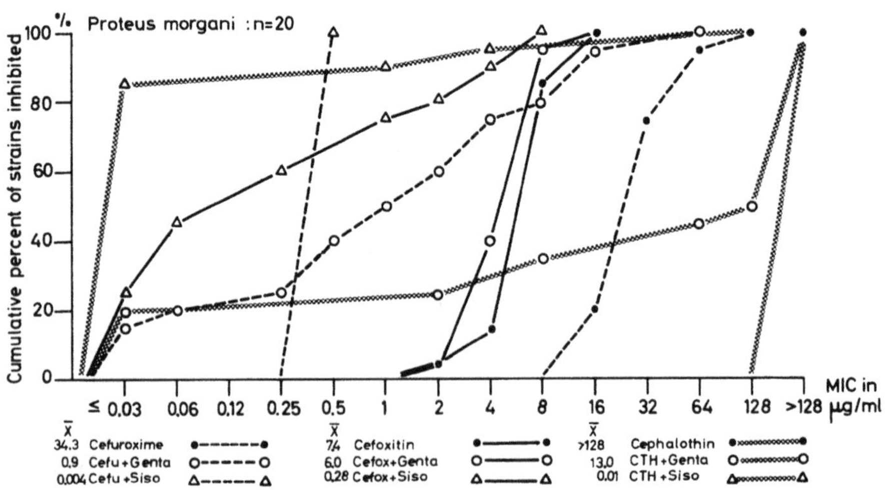

Fig. 19.5 - Combination of cephalosporins with aminoglycosides. In vitro activity - *P. morganii*.

combined synergistically against *E. coli* and *Klebsiella*; *b*) cefoxitin and sisomicin combined synergistically against *Enterobacter*, *Serratia* and *P. morganii*; and *c*) cefuroxime and gentamicin combined synergistically against *Klebsiella*, *Enterobacter*, *Serratia*, *P. morganii* and *P. vulgaris*.

Analysis of resistant strains The number of strains of the 8 species of Enterobacteriaceae tested that proved resistant to each drug and drug combination used in this study is shown in Table 19.4. Strains with a MIC of 64 mcg/ml or more were considered resistant. Cephalothin yielded the highest resistance rate (61.3%), followed by cefuroxime (40.6%) and the combination of cephalotin and gentamicin (40.6%). The resistance rate dropped to 19.3% with the combination of cephalothin and sisomicin, to 8.6% with cefoxitin and the combination of cefuroxime and gentamicin, to 4.7% with cefuroxime and sisomicin, and to 1.3% with cefoxitin and gentamicin. The combination of cefoxitin and sisomicin yielded the lowest resistance rate (0.7%), only 1 strain of *Enterobacter* proving resistant.

Discussion

The combination of a beta-lactam antibiotic and an aminoglycoside is generally indicated in the treatment of life-threatening infections and of infections characterized by resistant or moderately sensitive pathogens.

It is generally postulated that the mechanisms of action of the antibiotics in a combination must be different to produce synergistic effects. The beta-lactam-aminoglycoside combination ideally meets this requirement, since the beta-lactam antibiotic interacts with the bacterial cell wall and the aminoglycoside with protein metabolism.

Cefoxitin proved more effective than cefuroxime and cephalothin against the indole-positive *Proteus spp* and *Serratia*, whereas cefuroxime proved more effective than the other two cephalosporins against *E. coli* and *Klebsiella*.

Cefoxitin also proved effective against cephalothin- and cefuroxime-resistant strains of *Enterobacter*, *Serratia* and both *Proteus* species presumably because of its elevated stability against gram-negative bacteria producing beta-lactamase [7], [8]. The beta-lactamase resistance of cefuroxime lies between that of cefoxitin and cefamandol. The elevated effectiveness and enzymatic stability of cefoxitin in vitro and in vivo has been observed by a number of investigators [1], [3], [9], [10], [12].

The three cephalosporins each combined with sisomicin or gentamicin to act synergistically against a varying number of strains and

Table 19.3 Combination effects of cefoxitin, cefuroxime and cephalothin with sisomicin and gentamicin.

	Indifference	*Addition*	*Synergism*
E. coli (20)			
Cefoxitin/Sisomicin	–	7	13
Cefoxitin/Gentamicin	1	10	9
Cefuroxime/Sisomicin	–	1	19
Cefuroxime/Gentamicin	–	20	–
Cephalothin/Sisomicin	–	1	19
Cephalothin/Gentamicin	1	19	–
Klebsiella (20)			
Cefoxitin/Sisomicin	1	3	16
Cefoxitin/Gentamicin	1	9	10
Cefuroxime/Sisomicin	–	3	17
Cefuroxime/Gentamicin	2	4	14
Cephalothin/Sisomicin	2	1	17
Cephalothin/Gentamicin	4	–	16
Enterobacter (20)			
Cefoxitin/Sisomicin	4	4	12
Cefoxitin/Gentamicin	5	12	3
Cefuroxime/Sisomicin	–	5	15
Cefuroxime/Gentamicin	1	5	14
Cephalothin/Sisomicin	6	2	12
Cephalothin/Gentamicin	7	6	7
Serratia (20)			
Cefoxitin/Sisomicin	7	9	4
Cefoxitin/Gentamicin	7	11	2
Cefuroxime/Sisomicin	2	6	12
Cefuroxime/Gentamicin	2	11	7
Cephalothin/Sisomicin	8	3	9
Cephalothin/Gentamicin	16	1	3

species. Of all the combinations, that of cefuroxime and sisomicin proved the most effective, acting synergistically against over 50% of all strains of each of the 8 species tested. This combination also proved bactericidally more effective than the other combinations, except that cefuroxime and gentamicin proved most effective against *Serratia*.

The combination of cefoxitin and sisomicin proved the most effective against resistant strains. Of 7 strains that were resistant to cefuroxime and sisomicin, only 1 remained resistant to cefoxitin and sisomicin.

Confirming the findings of Thomas et al. [11], no combination of a cephalosporin and an aminoglycoside produced antagonistic effects.

Table 19.3 Combination effects of cefoxitin, cefuroxime and cephalothin with sisomicin and gentamicin (cont.)

	Indifference	*Addition*	*Synergism*
Proteus mirabilis (20)			
Cefoxitin/Sisomicin	13	7	–
Cefoxitin/Gentamicin	9	11	–
Cefuroxime/Sisomicin	1	7	12
Cefuroxime/Gentamicin	1	12	7
Cephalothin/Sisomicin	2	17	1
Cephalothin/Gentamicin	5	15	–
Proteus morganii (20)			
Cefoxitin/Sisomicin	2	4	14
Cefoxitin/Gentamicin	14	6	–
Cefuroxime/Sisomicin	–	–	20
Cefuroxime/Gentamicin	1	9	10
Cephalothin/Sisomicin	1	1	18
Cephalothin/Gentamicin	10	3	7
Proteus vulgaris (20)			
Cefoxitin/Sisomicin	13	7	–
Cefoxitin/Gentamicin	19	1	–
Cefuroxime/Sisomicin	–	3	17
Cefuroxime/Gentamicin	4	3	13
Cephalothin/Sisomicin	–	6	14
Cephalothin/Gentamicin	2	13	5
Proteus rettgeri (10)			
Cefoxitin/Sisomicin	5	4	1
Cefoxitin/Gentamicin	6	3	1
Cefuroxime/Sisomicin	1	1	8
Cefuroxime/Gentamicin	2	1	7
Cephalothin/Sisomicin	1	1	8
Cephalothin/Gentamicin	1	3	6

In light of the results of this study, it seems reasonable to consider the use of the combination of cefuroxime or cefoxitin and sisomicin in the treatment of severe infections, especially those caused by resistant or moderately sensitive bacteria. The real value of therapy based on a combination of one of the new cephalosporins and an aminoglycoside still requires thorough clinical assessment.

Acknowledgements

We wish to express our thanks to Ms. Johanna Fedder for her invaluable help in the experiments and to Ms. Carola Bauerschmidt for her excellent collaboration in preparing the diagrams.

Table 19.4 *Resistant strains* (MIC $\geq$ 64 μg/ml) *of Enterobacteriaceae against cephalothin (CTH), cefuroxime (CFU) and cefoxitin (CFX) and their combinations with gentamicin (Ge) and sisomicin (Si).*

Organisms	*n*	*CTH*	*CTH/Ge*	*CTH/Si*	*CFU*	*CFU/Ge*	*CFU/Si*	*CFX*	*CFX/Ge*	*CFX/Si*
E. coli	20	—	—	—	—	—	—	—	—	—
Klebsiella	20	6	3	3	2	—	—	2	—	—
Enterobacter	20	17	9	8	14	3	3	9	2	1
Serratia	20	20	17	9	18	2	1	2	—	—
Prot. mirab.	20	2	2	1	2	1	1	—	—	—
Prot. morganii	20	20	13	1	5	1	—	—	—	—
Prot. vulg.	20	20	14	5	20	6	2	—	—	—
Prot. rettg.	10	7	3	2	—	—	—	—	—	—
Sum abs.	150	92	61	29	61	13	7	13	2	1
%		61.3	40.6	19.3	40.6	8.6	4.7	8.6	1.3	0.7

References

[1] BRUMFITT, W., KOSMIDIS, J., HAMILTON-MILLER, J. M. T., and GILCHRIST, J. G., (1974) "Cefoxitin and cephalothin: Antimicrobial activity, human pharmacokinetis and toxicology", *Antimicrob. Ag. Chemother.*, **6**, 290-299.

[2] BULGER, R. J. and KIRBY, W. M. M. (1963), "Gentamicin and ampicillin: synergism with other antibiotics", *American. J. of Med. Sci.*, **246**, 717.

[3] FREITAG, V. (1978) "In-vitro-Aktivität von Cefoxitin-Natrium im Vergleich mit anderen Antibiotika und Trimethoprim-Sulfamethoxazol", *Med. Klin.*, **73**, 423-427.

[4] GAYA, H., BROWN, E. M., FRIEDMAN, P., COX, S. E. M. (1977) "Comparative In-vitro Activity of Cefuroxime and its Interactions with Aminoglycoside Antibiotics", *Int. Conf. on Cefuroxime, Proceedings of the Royal Society of Medicine*, **70**, supplement, **9**, 51-55.

[5] SCHASSAN, H. H. (1976) "Die antimikrobielle Wirksamkeit von Sisomicin. II. Untersuchungen zum Synergismus von Sisomicin und Beta-Lactam-Antibiotika", *Infection Suppl.*, **4**, 324-330.

[6] SCHASSAN, H. H., SUKARDI, ANICCA (1977) "In Vitro Microbiological Evaluation of Cefuroxime and its Combinations with Gentamicin and Sisomicin", *Proc. roy. Soc. Med.*, **70**, Suppl. 9, 200-202.

[7] SCHASSAN, H. H. (1978) "Cefoxitin, das erste Cephamycin-Antibiotikum: Enzymstabilität und mikrobiologische Standortermittlung im Vergleich zu anderen Beta-Laktam-Antibiotika", *Fortschritte der Medizin*, **96**, 1510-1517.

[8] SCHASSAN, H. H. (1978) "Eine neue Methode zur Prüfung der Beta-Laktamase produktion resistenter Enterobacteriaceaestämme", *Zbl. Bakt. Hyg., I. Ass. Orig.*, **242**, 347-354.

[9] SHAH, P. M., ZWISCHENBRUGGER, H. and STILLE, W. (1976) "In-vitro Aktivität von Cefoxitin, einem neuen Cephalosporin", *München. med. Wschr.*, **118**, 1469-1472.

[10] SHAH, P. M., HELM, E. B. and STILLE, W. (1978) "Behandlung von Cephalothin- und Cefazolin-resistenten Infektionen mit Cefoxitin", *Münch. med. Wschr.*, **120**, 375-376.

[11] THOMAS, F. E., JACKSON, R. T., and ALFORD, R. H. (1976), "Combined Effects of Cephalosporins and Aminoglycosides against Cephalothin-Gentamicin-resistant Klebsiella", 16th Intersci. Conf. on *Antimicrob. Ag. Chemother.*, Abstracts of papers no. 5, Chicago.

[12] WALLICK, H., HENDLIN, D. (1974), "Cefoxitin, a new semi-synthetic cephamycin antibiotic: Susceptibility studies", *Antimicrob. Ag. Chemother.*, **5**, 25-32.

20

Dibekacin, an aminoglycoside antibiotic in the combined treatment of brucellosis

A. G. Paradelis, G. Stathopoulos, A. Grigoriadou-Edipides, G. Salpigides, J. Papapanagiotou

Departments of Microbiology, Hygiene and Experimental Pharmacology, Medical School, Aristolelian University of Thessaloniki, Greece

Summary

Tetracyclines remain the most active chemotherapeutic agents for the treatment of Brucellosis. Dibekacin, a new aminoglycoside antibiotic, was found to be equally effective (MIC 0.312-0.625 μg/ml) as oxytetracycline (MIC 0.312-0.625 μg/ml) against 23 *Brucella melitensis* strains. Dibekacin is significantly more effective than streptomycin (MIC 2.500 μg/ml) and may be useful in the combined treatment to prevent Brucellosis relapses or for the treatment of Brucella reinfection.

Introduction

Brucellosis continues to present therapeutic problems for the mediterranean and some other countries despite the fact that Brucellae are still susceptible to a variety of antimicrobial chemotherapeutic agents [1]. Whilst acute Brucellosis responds well to tetracyclines or to the combination of tetracyclines and streptomycin [2], there is a number of patients whose disease proceeds to a chronic state, due to inadequate treatment or to reinfection.

The effective control of Brucellosis in the animal population has resulted in the reduction of the number of human infections [3]. However, Brucellosis continues to be a significant occupational hazard for many countries.

In the present paper we present data of the sensitivity of Brucella melitensis strains, all clinical isolates, against dibekacin, a new aminoglycoside antibiotic, oxytetracycline, ampicillin, streptomycin, kanamycin, gentamicin and amikacin.

Methods and materials

Twenty-three Brucella melitensis strains were isolated from blood cultures of patients admitted to the Hospital for Infectious Diseases of Thessaloniki (Greece), with acute clinical manifestations of Brucellosis.

All strains were tested for susceptibility to oxytetracycline, ampicillin, and five aminoglycoside antibiotics, i.e. dibekacin, streptomycin, kanamycin, gentamicin and amikacin, using a twofold serial dilution of the antibiotics in Brucella broth (BBL).

The inoculum was 0.1 ml of a suitable dilution of an overnight culture in Brucella broth to give a final concentration in the test dilution of approximately 10^5 colony-forming units (cfu). The minimal inhibitory concentration value (MIC) was expressed in terms of the lowest concentration of the antibiotic, which totally inhibited visible growth after incubation for 24 hours at 37 °C. All MIC values which are mentioned in the text are expressed in terms of the corresponding base of antibiotics. We used the following antibiotics: Oxytetracycline hydrochloride, Ampicillin sodium, Dibekacin sulfate, Streptomycin sulfate, Kanamycin sulfate, Gentamicin sulfate and Amikacin sulfate.

Results

The MIC values of streptomycin, kanamycin, dibekacin, gentamicin, amikacin, oxytetracycline and ampicillin are shown in Table 20.1. The MIC values of dibekacin are significantly lower than those of the other four aminoglycoside antibiotics and ampicillin and similar to those of oxytetracycline.

Discussion

Early treatment of Brucellosis first with sulphonamides and then with streptomycin, was only partially successful. Combined treatment with both chemotherapeutic agents was more so [1].

The treatment of Brucellosis with tetracyclines produces excellent results and is generally accepted as the treatment of choice. Brucella melitensis, suis and abortus still remains sensitive to tetracyclines. Excellent results are usually obtained in acute Brucellosis with a total daily dose of 2 g of any tetracycline for 21 days. Streptomycin, 0.5 g intramuscularly every 12 hours, is occasionally given at the same time as tetracycline [2]. The above combined treatment results in a lower incidence of relapse than that observed with a tetracycline alone.

Table 20.1 ***Susceptibility of 23 Brucella melitensis strains, by the tube dilution method in Brucella broth (BBL) to Oxytetracycline, Ampicillin, Dibekacin, Streptomycin, Kanamycin, Gentamicin and Amikacin.*** (MIC µg/ml)

Number of strain	*Oxytetracyclin*	*Ampicillin*	*Dibekacin*	*Streptomycin*	*Kanamycin*	*Gentamicin*	*Amikacin*
1	0.625	0.625	0.312	2.500	2.500	1.250	1.250
2	0.625	1.250	0.312	2.500	2.500	1.250	1.250
3	0.312	0.625	0.312	2.500	2.500	1.250	1.250
4	0.625	0.625	0.312	2.500	2.500	1.250	1.250
5	0.625	2.500	0.625	2.500	2.500	1.250	2.500
6	0.625	1.250	0.312	2.500	2.500	1.250	2.500
7	0.312	1.250	0.312	2.500	2.500	1.250	1.250
8	0.625	2.500	0.625	2.500	2.500	2.500	2.500
9	0.625	2.500	0.312	2.500	2.500	2.500	2.500
10	0.312	1.250	0.312	2.500	2.500	1.250	2.500
11	0.625	2.500	0.625	2.500	2.500	1.250	2.500
12	0.625	1.250	0.312	2.500	2.500	1.250	1.250
13	0.312	1.250	0.312	2.500	2.500	2.500	2.500
14	0.625	2.500	0.625	2.500	2.500	2.500	2.500
15	0.625	1.250	0.312	2.500	2.500	1.250	2.500
16	0.312	1.250	0.312	2.500	2.500	2.500	2.500
17	0.625	1.250	0.625	2.500	2.500	2.500	2.500
18	0.312	1.250	0.312	2.500	2.500	2.500	2.500
19	0.625	5.000	0.312	2.500	2.500	2.500	2.500
20	0.312	5.000	0.312	2.500	5.000	2.500	2.500
21	0.625	5.000	0.625	2.500	5.000	2.500	2.500
22	0.312	1.250	0.312	2.500	5.000	2.500	2.500
23	0.312	1.250	0.312	2.500	2.500	2.500	2.500

Relapses are not the result of the development of resistance by Brucella strains but in the great majority are real reinfections, since patients return to their infectious environment. Relapses usually respond to retreatment [2].

Co-trimoxazol, a combination of trimethoprim and sulphamethoxazol, has also proved effective in acute Brucellosis, though in a number of patients recovery has been incomplete and in some cases patients continued to have positive blood cultures, despite the clinical improvement [4], [5], [6], [7].

Sensitivity tests on bovine isolation of Brucella abortus showed that 47% of the strains were resistant to co-trimoxazol, 1.1% to tetracycline and 4.7% to streptomycin, and hence co-trimoxazol may be effective only in infections with certain strains [8]. Gastric discomfort and an extensive rash developing during the treatment of Brucellosis with co-trimoxazol have lead to discontinuation of treatment [7] and a few examples of toxic effects on the bone marrow have also been described [9].

The search continues for better drug combinations which should be easy to administer, with a low incidence of side effects, with efficacy against acute Brucellosis and sufficient to prevent relapses.

At the present, in addition to searching for new and effective antibiotics useful in treating infections caused by resistant bacteria, the chemical transformation of known antibiotics to more useful agents are under study. Dibekacin (3′, 4′-dideoxykanamycin B) has been derived from kanamycin B by such chemical transformation [10].

The antibacterial activity of dibekacin (MIC 0.312-0.625 μg/ml) was found to be equal to that of oxytetracycline (MIC 0.312-0.625 μg/ml) and much greater than that of ampicillin (MIC 0.625-5.000 μg/ml) and the other four tested aminoglycoside antibiotics (MIC 1.250-5.000 μg/ml) against 23 clinical isolates of Brucella melitensis.

In conclusion, our data obtained from *in vitro* and *in vivo* observations suggest that dibekacin, due to its greater antimicrobial activity and fewer side effects, in comparison with the other old and new aminoglycoside antibiotics, may be a useful drug for the substitution of streptomycin in the combined treatment of acute Brucellosis. The combination of a tetracycline and dibekacin may result in a lower incidence of clinical and bacteriological relapses than that observed with a tetracycline alone.

References

[1] Garrod, L. E., Lambert, H. P. and O'Grady, F. (1973), *Antibiotic and Chemotherapy*, 4th ed. p. 321, Churchill Livingston, Edinburgh-London.

[2] Grossman, M. and Jawetz, E. (1977) in *Medical Diagnosis and Treatment* (A. M. Krupp and M. J. Chatton eds), 16th Annual Revision, p. 819, Lange Medical Publication, Los Altos, California.

[3] Busch, L. A. and Parker, R. L. (1972) "Brucellosis in the United States", in *J. Inf. Diseases*, **125**, 289-294.

[4] Lal, S., Modawal, K. K., Fowle, A. S. E., Peach, B. and Popham, R. D. (1970) "Acute brucellosis treated with trimethoprim and sulphamethoxazol", in *Br. med. J.*., **3**, 256-258.

[5] Giunchi, G., Rosa, F. de and Fabiani, F. (1971) "Trimethoprim-sulphamethoxazol in the treatment of acute human brucellosis", in *Chemotherapy*, **16**, 332-335.

[6] Hassan, A., Erian, M. M., Farid, Z., Hathout, S. D. and Sorensen, K. (1971) "Trimethoprim-sulphamethoxazol in acute brucellosis", in *Br. med. J.*, **3**, 159-161.

[7] Daikos, G. K., Papapolyzos, N., Marketos, N., Mochas, S., Kastanakis, S. and Papasteriadis, L. (1973), in *Trimethoprim-sulphamethoxazol in Bacterial Infections* (Bernstein and Slater eds), pp. 201-205, Churchill Livingston, Edinburgh-London.

[8] Williams, E. (1973) "Brucellosis", in *Br. med. J.*, **1**, 791-793.

[9] Garrod, L. E. (1976) in *Drug treatment* (G. S. Avery ed) p. 845, Adis Press.

[10] Umezawa, H., Umezawa, S., Tsuchiya, T. and Okazaki, Y. (1971) "3', 4'-Dideoxy-kanamycin B active against kanamycin-resistant E. coli and Ps. aeruginosa", in *J. Antibiotics*, **24**, 485-487.

21

Proteus infections in the urological field: a therapeutic problem. Preliminary report

R. Tenaglia, V. Gentile

Clinica urologica, Università di Roma, Italy

A. Marocchi

Patologia speciale medica e Metodologia clinica II, Università di Roma, Italy

P. Criniti, F. Di Silverio

Clinica urologica, Università di Chieti, Italy

Treatment of urinary infections is a complex problem for the physician. In a specialized Urology Department the exact line of treatment must be defined bearing in mind the ethiopathogenesis, the wide choice of drugs available and the condition of the patient (kidney and liver function tests).

Urinary infections should be limited; preventive measures are: scrupulous sterilization of the hospital environment, and of the instruments (both surgical and diagnostic), hygiene of the hospital personnel.

The increase in bacterial resistance and the high incidence of recurrence require a greater use of antibiotics and of antibiotics combinations. The indiscriminate administration of these drugs by physicians is solicited by patients with a poor knowledge of sanitary problems.

All possible infections sources must be known; they are, in our field: the surgical operation, any endoscopial instrumental investigation, bladder and renal drainage principally.

It would be useful to determine exactly the host defences and the urinary dosage of the 5-fraction of latticodehydrogenase in order to distinguish between high tract urinary infection and low tract urinary infection or, in other words, between infection of the container and of the contents.

However when we are faced with chronic bacteriuria with important urinary symptoms we must try to obtain the total cure of the disease.

A recent research carried out in our clinic showed that *Proteus* is the most frequently found among urinary bacteria. Therefore, we have shown particular interest in this micro-organism which is, in our clinical experience, one of the most difficult to defeat.

This micro-organism of the Enterobacteriaceae family has a highly variable sensitivity to antibiotics, rarely of diagnostic significance. Antigens are usually present and differences in the urological specificities form the basis for the development of a large number of serotypes within each species. Virulent and temperate phages are known for all species. *Proteus vulgaris* is found in fecal matter of many animals, it is less frequently found in clinical specimens than *Proteus mirabilis*. The habitat of *Proteus morganii* is not known; it has been frequently found in human feces. The same may be said for *Proteus rettgeri*: many strains have been isolated from a variety of human clinical specimens and also isolated from chicken feces.

Thorough studies on this germ have led to a good knowledge of its biological activity and of its sensitivity to drugs. Consequently we tried to find the relationship between *Proteus* found in 25 urocultures, and the activity of some of the best known antibiotics *in vitro* and *in vivo*.

As can be seen from Table 21.1, we have controlled the sensitivity to 9 antibiotics.

Table 21.1 shows that gentamicin develops an excellent action; colistin alone, on the contrary, has a very low effect. Then, we combined a sulphonamide with colistin and obtained very good results: 13 cases of S standard, 12 I standard and no case of R standard, which means that in this way we went from 60% to 100% of an appreciable positive answer.

To arrive at these results we attempted to combine colistin with sulphonamides. We used different dosages of those drugs. As can be seen, the best combination was obtained between colistin 10 mcg using the antibiotic disk diffusion technique, and sulfametizole 1 mg. In this way we have the complete inhibition of all strains. Sulphonamides inhibit folic acid synthesis, whilst colistin seems to act by joining the phosphate group of the cell wall.

It has recently been shown that *Proteus* strains are completely inhibited by synergistic combination of colistin together with a number of sulphonamides. Electron microscopy shows that the cells are normal until optical density changes occur, when they begin to

Table 21.1 Proteus *sensitivity to antibiotics. Number of cultures: 25.*

Antibiotics	*Sensitivity* *			
	S	*I*	*R*	*S+I* (*% values*)
Gentamicin	24	0	1	96,00
Tobramicin	12	0	5	70,58
Nalidixic ac.	10	12	2	91,66
Cotrimoxazol	10	1	14	52,00
Caf	2	2	21	16,00
Colistin	2	13	10	60,00
Kanamycin	6	4	15	40,00
Furedan	10	4	11	56,00
Colistin + Sulfemetazolum	13	12	0	100,00

* S = Sensitive; I = Intermediate sensitive; R = Resistant.

develop projection on the outer lipoprotein and lipolysaccharide membrane. The ultrastructural changes induced in the normally colistin-resistant *Proteus vulgaris* by the synergistic combination of sulphadiazine and colistin are very similar to those produced by colistin alone, in colistin sensitive *E. coli*.

The lesions of the mucopeptide layers seem to occur as a result of the synergistic response: presumably at these points internal osmotic pressure becomes great enough to push the cytoplasm outwards so as to form spherical protrusions. It is possible to postulate that growth in sulphadiazine in some way exposes or sensitizes the lipopolyssaccharide layer to the action of colistin, possibly weakening the mucopeptide at the same time so that the colistin can reach the membrane, disorganizing it and finally combining with DNA.

After our microbiological tests (Table 21.2), our clinical experience with colistin and sulphamethizole therapy in urinary tract infections are described in Table 21.3.

The number of patients screened for study is 21. Twelve of these (57,1%) had undergone endoscopical instrumental tests either for diagnostic or for therapeutic purposes; the same patients had had or still had a catheter. In fact, infections due to use of catheters seem to represent more than one third of the infections found in hospitalized patients. The risk of urinary infections by a temporary use of catheter is 1-2%, but rapidly increases up to 90-95% when the catheter itself is used for more than 3 days. A catheterization according to the closed-system reduces the incidence of infection. Six patients had suffered from renal and/or urethral stones. 1 from a renal cyst, another from right renal ptosis, and another one from hydronephrosis with dysuria and haematuria. We adopted the following therapy with a single daily dosage:

h 7-8 p.m. 500 mg sulfamethizole tablets
h 9-10 p.m. 500.000-1.000.000 Colistin U.
according to the body-weight
Duration of treatment: 10-15 days

The results of our urinoculture controls (Table 21.3) showed: sterile urine in 16 patients. bacteriuria still present in 2 patients, bacteriuria reduced in 2 patients, bacteriuria still present in one patient but not caused by *Proteus*.

On the basis of our microbiological and clinical data we can affirm that the combination of colistin with sulfamethizole has a very high therapeutic index in urinary tract infections sustained by *Proteus* spp.

In our clinical experience we have never observed signs of toxicity.

Table 21.2 Proteus *sensitivity to colistin and to colistin + sulfonamides* *.

Patients	*COL (10 mcg)*	*COL 10 mcg + SMZ 1 mg*	*COL 10 mcg + SMZ 0,25 mg*	*COL 2 mcg + SMZ 1 mg*	*COL 2 mcg + SMZ 0,25 mg*	*COL 10 mcg + SDZ 1 mg*	*COL 2 mcg + SDZ 1 mg*
FF	I	S	I	S	I	S	S
TA	I	I	R	R	R	R	R
DPA	I	I	R	R	R	R	R
BM	R	S	S	S	I	S	I
TA	R	S	S	I	S	S	I
AV	R	S	S	S	S	S	S
CMT	I	I	I	I	R	I	I
SM	R	S	S	I	I	S	I
BA	I	S	S	S	I	S	S
PA	R	I	I	I	I	I	I
PE	I	S	S	I	S	S	S
DGI	R	I		R			
TL	R	S		I			
FA	R	S		I			
FCA	I	I		R			
CE	I	I		R			
FM	I	S		I			
DF	R	S		I			
DLN	I	I		I			
DGT	R	I		R			
VC	I	I		I			
GG	I	I		R			
DOE	I	I		I			
DP	I	S		I			
SO	I	S		I			

* S = Sensitive; I = Intermediate sensitive; R = Resistant.

Table 21.3 Our experience with SM+COL therapy in urinary tract infections.

Patients	Sex	Diagnosis	Uroculture	
			Initial	Control
BV	M	Bladder papilloma	*Proteus rettgeri*	Negative
DGI	F	Cystitis, dysuria	*Proteus mirabilis*	Positive
DSV	M	R. calix stones	*Proteus mirabilis*	Negative
MM	F	Dysuria, hydronephrosis	*Proteus vulgaris*	Negative
VP	M	Prostate cancer	*Proteus mirabilis*	Negative
SM	F	Ureteral stone	*Proteus rettgeri*	Negative
CM	M	Prostate adenoma	*Proteus vulgaris*	Negative
FMA	F	Ureteral stone	*Proteus mirabilis*	Positive
SA	M	Neurogenic bladder	*Proteus mirabilis*	Negative
FF	F	L. calix stone	*Proteus mirabilis*	Negative.
FE	M	Ureteral stenosis	*Proteus rettgeri*	Positive
TG	M	Bladder carcinoma	*Proteus mirabilis*	Negative
DSG	F	L. renal cyst	*Proteus mirabilis*	Negative
DMA	F	R. renal ptosis	*Proteus mirabilis*	Negative
DVA	M	Bladder carcinoma	*Proteus mirabilis*	Positive
MA	F	Ureteral stone	*Proteus mirabilis*	Negative
DLN	M	Bladder carcinoma	*Proteus mirabilis*	Negative
FC	M	Ureteral stenosis	*Proteus rettgeri*	Negative
MC	M	Strangury, pyuria	*Proteus mirabilis*	Negative
BP	F	L. renal stone	*Proteus mirabilis*	Positive
GG	F	Chronic cystitis	*Proteus morganii*	Negative

22

Retrospective study on clinical nephrotoxicity induced by the combination of cephalothin and gentamicin or tobramycin
A preliminary report

F. Paradisi, G. Caruso, P. Cristiano, R. Cioffi, C. D'Onofrio

Istituto di Clinica Medica
Nuovo Policlinico - Via S. Pansini - 80129 Napoli

Summary

A retrospective study of hospital records on 166 patients who were treated between 1974 and 1978 with a combination of cephalothin and gentamicin or tobramycin for severe gram-negative infections showed that: (*a*) the combination of cephalothin and gentamicin or tobramycin is probably more nephrotoxic than cephalothin, gentamicin or tobramycin alone; (*b*) the combination of cephalothin and gentamicin is probably more nephrotoxic than that of cephalothin and tobramycin; and (*c*) the nephrotoxicity probably induced by cephalothin and gentamicin or tobramycin is apparently correlated more with age and impaired renal function before treatment than with duration of treatment and normal renal function before treatment. The findings on nephrotoxicity induced by the combination of cephalothin and gentamicin or tobramycin are qualified as probable because all patients had received other drugs and/or because they were affected by severe life-threatening infections that may in themselves induce renal insufficiency.

Introduction

Nephrotoxicity and ototoxicity are the major limiting factors in the clinical use of aminoglycoside antibiotics, a group of antibiotics that includes streptomycin, neomycin, kanamycin, gentamicin, tobramycin, amikacin and sisomicin. Of these antibiotics, amikacin and sisomicin have only recently been introduced and are still undergoing clinical evaluation, and gentamicin and tobramycin are extensively used in the treatment of severe infections caused by gram-negative bacteria.

As regards gentamicin, investigators have found that this drug induces nephrotoxicity in about 2% of the patients receiving the drug, and that nephrotoxicity is usually indicated by a rise of the blood urea nitrogen (BUN) level [1], [2], [3]. In reviewing the case records of 3370 patients who had received gentamicin treatment, Hewitt [4] estimated that the drug had possibly or probably induced nephrotoxicity in 140 (4.1%).

As regards tobramycin, experimental studies in rodents [5] and in rabbits [6] showed that tobramycin was less nephrotoxic than gentamicin, and a review of case records showed that tobramycin induced nephrotoxicity in only 1% of the patients receiving this drug [7].

Combinations of gentamicin or tobramycin and cephalothin are frequently used in the treatment of severe infections caused by gram-negative bacteria. A number of reports have suggested that therapy based on the combination of gentamicin and cephalothin is particularly nephrotoxic [8], [9], [10]. This is a little surprising since cephalothin has only been occasionally reported to induce renal failure [11], [12]. Further, five times the usual therapeutic dosage of cephalothin is required for the experimental production of mild tubular necrosis [12], and single doses of cephalothin at 4000 mg/kg s.c. are needed to induce tubular damage in the rabbit [13]. Be that as it may, we cannot ignore the fact that combinations of cephalothin and an aminoglycoside antibiotic apparently seriously increase the risk of nephrotoxicity. Unfortunately we are often obliged to administer a combination of cephalothin and gentamicin or tobramycin to patients with normal or sometimes even abnormal renal function who are affected by severe infections caused by gram-negative bacteria.

To obtain further information on the incidence of nephrotoxicity induced by combinations of cephalothin and gentamicin or tobramycin, we reviewed the hospital records of 166 patients of the New Polyclinic in Naples who received one or other of these combinations between 1974 and 1978, a period that witnessed the introduction and subsequent widespread use of gentamicin and tobramycin.

Analysis of case records

In reviewing the case records of patients who had received cephalothin + gentamicin or cephalothin + tobramycin treatment we:

1. Excluded the case records of patients who had died during treatment or who had been treated for less than 3 days.
2. Considered a 5 mg/100 ml increase of BUN levels as indicative of

deterioration of renal function. We ignored changes in serum creatinine levels and creatinine clearance rates because these indices were not recorded in all cases and inclusion of data on these indices in our review would not have permitted valid statistical analysis.

With respect to the 166 case records reviewed, we extracted the following information on each patient:

a) Type of treatment (cephalothin + gentamicin or cephalothin + tobramycin).
b) Age (years).
c) Duration of treatment (days).
d) Gentamicin or tobramycin dose (mg).
e) Cephalothin dose (g).
f) BUN value (mg/100 ml) before and after treatment.

All data respecting the two treatment groups and the following groups within each treatment group were summed and the means (± s.e.) calculated:

a) Group I - Patients with elevated BUN level before treatment and with a 5 mg/100 ml or more increase of BUN level after treatment.
b) Group II - Patients with normal BUN level before treatment and with a 5 mg/100 ml or more increase of BUN level after treatment.
c) Group III - Patients with elevated BUN level before treatment and with no increase of BUN level after treatment.
d) Group IV - Patients with normal BUN level before treatment and with no increase of BUN level after treatment.

Since frequency distribution did not follow normal distribution, the Komolgorov and Smirnov non-parametric test [14] was used to compare BUN values before and after treatment. A value of $P < 0.05$ was accepted as significant.

Results

The data obtained from the case records and their statistical assessment are summarized in Tables 22.1, 22.2, 22.3 and 22.4, and in Figure 22.1.

Comment

The combination of cephalothin and gentamicin or tobramycin appears to be more nephrotoxic than any of the three drugs alone.

After reviewing the case records of 1450 patients who had received gentamicin or gentamicin + cephalothin therapy, Hewitt [4] found drug-related nephrotoxicity in 3 patients who had received

Table 22.1 General data.

	Gentamicin + Cephalothin		*Tobramycin + Cephalothin*	
Number of patients	130		36	
Average age (years)	55.1 min 7 max 81		50.1 min 12 max 75	
Male	75 = 57.69%		17 = 47.22%	
Female	55 = 42.30%		19 = 52.77%	
Average length of treatment (days)	9 min 3 max 25		8.277 min 3 max 25	
Average dose (mg)	Gentamicin 134.46 min 40 max 240		Tobramycin 196.94 min 50 max 300	
Average dose (g)	Cephalothin 8.49 min 3 max 12		Cephalothin 8.38 min 2 max 12	
Rise of BUN 5 mg% with abnormal initial value	16 patients = 12.3%		2 patients = 5.5%	
Rise of BUN with normal initial value	25 patients = 19.2%		6 patients = 16.6%	
Unchanged abnormal initial value	13 patients = 10%		3 patients = 8.3%	
Unchanged normal initial value	76 patients = 58.4%		25 patients = 69.4%	
	Initial	Final	Initial	Final
BUN average (x)	47.7	65.7	43.5	51.9
Standard error	3.2848	5.15509	5.9733	6.8198
Significance (Komolgorov and Smirnov test)	0.05		Not significant	

gentamicin alone and in 4 patients who had received gentamicin + cephalothin.

In patients receiving gentamicin therapy, the rate of gentamicin-induced nephrotoxicity is generally accepted as 4% [1], [2]. We have less information on the nephrotoxicity of tobramycin, but it appears likely that this drug is less nephrotoxic than gentamicin. In patients receiving tobramycin treatment, Jackson [7] in fact reports that the tobramycin-induced nephrotoxicity rate is about 1%.

In our review, we found increased BUN levels in 31.5% of the patients given cephalothin and gentamicin and in 16.66% of the patients given cephalothin and tobramycin. Our findings differ from those of Fanning et al. [15] who, after reviewing the hospital records

Table 22.2

Tobramycin + Cephalothin				*Gentamicin + Cephalothin*			
Patients with initial high BUN levels	→ 5	↗ Number of patients with rise of BUN levels	→ 2 (40%)	Patients with initial high BUN levels	→ 29	↗ Number of patients with rise of BUN levels	→ 16 (55.17%)
		↘ Number of patients without rise of BUN levels	→ 3 (60%)			↘ Number of patients without rise of BUN levels	→ 13 (44.83%)
Patients with initial normal BUN levels	→ 31	↗ Number of patients with rise of BUN levels	→ 6 (19.35%)	Patients with initial normal BUN levels	→ 101	↗ Number of patients with rise of BUN levels	→ 25 (24.75%)
		↘ Number of patients without rise of BUN levels	→ 25 (80.65%)			↘ Number of patients without rise of BUN levels	→ 76 (75.25%)

Table 22.3 *Particular data.*

	Tobramycin + Cephalothin								*Gentamicin + Cephalothin*							
	I Group		*II Group*		*III Group*		*IV Group*		*I Group*		*II Group*		*III Group*		*IV Group*	
Patients	2		6		3		25		16		25		13		76	
Average age (years)	68 min 62 max 74		57.16 min 42 max 68		44.33 min 28 max 57		47.7 min 12 max 75		63.35 min 48 max 78		61 min 23 max 80		65.69 min 44 max 80		50.88 min 6 max 80	
Male	1 = 50%		4 = 66.6%		1 = 33.3%		11 = 44%		9 = 56.25%		13 = 52%		10 = 76.32%		41 = 53.94%	
Female	1 = 50%		2 = 33.3%		2 = 66.6%		14 = 56%		7 = 43.75%		12 = 48%		3 = 23.07%		35 = 46.05%	
Average length of treatment (days)	6.5 min 5 max 8		7.8 min 5 max 10		11.3 min 5 max 20		8.16 min 3 max 25		9.93 min 3 max 20		8.24 min 3 max 18		9.69 min 5 max 14		8.80 min 3 max 25	
Average dose of tobramycin (mg)	125 min 50 max 200		211.6 min 150 max 300		123.3 min 70 max 200		202.8 min 50 max 300		117.5 min 40 max 160		149.2 min 80 max 240		132.3 min 80 max 160		134.2 min 80 max 240	
Average dose of cephalothin (g)	10 min 8 max 12		7.66 min 4 max 12		6.33 min 3 max 8		8.48 min 2 max 12		8.37 min 4 max 12		8.88 min 6 max 12		8.15 min 4 max 12		8.44 min 4 max 16	
	Initial	Final	Initial	Final	Initial	Final	Initial	Final	Initial	Final	Initial	Final	Initial	Final	Initial	Final
BUN average (x)	163.5	184.5	40.3	76	84.66	61.6	29.32	34.24	107.75	164.31	40.84	111.48	74.69	60.30	35.46	34.92
Standard error	25.8093	35.0017	2.1438	6.994	16.97	23.9121	2.0141	1.7272	14.8688	19.4160	1.4438	3.4763	9.5563	10.8081	1.2801	1.3165
Significance (Komolgorovand Smirnov test)	Not significant		0.01		Not significant		Not significant		0.05		0.001		Not significant		Not significant	

I Group = Patients with abnormal BUN initial value and with rise 5 mg %.
II Group = Patients with normal BUN initial value and with final azotemia.
III Group = Patients with normal BUN initial value and with unchanged final value.
IV Group = Patients with normal BUN initial and final values.

Table 22.4 Average age, length of treatment and dosage in the patients of different groups.

	Antimicrobial association	*Total data*	*Partial data: I Group*	*Partial data: II Group*	*Partial data: III Group*	*Partial data: IV Group*
Age (years) (average ± SE)	Gentamicin + Cephalothin	55.12 ± 1.33	63.25 ± 2.34	61.0 ± 2.04	54.76 ± 5.08	50.96 ± 2.01
	Tobramycin + Cephalothin	50.13 ± 2.72	68.0 ± 25.45	57.16 ± 3.74	44.33 ± 9.11	47.48 ± 3.57
Length of treatment (days) (average ± SE)	Gentamicin + Cephalothin	8.97 ± 0.40	9.93 ± 1.46	8.64 ± 0.82	9.69 ± 0.83	8.79 ± 0.52
	Tobramycin + Cephalotin	8.27 ± 0.79	6.5 ± 1.06	7.83 ± 0.87	11.33 ± 3.76	7.8 ± 1.10
Dosage (average ± SE)	Gentamicin + (mg/day)	134.46 ± 3.69	117.5 ± 9.66	147.2 ± 4.36	132.3 ± 10.09	134.21 ± 5.08
	Cephalothin (g/day)	8.49 ± 0.25	8.37 ± 1.06	8.88 ± 0.36	8.15 ± 0.55	8.05 ± 0.44
	Tobramycin + (mg/day)	169.94 ± 12.88	125 ± 53.03	211.66 ± 27.94	166.66 ± 27.35	202.8 ± 15.68
	Cephalothin (g/day)	8.38 ± 0.49	10 ± 1.41	7.6 ± 0.99	8 ± 0	8.48 ± 0.64

of 1070 patients who had received gentamicin or cephalothin alone or in combination, concluded that the combination of the two drugs produced no increase of nephrotoxicity. On the other hand, our findings agree with those of Gaya et al. [16] who found a higher incidence of elevated serum creatinine in patients receiving cephalothin and gentamicin than in patients receiving cephalothin and carbenicillin. Our findings are also similar to those of Klastersky et al. [17] with respect to 10 patients, out of a total of 48 patients, who had received a combination of cephalothin and tobramycin.

The combination of cephalothin and gentamicin appears to be more nephrotoxic than that of cephalothin and tobramycin (Tables 22.1, 22.2, 22.3). The combination of cephalothin and gentamicin in fact induced a rise of BUN values in 24.75% of patients with normal BUN values before treatment and in 55.17% of patients with elevated BUN values before treat. The corresponding values in the cephalothin and tobramycin treatment group were 19.35% and 40%. In considering the statistical difference between BUN values before and after treatment, it is only fair to add that there were only two patients in the cephalothin and tobramycin treatment group with elevated BUN levels before treatment (Table 22.3).

Confirming the findings of Cabanillas et al. [18], our findings show that nephrotoxicity in both treatment groups appears to be related to age (Table 22.4). As Friedman et al. [19] pointed out, the possibility of decreased renal reserve due to advanced age cannot be disregarded. Our findings indicate that both combinations are more nephrotoxic in patients with renal insufficiency before treatment than in patients with normal renal function before treatment. Whereas this difference is statistically significant in the cephalothin and gentamicin treatment group, it is not in the cephalothin and tobramycin treatment group (Table 22.3).

Finally, duration of treatment does not seem to have any significant influence on the onset of renal insufficiency. It should, however, be pointed out that the duration of treatment in our records ranged from 6½ to 11 days and that our records did not include any patients who had undergone long-term therapy.

Conclusion

Comparison of our findings with those of other retrospective studies involving the same objectives and antibiotics is complicated by a number of factors such as varying criteria for assessing nephrotoxicity, varying methods of analyzing data, and varying dosage schedules, depending on the condition of renal function. In our study, for example, the mean dosage of both combinations of antibiotics

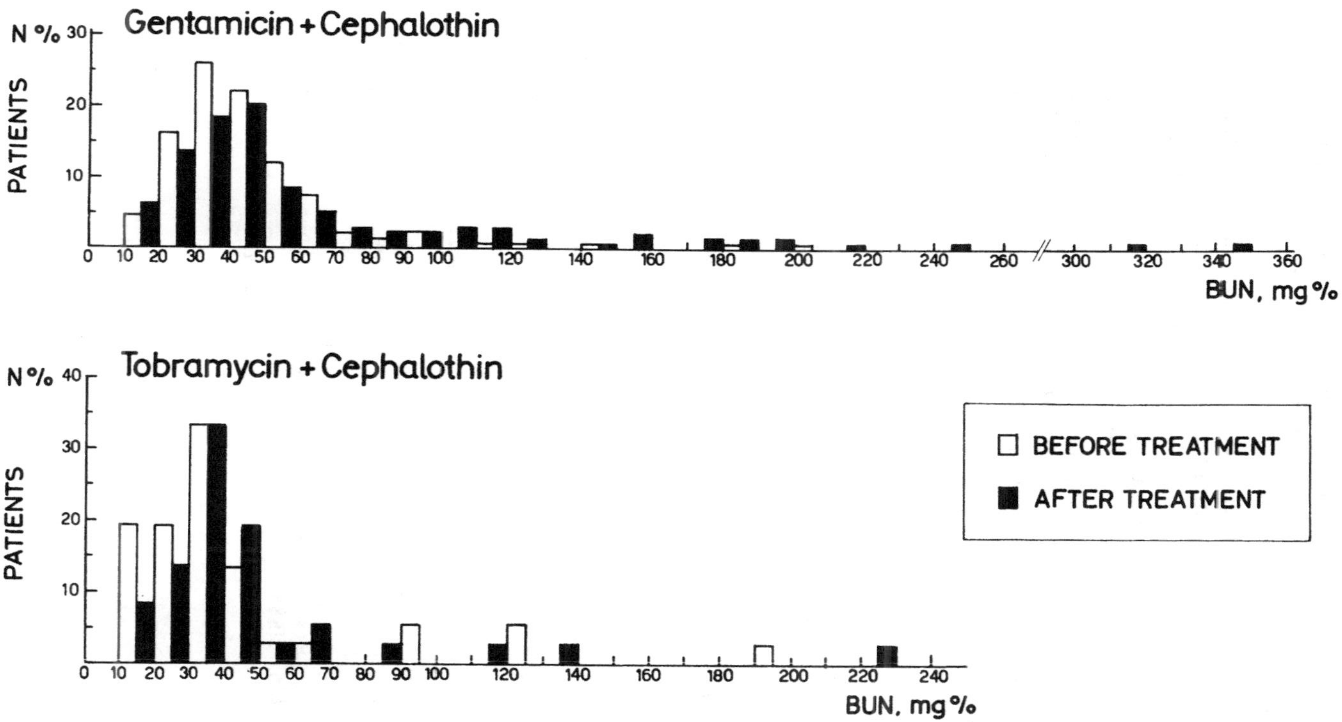

Fig. 22.1 - Frequency distribution of BUN levels in patients before and after treatment with gentamicin + cephalothin or tobramycin + cephalothin.

was appreciably lower in Group I patients (i.e. patients with elevated BUN levels before treatment and rising BUN levels during treatment) than in Group II, III and IV patients (Table 22.4).

Further, our findings inevitably reflect to a greater or lesser extent the many problems that affect all retrospective multicenter studies (availability of data, homogeneity of treatment, etc.).

Despite these complications and difficulties, two basic conclusions clearly emerge from our findings: the first, that the combination of cephalothin and gentamicin or tobramycin is probably more nephrotoxic than cephalothin, gentamicin or tobramycin alone; the second, that the combination of cephalothin and gentamicin is probably more nephrotoxic than that of cephalothin and tobramycin.

References

[1] Jackson, G. G. (1967) *Practitioner*, **198**, 855-859.
[2] Abramowicz, M. and Edelman, C. (1965) *Clin. Pediatr.*, **7**, 389-394.
[3] Kahn, T. and Stein, R. M. (1972) *Lancet*, **1**, 489-491.
[4] Hewitt, W. L. (1974) *Postgrad. Med. J.*, **50** (Supplement) 55-59.
[5] Welles J. et al. (1973) *Toxicol. Appl. Pharmacol.*, **25**, 398-399.
[6] Yamasuku, F. et al. (1975) *Chemotherapy*, **(Japan)**, **23**, 945-956.
[7] Jackson, G. G. (1977) *Clinical Therapeutics*, **1**, 200-209.
[8] Schultze, R. G., Winters, R. E. and Kauffman, H. (1971) *J. Infect. Dis.*, **124**, S 145-151.
[9] Kleinknecht, D., Ganeval, D. and Droz, D. (1973) *Lancet*, **1**, 1129-1130.
[10] Fillastre J. P., Laumonier, R., Humbert, G. et al. (1973) *Brit. Med. J.*, **2**, 396-402.
[11] Perkins, R. L., Apicella, M. A., Lee, I. S. et al. (1968) *J. Lab. Clin. Med.*, **71**, 75-79.
[12] Thomas, B. L. (1969) *New Engl. J. Med.*, **280**, 505-510.
[13] Venuto, R. C., Stein, J. H. and Ferris, T. F.: *unpublished data*.
[14] Wilkes, S. (1962) *Mathematical statistics*, pag. 454-459, Wiley, New York.
[15] Fanning, W. L. et al. (1975) *Abstracts of the Fiftheenth Interscience Conference on Antimicrobial Agents and Chemotherapy*, Washington, September 24-26.
[16] Gaya, H. et al. *Ibidem*.
[17] Klastersky, J. et al. (1975) *Antimicrob. Ag. Chemother.*, **7**, 640-645.
[18] Cabanillas, F., Burgos, R. C., Rodriguez, R. C. and Baldizon, C. (1975) *Arch. Int. Med.*, **135**, 850-852.
[19] Friedman, S. A., Raizner, A. E., Rosen, H. et al. (1972) *Ann. Int. Med.*, **76**, 41-45.

23

Classification of trimethoprim resistance plasmids

G. Manicardi, M. Bondi, L. Ferrari, R. Neglia, G. P. Quaglio, U. Fabio

Istituto di Igiene e Microbiologia
Via G. Campi, 287 - 41100 Modena - Italy

Plasmids conferring resistance to trimethoprim (TMP) (R factors) were first described in the United Kingdom by Datta and Hedges [1], Fleming et al. [2], Hedges et al. [3], and Jobanputra and Datta [4]. More recently, they have been described in Italy by Romero and Perduca [5] and in France by Acar et al. [6]. The plasmids isolated have presented interesting differences according to the different isolation areas.

To define more precisely the epidemiology of the plasmids emerging under the selective pressure of TMP (a drug increasingly used in therapy, in combination with a sulphonamide), we studied the main characters of a number of R factors found in patients admitted to the Modena Hospital. Trimethoprim-resistant bacteria were isolated from pathological specimens (mainly from urine, but also from blood, feces and sputum) in our microbiological laboratory and were identified by reference to the tables of Edward and Ewing [7].

The TMP resistance of the strains isolated was tested on Mueller-Hinton medium (Difco) by a disk diffusion method in which the size of the inhibition zone produced by the test strain was compared with that produced by a standard susceptible strain. The MIC of TMP was determined by the spot method, with inoculation of 10^4 bacteria on plates of Mueller-Hinton medium containing serial concentrations of TMP.

Transfers of R factor from wild strains and between lines of *Escherichia coli* K-12 were made as described by Datta and Hedges [1] and by Coetzee et al. [8]. The *E. coli* K-12 strains used were as follows: J53 ($pro^-met^-Nal^R$)F^-; J62 ($pro^-his^-trp^-lac^-S^R$ F^-; 309 (P13) ($cys^-lys^-Rif^R$)Hfr; and HfrC (met^-S^R)Hfr. Classification of R factors by fi character was performed as described by Datta and Hedges [1], with the use of MS_2 phage. Tests for compatibility were carried out as reported by Coetzee et al. Standard plasmids representative of the known compatibility groups are listed in Table 23.1.

Preliminary data have been reported at the 10th International Congress of Chemotherapy [9].

Over a two-year period (1976-77), we tested 1508 isolates belonging to various genera of the Enterobacteriaceae family, and among these we found 142 strains resistant to TMP. The generic distribution of all the isolates and of the 142 that were resistant is shown in Table 23.2. The MICs of TMP for all 1508 strains ranged from 10 to 2,000 μg/ml, and 28% of the 142 resistant strains were resistant to TMP at concentrations of 1,000 μg/ml (among these were all the strains belonging to the genera *Escherichia* and *Salmonella*). TMP resistance factors were isolated from 25 of the 142 resistant strains as follows: from 14 of 16 *Escherichia* strains, from 5 of 6 *Salmonella* strains, from 3 of 15 *Citrobacter* strains, from 2 of 40 *Proteus* strains, and from 1 of 4 *Enterobacter* strains. All of the TMP resistance factors isolated conferred resistance to concentrations higher than 1,000 μg/ml.

As has often been reported, we frequently found associated TMP and sulphonamide resistance. These resistances were co-transferred in 21 of the 25 plasmids we isolated.

This is important because therapeutic preparations of TMP are

Table 23.1 Standard R factors used in tests for compatibility.

Standard plasmids	*Compatibility group*	*fi*	*Genetic markers **
RA1	A	–	T Su
R4Oa	C	–	AKSu
R27	H	–	T
R144	Ia	–	TK
R621a	Iy	–	T
R391	J	–	K
R387	K	–	CS
R471a	L	–	A
R446b	M	–	ST
RN3	N	–	STSu
R16	O	–	ASTSu
RP4	P	–	ATK
R478	S	–	CTK
RtSI	T	–	K
RSa	W	–	CSKSu
R6K	X	–	AS
RP1CM	Y	–	C
R386	FI	+	T
R1	FII	+	ASCKSu
R124	FIV	+	T

* Symbols designate resistance to: A, ampicillin; T, tetracycline; C, chloramphenicol; K, kanamycin; S, streptomycin; Su, sulphonamides.

Table 23.2 Characteristics of TMP resistance factors isolated in Modena.

N. strains	*N. res.* *	*Identifi-cations* **	*Resistance transfer-red to K-12* ***	*F factor*	*fi*	*Group*
460	16	*E. coli*	SKTpSu	pMM001	–	N
		E. coli	SKTpSu	pMM002	–	N
		E. coli	STp	pMM003	–	Ia
		E. coli	TTpSu	pMM004	–	N
		E. coli	Tp	pMM005	–	N
		E. coli	TpSu	pMM006	+	FI
		E. coli	TpSu	pMM008	–	Ia
		E. coli	TpSu	pMM009	–	N
		E. coli	ASCTpSu	pMM010	+	FII
		E. coli	TpSu	pMM011	–	N
		E. coli	CTpSu	pMM016	+	FII
		E. coli	STpSu	pMM018	–	L
		E. coli	SCTpSu	pMM019	–	Ia
		E. coli	CTpSu	pMM020	+	FII
400	6	*S.s-paul*	TpSu	pMM101	–	N
		S.s-paul	TpSu	pMM102	–	N
		S.t-m	CTp	pMM103	–	N
		S.t-m	SCTpSu	pMM106	–	N
		S. agona	SCTpSu	pMM107	–	N
195	15	*C. freundii*	TpSu	pMM201	–	N
		C. freundii	TpSu	pMM202	–	N
		C. freundii	TpSu	pMM203	–	N
32	4	*E. aerog.*	STpEu	pMM301	–	N
421	101	*P. mirab.*	TpSu	pMM401	–	L
		P. mirab.	TpSu	pMM402	–	N

* Resistant to TMP at a conc. of 10 μg/ml.
** Resistant strains from which R factors were isolated. The MIC for all of these strains was > 1,000 μg/ml.
*** A, ampicillin; C, chloramphenicol; K, kanamycin; S, streptomycin; Su, sulfonamides; T, tetracycline; Tp, trimethoprim.

always made in combination with a sulphonamide (to take advantage of the synergistic antibacterial effects of the two drugs and to prevent, at the same time, the emergence of mutant bacteria resistant to either drug). The TMP resistance factor was transferred alone only in one strain of *E. coli*. Of 25 plasmids examined, 21 belonged to the fi$^-$ type and 4 to the fi$^+$ type. With a few exceptions, all of the TMP R factors isolated in the United Kingdom have been of the fi$^-$ type, whereas Romero and Perduca [5] in Italy and Acar et al. [6] in France found some plasmids of the fi$^+$ type. Among our 22 fi$^-$ plasmids 15 were classified in the N group, 4 were placed in the Ia group and 2 in the L group. Among the 4 plasmids of the fi$^+$ type, 3 were classified in the

FII group and 1 in the FI group. We found no plasmids belonging to the W group, as reported, with a few exceptions, by Hedges et al. [3] and Jobanputra and Datta [4].

References

[1] Datta, N., and Hedges, R. W. (1972) "Trimethoprim resistance conferred by W plasmids in Enterobacteriaceae", *J. Gen. Microbiol.*, **72**, 349-355.

[2] Fleming, M. P., Datta, N. and Grüneberg, R. N. (1972) "Trimethoprim resistance determined by R factors", *Br. Med. J.*, **1**, 726-728.

[3] Hedges, R. W., Datta, N. and Fleming, M. P. (1972) "R-factors conferring resistance to trimethoprim but not sulphonamides", *J. Gen. Microbiol.*, **75**, 573-575.

[4] Jobanputra, R. S. and Datta, N. (1974) "Trimethoprim R factors in enterobacteria from clinical specimens", *J. Med. Microbiol.*, **7**, 169-177.

[5] Romero, E., and Perduca, M. (1976) "Compatibility group of R factors for trimethoprim resistance isolated in Italy", *Proc. Meet. Bacterial Resistance*, vol. 1, pp. 24-26.

[6] Acar, J. F., Goldstein, F. W., Gerbaud, G. R. and Chabbert, Y. A. (1977) "Plasmides de resistance ou trimethoprim: transféribilité et groupes d'incompatibilité", *Ann. Microbiol.* (Paris), **128 A**, 41-47.

[7] Edwards, P. R. and Ewing, W. H. (1972) *Identification of Enterobacteriaceae*, 3rd ed., Burgess Publishing Co., Minneapolis, Minn.

[8] Coetzee, J. N., Datta, N. and Hedges, R. W. (1972) "R factors from *Proteus rettgeri*". *J. Gen. Microbiol.*, **72**, 543-552.

[9] Manicardi, G., Bondi, M., Quaglio, G. P. and Fabio, U. (1978) "Transferable Trimethoprim Resistance in Some Genera of the Enterobacteriaceae Family", *Proceedings of the 10th International Congress of Chemotherapy*, in *Current Chemotherapy*, **25**, 453.

24

Double-blind clinical study of a rifampicin-trimethoprim combination and of rifampicin alone in urinary tract infections

A. Visconti, U. Marini, F. Azzollini

Ospedale S. Carlo Borromeo - Via Pio II, 3 - Milano - Italy

E. Robotti

Clinical Research Dept. - Lepetit S.p.A.
Via Durando, 38 - Milano - Italy

Summary

Thirty-one UTI patients with significant bacteriuria were randomly allocated under double-blind conditions to either rifampicin plus trimethoprim (450 plus 120 mg, *b.i.d.*) or to rifampicin alone (450 mg, *b.i.d.*) for 10 consecutive days. The success rate at the end of treatment was significantly higher in the first group. Clearcut emergence of resistance from originally sensitive strains was documented in four cases with rifampicin alone; only one doubtful case was observed in the group treated with the combination.

Introduction

Rifampicin (RMP) is a broad spectrum bactericidal antibiotic which reaches high and dose-dependent urinary concentrations; it does not have cross-resistance with any other antibiotic except rifamycin SV [1]. Its activity against urinary pathogens has been recently confirmed with a large series of hospital isolates [2], and its therapeutic use in urinary tract infections (UTI) has been reported in the literature [3-12].

A dose regimen of 300 mg RMP 12-hourly gave a low success rate in UTI, due to the selection of resistant mutants [3]. The bacterial count in the urine decreased in the first few days, but then increased strikingly in those cases that failed to respond [4]. In the failures, resistance to RMP appeared on day 3 to 7 of treatment and reached levels much higher than the urinary concentrations of RMP [5], [6].

The selection of resistant strains was repeatedly confirmed, but it was not an absolute rule: the persistence of RMP-sensitive strains after a full course of treatment was documented in a few cases with failure [7], [8].

A higher success rate was obtained, in chronic refractory UTI free from stones or anatomical abnormalities, by increasing the dose of RMP from 450 mg *b.i.d.* (success rate 33%) to 600 mg *b.i.d.* (success rate 75%).

The combination of RMP with other antibacterial agents, such as cotrimoxazole [11], or tetracycline, gentamycin or colistin [12], has been used in order to overcome the problem of resistant mutants, and has given satisfactory results in difficult cases of chronic UTI with multiple resistant strains and an impaired defence mechanism in the host.

The addition of trimethoprim (TMP) to RMP can prevent selection of RMP-resistant strains *in vitro* [13], [14]. Moreover, the combination has shown a certain degree of synergism against different organisms, both *in vitro* and *in vivo* [13-18]. The present study was undertaken to compare the RMP-TMP combination (Rifaprim®) with RMP alone.

Patients and treatments

Thirty-one hospitalized patients (14 men and 17 women), aged from 37 to 83 (mean 69), with cystitis or cystopyelitis, entered the study; all had significant bacteriuria, i.e. $\geqslant 10^5$ organisms/ml of urine. Almost all patients had one or more symptoms of UTI, such as fever, flank pain, or dysuria. Half of them had been treated with other antibacterial agents recently, but these drugs had been discontinued at least 48 hours before starting the experimental treatment. There was no contraindication to the experimental drugs, and creatinine clearance was always greater than 30 ml/minute. The majority of these patients had one or more concomitant diseases (such as diabetes, hypertension, heart failure or cerebral vascular insufficiency), or complicating conditions (such as prostatic hypertrophy or other morphological abnormalities of the urinary tract).

Bacterial sensitivity was tested in Wellcotest agar with lysed horse blood. A diameter of at least 12 mm for the RMP disc and of at least 10 mm for the TMP disc [corresponding to MIC of about 50 and 12 mcg/ml (Berti, Lepetit files)] was considered to indicate that the organism was sensitive.

Patients were randomly allocated by a double-blind procedure to the treatments. One group received two tablets of Rifaprim® (RMP 225 mg plus TMP 60 mg in each tablet) twice daily for 10 days, and the other group received identical tablets containing RMP only. Urine cultures were made on the third day of treatment, on the day after the end of the antibacterial treatment, and one week later. Safety tests (hematology and blood chemistry) were carried out before and after treatment.

Results

In the Rifaprim® group there was one drop-out on the third day of treatment because of jaundice (the patient was also being treated with an anabolic steroid), which disappeared within 3 days after discontinuation of the experimental treatment. Two additional Rifaprim® patients could not be evaluated for efficacy because they did not comply with the protocol; and there were also three such patients in the other group. Therefore the patients available for the analysis of efficacy are 12 in the Rifaprim® group and 13 in the RMP group.

The bacteriological results in the Rifaprim® group are presented in Table 24.1. On the third day of treatment urine cultures were negative in all patients except case No. 11, who showed persistence of an *E. coli* sensitive to RMP and TMP, and who became negative on continuing treatment. On the day after the end of treatment, two strains of *E. coli*, RMP and TMP resistant, were cultured from one patient (No. 8, cystopyelitis and gastric cancer). The sensitivity pattern of these strains was definitely different from that of the original strain, and we suspect that this was a different serotype of *E. coli*. Negative urine cultures were found in the 9 cases which were controlled one week after the end of treatment.

The bacteriological results in the RMP group are presented in Table 24.2. On the third day of treatment only 6 out of 13 patients had negative cultures. One patient had a positive culture for *E. coli*, but the organism was still sensitive to RMP and had disappeared by the next control (No. 5); the other six patients had resistant strains, which were still present in the next control except in one case (No. 13). On the day after the end of treatment, selection of RMP-resistant organisms from the previously sensitive population was documented for two *Proteus rettgeri* infections (case No. 2 and 3), and for two *E. coli* infections (case No. 7 and 9); the sensitivity patterns of these organisms to the other antibacterials tested had remained essentially the same, with the possible exception of *E. coli*, in case No. 7. On the day after the end of treatment there were three additional failures: case No. 4 due to reinfection; case No. 11 due to the persistence of an originally resistant strain; case No. 12 due to the persistence of the same strain, still sensitive to RMP.

Table 24.1 *Group treated with Rifaprim®. (Clinical diagnosis: CP = cystopyelitis; C = cystitis. Microbiological data: S = sensitive; R = resistant; NA = not available, although required by the protocol).*

			Before treatment			*Urine culture*			
Case n.	*Diagnosis*	*Body temp. (°C)*	*Organism (s)*	*Sensitivity RMP*	*Sensitivity TMP*	*3rd day of treatm.*	*On the day after the end*	*1 week after the end*	*Result*
1	C	<37.0	*E. coli*	S	S	–	–	NA	Success (on the day after the end)
2	C	<37.0	*E. coli*	S	S	–	–	–	Success
			Ps. aeruginosa	S	R	–	–	–	
3	C	35.7	*E. coli*	S	S	–	–	–	Success
4	CP	38.2	*E. coli*	S	S	NA	–	NA	Success (on the day after the end)
5	CP	38.5	*Pr. morganii*	S	S	–	–	–	Success
			Klebsiella spp	S	R	–	–	–	
6	CP	37.8	*E. coli*	S	S	–	–	–	Success
7	CP	39.0	*E. coli*	S	S	–	–	–	Success
8	CP	38.4	*E. coli*	S	S	–	+		Failure *
9	CP	39.0	*E. coli*	S	R	–	–	–	Success
10	CP	38.0	*E. coli*	S	S	–	–	–	Success
11	CP	39.0	*E. coli*	S	S	+	–	–	Success **
12	CP	40.0	*E. coli*	S	S	–	–	–	Success
			Pr. mirabilis	S	R	–	–	–	

* Selection of two strains resistant to both RMP and TMP (the sensitivity pattern was different from that of the original strain).
** Sensitivity to both RMP and TMP was maintained on the 3rd day, and the organism was eradicated.

Table 24.2 *Group treated with RMP. (Clinical diagnosis: CP = cystopyelitis; C = cystitis. Microbiological data: S = sensitive; R = resistant; NA = not available, although required by the protocol).*

Case n.	Diagnosis	Body temp. (°C)	*Before treatment* Organism (s)	Sensitivity to RMP	*Urine culture* 3rd day of treatm.	On the day after the end	1 week after the end	Results
1	CP	<37.0	*E. coli*	S	–	–	–	Success
2	C	<37.0	*Pr. rettgeri*	S	+	+		Failure *
3	CP	38.0	*Pr. rettgeri*	S	+	+		Failure *
4	C	37.8	*Pr. vulgaris* *Klebsiella* spp.	R S	+	+		Reinfection with *Pr. mirabilis* resistant to RMP
5	C	37.5	*E. coli*	S	+	–	NA	Success (on the day after the end)
6	CP	38.0	*Enterobacter* spp.	S	–	–	–	Success
7	CP	38.5	*E. coli*	S	–	+		Failure *
8	CP	38.0	*Klebsiella* spp. *E. coli*	S S	– –	– –	– +	Failure **
9	CP	38.5	*E. coli*	S	–	+		Failure *
10	CP	38.4	*Klebsiella* spp.	S	–	–	+	Reinfection with *E. coli* res. to RMP and *Pr. mirabilis* sens. to RMP
11	CP	38.2	*Pr. rettgeri* *Ps. aeruginosa*	R S	+ –	+ –		Failure
12	CP	37.8	*Pr. mirabilis*	S	+	+		Failure **
13	C	38.5	*Pr. rettgeri*	S	+	–	–	Success

* Selection of strains resistant to RMP.

** The strains isolated in the last examination were still sensitive to RMP.

Clinical improvement and subsidence of fever was observed in all cases during the first day of treatment. After the end of treatment symptoms were absent in all treated patients except in one on RMP who still had dysuria. Both treatments were well tolerated, and there was no clinically significant change in the laboratory values except for the increased bilirubinemia in the previously mentioned drop-out.

Conclusions

Rifaprim® gave a higher success rate than did RMP alone (11/12 versus 6/13) on the day after the end of treatment, $P < 0.025$). Selection of RMP-resistant strains was observed in 4/13 RMP patients; only one doubtful case was observed in the Rifaprim® group. These results agree with those reported from the multicenter study in which we took part [19], and appear to confirm that Rifaprim® offers advantages over RMP alone in UTI patients.

In particular, it should be noticed that four strains in the Rifaprim® group which had been classified RMP-sensitive but TMP-resistant before treatment were eradicated. This fact is in agreement with the findings that subinhibitory concentrations of TMP can prevent the emergence of RMP resistance [13], [14]. Other explanations can be envisaged: the high concentration of TMP in the urine, in contrast with the rather conservative criterion for TMP sensitivity as detected by the disc method we adopted; the reduction of the bacterial population due to the cidal effect of RMP. On the basis of these considerations we plan to explore the potential of Rifaprim® against multiresistant *Proteus* strains, which in our experience are usually sensitive to RMP only.

Aknowledgments

We wish to thank Miss Silveria Zanoli for her technical assistance.

References

[1] Binda, G., Domenichini, E., Gottardi, A., Orlandi, B., Ortelli, E., Pacini, B. and Fowst, G. (1971) *Arzneim. Forsch. (Drug Res.)*, **21**, 1907-1977.

[2] Sanna, A. (1976) in: *Atti Società Italiana di Chirurgia, 78° Congresso*, Piccin, Padova, vol. I, pp. 40-63.

[3] Atlas, E. and Turck, M. (1968) *Am. J. Med. Sci.*, **256**, 247-254.

[4] Valentino, R. V. (1968) *Minerva Urol.*, **20**, 33-41.

[5] Brickner, P. W. (1969) *J. Clin. Pharmacol.*, **9**, 243-250.

[6] McMurdoch, J., Speirs, C. F., Wright, N., Wallace, E. T. (1969) *Lancet*, **1**, 1094.
[7] Trafton, H. M. and Lind, H. E. (1970) *Clin. Med.*, **77**, 30-33.
[8] Oosterlinck, W. and De Sy, W. (1972) *J. Urol. Nephrol.*, **3**, 233-240.
[9] Porpáczy, P. (1970) in: *Proceedings of the 6th International Congress of Chemotherapy*, University of Tokio Press, Tokio, vol. 1, pp. 895-896.
[10] Rocha Posada, H., Leuro, F. E., Sanches, A. S. (1972) *Arzneim. Forsch.* (*Drug Res.*), **22**, 1188-1190.
[11] Bourret, J., Girard, R., Deidier, Ch. (1974) *Lyon. Méd.*, **231**, 1013-1021.
[12] Acar, J. F. and Brisset, J. M. (1975) in: *Clinical Use of Combinations of Antibiotics* (J. Klastersky and E. Jawetz, ed.), Hodder and Stoughton, London, pp. 126-134.
[13] Kerry, D. W., Hamilton-Miller, J. M. T., Brumfitt, W. (1975) *J. Antimicr. Chemother.*, **1**, 417-427.
[14] Arioli, V., Berti, M., Carniti, G., Rossi, E., Silvestri, L. G. (1977) *J. Antimicr. Chemother.*, **3**, 87-94.
[15] Cluzel, R., Sirot, J., Cluzel, M., Joly, B. (1972) *Pathol. Biol.*, **20**, 871-879.
[16] Farrell, W., Wilks, M., Drasar, F. A. (1977) *J. Antimicr. Chemother.*, **3**, 459-462.
[17] Grüneberg, R. N. and Emmerson, A. M. (1977) *J. Antimicr. Chemother.*, **3**, 453-457.
[18] Hamilton-Miller, J. M. T., Kerry, D. W., Brumfitt, W. (1977) *J. Antimicr. Chemother.*, **3**, 193-194.
[19] Palminteri, R. and Sassella, D. (1979) *Chemotherapy*, **25**, 181-188.

25

Bacteriological studies on rifampicin and trimethoprim

A. M. Emmerson

Department of Microbiology, Whittington Hospital, Highgate Hill, London N19 3UA, England

R. N. Grüneberg

Department of Microbiology, University College Hospital, Gower Street, London WC 1, England

Summary

The interactions of Rifampicin (RIF) and Trimethoprim (TMP) have been studied against 100 urinary pathogens of various species and sensitivities using 'chequer board' titrations on solid media. Synergistic activity was demonstrated in 16 strains, which is likely to be clinically significant in 11 cases. In 14 of these strains synergy occurred in strains either singly or doubly resistant to TMP and RIF. No antagonism was demonstrated with the combination.

Resistance to Rifampicin was determined in 3,299 unselected urinary pathogens tested from the years 1975, 1976 and 1977. Overall resistance to Rifampicin was rare and less than 1% between the years 1975-77. Moderately increased resistance was observed in less than 6% of strains over the same period. This 'in vitro' evidence suggests that the combination of RIF/TMP is a promising therapeutic combination for use in urinary tract infection.

Introduction

Rifampicin in combination with isoniazid and ethambutol is the first choice for the treatment of tuberculosis. Rifampicin however has an extremely broad antimicrobial spectrum and has been used for the treatment of non-tuberculous infections. Resistance may occur during treatment of bacterial infections due to a high mutation rate which can be reduced by using combinations with trimethoprim, erythromycin, novobiocin or tetracycline. Since trimethoprim acts on the same metabolic pathway as rifampicin and equally has a broad spectrum it seems an obvious choice to combine the two. Since any consideration of potential interactions between drugs should take Into account the nature of their clinical applications, it seems that one probable applications of a combination of Rifampicin (RIF) and Trimethoprim (TMP) would be in the treatment of urinary tract infections (UTI).

In order to cover likely clinical situations we have tested various combinations of these two drugs against strains of varying sensitivity. Since it is not known how frequently RIF resistance occurs in urinary pathogens, this has been determined from 1,000 unselected urinary isolates examined for each of the years 1975, 1976 and 1977.

Methods (1)

In general the methods used were those of Kerry et al. [1] Freshly prepared solutions of RIF and TMP were incorporated in various combinations in Wellcotest agar. Strains were purified and identified using standard bacteriological methods and diluted to deliver an inoculum of 200-300 organisms. Plate inoculation was made using a Multipoint inoculator (Denley Instruments Ltd). The important features of this method are shown in Table 25.1.

Fresh urinary isolates of *Esch. coli, Proteus mirabilis, Klebsiella aerogenes, Strep. faecalis* and *Staph. albus* were selected. In each case five strains were used with the following sensitivity patterns:

a) organisms sensitive to both RIF and TMP;
b) organisms resistant to RIF and sensitive to TMP;
c) organisms sensitive to RIF and resistant to TMP;
d) organisms resistant to both RIF and TMP.

The methods used for selection were those previously described [2] which resulted in 100 strains available for study. The range of MIC's for these organisms are shown in Table 25.2. 'Chequer Board' titrations were performed on solid medium in such a way that each strain was tested against 144 different combinations of RIF and TMP.

Methods (2)

For the second part of this study more than 1,000 unselected urinary isolates were examined for the years 1975, 1976 and 1977. Approximately equal numbers were derived from general practice and hospital patients.

Antibiotic disc sensitivity tests were carried out on Oxoid Diagnostic sensitivity test agar (DST) with 5% added lysed blood, using Stokes' method [3]. The sensitive control organism was *E. coli*

Table 25.1 Minimum Inhibitory Concentration (MIC).

Definition	Lowest concentration which prevents all growth
Method	Plate incorporation method - Wellcotest Agar
Inoculum	200-300 organisms *via* inoculator
Incubation	37 °C overnight
Rifampicin	Range 0.001-1 μg/ml and 0.06-64 μg/ml
Trimethoprim	Range 0.03-32 μg/ml

Table 25.2 The range of MIC's of organisms tested for sensitivity to TMP and RIF. MIC (μg/ml).

	TMP		*RIF*	
Organism	*Sensitive*	*Resistant*	*Sensitive*	*Resistant*
Esch. coli	0.125-0.5	≥ 32	4-16	≥ 64
Proteus mirabilis	0.5 -2.0	≥ 32	2-4	≥ 64
Klebs. spp.	0.25 -0.5	≥ 32	4-8	≥ 64
Strep. faecalis	0.125-0.5	≥ 16	0.5-2	≥ 64
Staph. albus	0.125-0.5	≥ 16	0.004-0.008	≥ 64

NCTC 10418. Discs contained 30 μg rifampicin, 100 μg sulphonamide, 1.25 μg trimethoprim and 25 μg cotrimoxazole. The important features of this technique are shown in Table 25.3.

Minimum inhibitory concentrations (MIC's) were determined for all strains showing a reduced zone to a 30 μg rifampicin disc. The MIC was determined in the manner mentioned early in the paper.

Results (1)

Table 25.4 shows an example of the chequer board titration of 12 dilutions of RIF tested in all combinations with 12 dilutions of TMP against an *E. coli* strain resistant to RIF (MIC > 64 μg/ml) and TMP (MIC > 32 μg/ml).

Using the Fractional Inhibitory Concentration Index (FIC Index) [4] as the criterion for measuring synergy 16 out of the 100 strains

Table 25.3 Antibiotic disc sensitivity tests.

Media	Oxoid DST agar + 5% lysed blood.
Method	Stokes' controlled method.
Controls	Esch. coli NCTC 10418. Ps. aeruginosa NCTC 10490.
Discs	Rifampicin 30 μg, Sulphonamide 100 μg Trimethoprim 1.25 μg and Co-trimoxazole 25 μg

Table 25.4 "Chequer-Board" Titration (μg/ml)

Rifampicin	A	B	C	D	E	F	G	H	I	J	K	L
									Trimethoprim			
1	0	·03	·06	·125	·25	·5	1	2	4	8	16	32
2	·06	+	+	+	+	+	+	+	+	+	+	+
3	·125	+	+	+	+	+	+	+	+	+	+	+
4	·25	+	+	+	+	+	+	+	+	+	+	+
5	·5	+	+	+	+	+	+	+	+	+	+	+
6	1	+	+	+	+	+	+	+	+	+	+	+
7	2	+	+	+	+	+	+	+	+	+	+	+
8	4	+	+	+	—	—	—	—	—	—	—	—
9	8	+	+	+	—	—	—	—	—	—	—	—
10	16	+	+	+	—	—	—	—	—	—	—	—
11	32	+	+	+	—	—	—	—	—	—	—	—
12	64	+	+	+	—	—	—	—	—	—	—	—

Esch. coli - MIC to Rifampicin > 64 μg/ml; MIC to Trimethoprim > 32 μg/ml
TMP - FIC = 0·25/32 = 0·01; RIF - FIC = 4/64 = 0·07; FIC Index = 0·08

tested showed varying degress of synergy between RIF and TMP. In 11 of these strains this synergy was likely to be of clinical significance as judged by clinically achievable serum levels following normal dosage schedules. We define this as up to 4 μg of TMP and up to 16 μg RIF per ml of blood.

Table 25.5 shows the synergy found in 5 strains doubly resistant to RIF and TMP. Table 25.6 shows the synergy found in 8 strains resistant to TMP and sensitive to RIF. Two strains of *Staph. albus* doubly sensitive showed moderate degrees of synergy which were not clinically significant. A further strain of *Staph. albus* resistant to TMP and sensitive to RIF showed moderate synergy which was clinically significant.

Table 25.5 Synergy shown by strains resistant *to* Rifampicin *and* Trimethoprim.

Organism	*FIC Index*	*Degree of synergy*	*Clinical significance*
Esch. coli	0.08	Strong	Yes
Strep. faecalis (2)	0.50	Moderate	No
Staph. albus	0.01	Strong	Yes
Staph. albus	0.53	Moderate	No

Table 25.6 Synergy shown by strains sensitive to Rifampicin *and* resistant to Trimethoprim.

Organism	*FIC Index*	*Degree of synergy*	*Clinical significance*
Proteus mirabilis	0.515	Moderate	Yes
Strep. faecalis (4)	0.50	Moderate	Yes
Strep. faecalis	0.53	Moderate	Yes
Staph. albus (2)	0.515	Moderate	Yes

Results (2)

The distribution of 3,299 urinary isolates is shown in Table 25.7. This Table shows that *E. coli* forms 76.2% of the general practice isolates but only 54.3% of isolates from Hospital Practice. A greater number of *Klebs-Enterobacter, Pseudomonas* and *Proteus* spp. were isolated from hospital practice. Table 25.8 shows the results of 3, 245 Rifampicin sensitivity tests. MIC's determination were made to confirm those strains entered in the Moderately Sensitive

Table 25.7 3,299 Urinary Isolates 1975-1977.

Organisms	*General Practice*			*Hospital*		
	1975	*1976*	*1977*	*1975*	*1976*	*1977*
Esch. coli	422	369	507	259	267	341
Proteus mirabilis	32	30	41	72	68	76
Klebs-Enterobacter	23	20	19	62	75	75
Enterococci	9	20	13	15	34	41
Staphylococci	41	43	73	24	21	32
Ps. aeruginosa	6	3	5	14	15	23
Others	7	7	13	14	28	40
Totals	540	492	671	460	508	628

Table 25.8 Rifampicin sensitivity in 3,245 unselected Urinary isolates.

Organism	*Sensitive* (MIC < 16 μg/ml)			*Mod. Sensitive* (MIC = 16 μg/ml)			*Resistant* (MIC ⩾ 32 μg/ml)		
	1975	*1976*	*1977*	*1975*	*1976*	*1977*	*1975*	*1976*	*1977*
Esch. coli	681	632	847	0	4	1	0	0	0
Proteus spp.	103	97	116	1	1	1	0	0	0
Klebs-Enterobact-er	53	64	59	28	29	30	4	2	5
Staphylococci	64	63	105	0	0	0	1	1	0
Ps. aeruginosa	17	10	19	2	8	2	1	0	7
Others	43	84	52	0	5	1	2	0	0
Totals	961	950	1198	31	47	35	8	3	12

Table 25.9 Percentage *of 3,299 unselected Urinary Isolates* Fully sensitive *to various drugs.*

	CO.T	*SU*	*TMP*	*RIF.*
General Practice (1,703 strains)				
1975	96.1	77.6	93.0	99.6
1976	94.7	75.4	92.1	100.0
1977	95.7	74.1	90.6	98.0
Hospital Practice (1,596 strains)				
1975	84.3	65.4	79.1	98.7
1976	81.9	58.3	77.8	98.4
1977	86.3	63.5	75.5	94.3

(MIC = 16 μg/ml) and the Resistant groups (MIC ⩾ 32 μg/ml).

Although only 4% of 274 *Klebsiella-Enterobacter* spp were resistant, 31.8% showed increased resistance with MIC's at 16 μg/ml.

Table 25.9 shows the percentages of 3,299 unselected urinary isolates fully sensitive (MIC < 16 μg/ml) to co-trimoxazole, sulphamethoxazole, trimethoprim and rifampicin. These figures show that whereas resistance to RIF in hospital practice is rare (< 6%) resistance to Sulphonamides is common (approximately 40%).

Discussion

The only combination of antimicrobial agents commonly used in the treatment of UTI is co-trimoxazole, a combination of TMP and sulphamethoxazole. If resistance to either component is common, as in the case of Sulphonamide (see Table 25.9), the effectiveness of the combination may be expected to decrease. Kerry et al. [1] suggested that because of their essentially similar actions, combinations of TMP and RIF might result in synergistic activity. We have demonstrated synergy in 16 out of the 100 well defined strains tested. The synergism seen was likely to be clinically important in 11 of the 16 strains tested according to the criteria presented. The important features demonstrated here suggests that apart from treating UTI caused by doubly sensitive organisms, there is good evidence to recommend treatment with RIF/TMP in cases of UTI caused by either singly or doubly resistant strains. Even if no synergy is demonstrable, there may be a case for using such a combination in order to delay the emergence of bacterial resistance as indicated by Kerry et al. [1].

At present rifampicin is not used routinely on its own, and has been reserved in most instances for the treatment of tuberculosis in combination with other drugs. The extent of rifampicin resistance among urinary pathogens is unknown but is now presented here. In general practice RIF resistance is very rare and ranges from 0.4% in 1975 to 2% in 1977. In hospital practice this ranges from 1.3-5.7%. This however includes those strains with decreased sensitivity and should only fully resistant strains be included then only 23 out of 3,245 (< 1%) strains were recorded in the years 1975-1977. As can be seen from this Table 25.8 the majority of moderately sensitive or resistant strains were in the *Klebs-Enterobacter* group. We were unable to demonstrate synergy in the 20 strains of *Klebsiella aerogenes* tested. Interestingly enough was the number of *Pseudomonas* strains which fell within the therapeutic range.

Whereas we have discussed degrees of resistance with reference to serum levels pharmacokinetic studies by others [5], [6], [7] show that urinary concentrations reach levels well in excess of the MIC's of most of the more resistant organisms.

In summary, our studies using 'chequer board' titrations and relevant studies of unselected urinary isolates strongly suggest the value of combining RIF and TMP in the treatment of UTI. We believe this is a responsible proposition since there is no evidence as yet to show that the use of Rifampicin for non-tuberculous conditions increases the likelihood that Rifampicin resistant strains of Mycobacteria will emerge.

Acknowledgements

The authors wish to express their thanks to Mrs. Anne Leakey, and Messrs A. Cremer and P. Lamport for their technical assistance. We are grateful to Messrs Lepetit and Co. and in particular to Dr. J.

Mc Gilchrist for their support of this project. We are also grateful to Dr E. Robotti for his constructive criticism and advice.

References

[1] Kerry, D. W., Hamilton-Miller, J. M. T. and Brumfitt, W. (1975), *J. Antimicr. Chemo.*, **1**, 417-427.
[2] Grüneberg, R. N., Emmerson, A. M. (1977), *J. Antimicr. Chemo.*, **3**, 453-457.
[3] Stokes, E. J. (1968), *Clinical Bacteriology*, 3rd ed., p. 179, Arnold, London.
[4] Study Group (1973), *J. Inf. Dis.*, **128** (Suppl.), 502-507.
[5] Hamilton-Miller, J. M. T. and Brumfitt, W. (1976), *J. Antimicr. Chemo.*, **2**, 181-188.
[6] Acocella, G., Scotti, R. (1976), *J. Antimicr. Chemo.*, **2**, 271-277.
[7] Emmerson, A. M., Grüneberg, R. N. and Johnson, E. S. (1978), *J. Antimicr. Chemo*, **4**, 523-531.

26

Comparison of a rifampicin-trimethoprim combination and cotrimoxazole in urinary tract infections

G. M. Savazzi

Clinica Medica II - Università di Parma - Italy

D. Sassella

Clinical Research Dept. - Lepetit S.p.A.
Via Durando, 38 - Milano - Italy

M. Toni

Istituto di Microbiologia - Università di Parma - Italy

Summary

A combination of rifampicin and trimethoprim (RMP-TMP), given in a single daily dose, was compared with a standard *b.d.* regimen of cotrimoxazole in a randomized clinical trial carried out on 70 patients with urinary tract infections. The success rates of the two combinations (80% and 72%), as well as their tolerabilities, were very similar. Treatment was discontinued in one patient in each group because of adverse reactions.

No rifampicin-resistant mutant was isolated from those RMP-TMP patients whose infections persisted.

Introduction

One of the suggested purposes of the use of antimicrobial agents in combination is the prevention of the emergence of bacterial resistance [1], [2], [3], particularly in the case of antibiotics towards which resistance is likely to emerge owing to high mutation rates [4].

In view of its broad spectrum of activity and its high urinary excretion [5], rifampicin (RMP) has been used for urinary tract infections (UTI) [5], [6], but quite frequently therapeutic failures due to the selection of RMP-resistant mutants have been observed [6], [7], [8], [9]. The combination of trimethoprim (TMP) with rifampicin significantly suppresses *in vitro* the emergence of resistant strains [10], [11]. Furthermore, the two drugs display varying degrees of synergism, over a wide range of ratios, against different bacterial species, especially Gram-negative ones, including the most common urinary pathogens and also "problem" organisms [10], [11], [12], [13], [14]. A previous clinical trial comparing RMP-TMP with RMP alone in UTI had shown a significantly higher success rate for the combination and a lower selection of RMP-resistant strains [15].

These results led us to investigate further the efficacy and tolerability of the RMP-TMP combination in patients with UTI. Cotrimoxazole was chosen as a comparison drug, in view of its well-established therapeutic value in UTI [16]. The dosage schedule of RMP-TMP used in this trial had been proved previously, through pharmacokinetic studies in healthy volunteers, to provide adequate serum and urinary levels for both drugs [17].

Patients and methods

Seventy in-patients with UTI (56 women and 14 men) were admitted to the study. Urinary tract infection was defined as the presence of significant bacteriuria (10^5 or more bacteria per ml) due to the same organisms in two consecutive mid-stream urine specimens collected within a day of each other. The criteria for exclusion were: age less than 15 years, pregnant or breast-feeding woman, presence of indwelling catheter or pyelostomy, previous intolerance to rifampicin or cotrimoxazole, impaired renal function with a creatinine clearance lower than 30 ml/min, and abnormalities in the liver function tests.

Patients fulfilling the admission criteria were assigned to treatment with either RMP-TMP or cotrimoxazole by a random list. Thirty-four patients received two sugar-coated tablets of RMP-TMP (each containing 300 mg of rifampicin and 80 mg of trimethoprim) once daily in the evening, and thirty-six patients received two tablets of cotrimoxazole (each containing 400 mg of sulphamethoxazole and 80 mg of trimethoprim) twice daily. Treatments were given for 7 days. The trial was carried out in open conditions: the requirements for a double-blind study could not have been fulfilled, even if dummy preparations were used, because of the reddish colour given to the urine by rifampicin. Sensitivity tests, however, were carried out under blind conditions.

Follow-up urine specimens for bacteriological examination were collected on the third to fifth day after the end of treatment, and, whenever possible, one month later. Quantitative urine cultures and organism identification were performed by routine methods. The sensitivities of the isolated pathogens were assessed by a disc diffusion method in Wellcotest Sensitivity Agar with 7.5% lysed horse blood added, using paper discs containing 30 mcg of RMP or 1.25 mcg of TMP or 25 mcg of sulphamethoxazole (SMZ). Organisms were considered to be sensitive when the diameter of the inhibition zone corresponded to an estimated MIC [18] lower than 50 mcg/ml for RMP, 12 mcg/ml for TMP and 200 mcg/ml for SMZ. The same value for SMZ had been used by Anderson et al. [19]. According to the laboratory routine, the sensitivity to cotrimoxazole was also tested with the combined TMP-SMZ disc, using the conventional sensitivity threshold (zone size corresponding to an MIC $\leqslant$ 0.8 mcg/ml for TMP and $\leqslant$ 16 mcg/ml for SMZ). Since quite often treatment was started before the results of the sensitivity tests were available, some patients with organisms resistant to the drugs under study were also included in the trial.

The results of therapy were classified as successes when urine cultures at follow-up were negative or revealed $\leqslant 10^4$ organisms/ml, and as failures when urine cultures revealed more than 10^4 organisms/ml either of the same or ofdifferent species as compared with the initial isolate. Because bacterial serotyping was not performed, it was not possible to differentiate between relapses and reinfections when organisms were of the same genus and species as the initial isolates.

A laboratory work-up consisting of routine urinalysis, hematologic examinations, liver and kidney function tests was performed before and at the end of treatment.

Results

Sixty-one patients (30 in the RMP-TMP group and 31 in the cotrimoxazole group) were considered to have complied to the protocol. Five among them showed minor deviations from the protocol, such as urine cultures performed on the first or second day following the end of treatment (two cases on RMP-TMP and one case on cotrimoxazole), or only at the one month follow-up (two cases on cotrimoxazole). All these patients were considered "valid cases" and taken into account in the evaluation of efficacy.

The remaining 9 cases were deleted from the evaluation of efficacy either because of discontinuation of the treatment due to adverse reactions (2 cases, one in each group) or because of major deviations from the protocol (lack of confirmation of significant bacteriuria in 4 cases and lack of any follow-up urine cultures in 3 cases).

Table 26.1 shows the characteristics of "valid cases" in the two

Table 26.1 Characteristics of patients (only patients evaluated for efficacy are included).

Characteristic	*RMP-TMP (N = 30)*	*Cotrimoxazole (N = 31)*
Mean age (range)	65 (29-85)	49 (15-83)
Males/Females	7/23	4/27
With UTI symptoms *	15	19
Complicated UTI **	6	7
Concomitant hypertension	9	8
Concomitant diabetes	4	4

* Dysuria and frequency in most cases; also flank pain and/or fever in about one third of patients in each group.
** Presence of stones and/or any urologic abnormality.

treatment groups, which appear to be homogeneous except for a slightly higher proportion of females in the cotrimoxazole group (not significant, χ^2test) and a lower mean age in the same group (not significant, *t* test).

Organisms isolated before treatment are listed in Table 26.2. *E coli* was the most common isolate in both groups, although the incidence of this organism as well as of the other pathogens was unevenly distributed among the two treatment groups. The difference was not statistically significant (χ^2 test). From all the reported mixed infections, two bacterial species were isolated: *E. coli* and *Proteus* spp. in 3 cases (2 of them were in the RMP-TMP group); *Klebsiella-Enterobacter* and *Proteus* spp. in 3 cases (one in the RMP-TMP group); *E. coli* and *Staphylococcus* spp. in 2 cases (both in the RMP-TMP group).

The overall results of treatment are listed in Table 26.3. No significant difference (χ^2 test) was found between the cure rates of the two combinations, either considering all the cases or subgroups including only cases with combination-sensitive organisms before treatment. Apart from these latter, 7 patients in the RMP-TMP group and 2 in the cotrimoxazole group had organisms sensitive to only one component of each combination (RMP in the former group and TMP in the latter): treatment was successful in 5/7 and 2/2 patients respectively. In general, the bacteriological results correlated reasonably well with the clinical results, although eradication of the offending pathogen did not always result in complete elimination of symptoms.

Only a small proportion of patients had urine cultures performed one month after the completion of treatment: the success rates were 4/7 and 12/15 respectively (difference not significant, Fisher exact probability test).

The overall cure rate on the third to fifth day after the end of treatment was broken down in Table 26.4 according to the infecting pathogen. It is apparent that infections due to species other than *E. coli* accounted for a success rate of 15/17 in the RMP/TMP group and of 6/10 in the cotrimoxazole group.

Among the 6 failures observed after RMP-TMP, there were two definite instances of reinfection; in the remaining 4 cases, the isolates were of the same species as before treatment (one strain of *Proteus mirabilis* and 3 strains of *E. coli*). None of these latter showed any change in sensitivity to either RMP or TMP (three of the original strains were doubly-sensitive; the fourth was sensitive only to RMP). Two of the eight failures of cotrimoxazole were due to definite reinfection; in the remaining 6 cases, organisms of the same species as before treatment were isolated (in 2 cases *Proteus rettgeri* and in 4 cases *E. coli*). In three of these 6 cases, the original strain was cotrimoxazole resistant. Of the remaining three cases, a strain of *E. coli* which was sensitive to cotrimoxazole, as assessed with the combined disc, but resistant to SMZ alone before treatment, was

Table 26.2 Distribution of organisms isolated before treatment (percentage in brackets).

Organism	RMP-TMP (N = 30)	Cotrimoxazole (N = 31)
E. coli	13 (43)	20 (64)
Proteus spp.	5 (17)	3 (10)
Klebsiella-Enterobacter	3 (10)	5 (16)
Other Gram-negatives *	2	–
Streptococcus faecalis	2	–
Mixed infections	5 (17)	3 (10)

* One *Pseudomonas* and one *Citrobacter* spp.

Table 26.3 Success rate on the 3rd to 5th day after the end of treatment (percentage in brackets).

	RMP-TMP		Cotrimoxazole **	
	Sterile urine	≤ 10^4 organisms/ml	Sterile urine	≤ 10^4 organisms/ml
all cases	19/30 (63)	24/30 (80)	18/29 (62)	21/29 (72)
Combination-sensitive organisms before treatment *	10/16 (62)	12/16 (75)	13/20 (65)	15/20 (75)
Complicated UTI	3/6	4/6	3/7	4/7

* Patients with organisms sensitive to both RMP and TMP in the RMP-TMP group and patients sensitive to cotrimoxazole, as assessed by the combined TMP-SMZ disc, in the cotrimoxazole group.

** Two other patients had urine cultures only at the one month follow-up (one success and one failure).

Table 26.4 Success rate on the 3rd to 5th day after the end of treatment according to the organism.

Organism	*RMP-TMP (N = 30)*	*Cotrimoxazole (N = 29 *)*
E. coli	9/13 (69%)	15/19 (79%)
Proteus spp.	3/5	1/3
Klebsiella-Enterobacter	3/3	3/4
Other Gram-negatives	2/2	–
Streptococcus faecalis	2/2	–
Mixed infections	5/5	2/3
Total	24/30 (80%)	21/29 (72%)

* Two other patients (one *E. coli* and one *Klebsiella* spp. infection) had urine cultures only at the one-month follow-up: the first was cured, the latter still had significant bacteriuria.

replaced by a strain resistant to both TMP and SMZ. However, its general sensitivity pattern was also different from the original strain, thus suggesting a probable reinfection.

Tolerability. Both combinations were well tolerated in most cases. Adverse reactions were observed in two patients. In the group on RMP-TMP, a patient with scleroderma and impaired renal function developed jaundice. Treatment was discontinued on the sixth day; total bilirubin had increased from 1.1 to 4.9 mg/100 ml. During the experimental treatment the patient had been given a blood transfusion because of anemia present on admission. Total bilirubin decreased to 1.6 mg/100 ml six days after withdrawal of the drug. In the group treated with cotrimoxazole, one patient had an erythematous itching rash on all four limbs. Treatment was discontinued on the second day. The reaction disappeared within 24 hours.

Apart from the bilirubin values in the patient with jaundice and a transient rise of SGOT and SGPT noticed in another patient on RMP-TMP, no clinically significant or definitely drug-related changes in laboratory values were observed at the end of treatment. The *t* test for paired comparisons failed to demonstrate any significant differences between the laboratory values observed before and after treatment in both groups.

Conclusion

The results of the present study show that a 7-day course of RMP-TMP given once daily was as successful as a standard *b.d.* regimen of cotrimoxazole in treating UTI in hospitalized patients, with cure rates of 80% and 72% respectively, as assessed on the third to fifth day after the completion of treatment. Both combinations were satisfactorily tolerated.

In our population, the relative incidences of the different urinary pathogens were not exactly the same in the two treatment groups, with a slightly lower frequency of *E. coli* in the RMP-TMP group than in the cotrimoxazole group. Although our numbers are too small to draw any significant inferences, the results obtained with RMP-TMP in infections caused by typical hospital isolates appear to be encouraging.

In our study, no RMP-resistant strain was isolated from those patients treated with RMP-TMP in whom the original pathogen probably persisted. Furthermore, treatment was also successful in five patients whose infecting organisms were sensitive to RMP, but were classified as resistant to TMP, as assessed by the paper disc method (estimated MIC $>$ 12 mcg/ml). These data appear to be in agreement with previously reported *in vitro* results which indicate that even subinhibitory doses or TMP may be sufficient to prevent the selection of RMP-resistant mutants [10].

In conclusion, in view of the satisfactory results obtained in hospital infections and of the lack of emergence of rifampicin-resistance, we think that the use of the combination rifampicin-trimethoprim in UTI may be sound, provided that the general sensitivity pattern of the infecting organisms is taken into account. Present results must be confirmed by longer term follow-ups.

Acknowledgements

We are very grateful to Dr. P. Somenzi for his excellent contribution in microbiological testing.

References

[1] Cohen, S. N. (1975) in *Clinical use of combinations of antibiotics* (J. Klastersky and E. Jawetz eds.), Hodder and Stoughton, London, pp. 1-10.

[2] Weinstein, L. (1975) in *The pharmacological basis of therapeutics* (L. S. Goodman and A. Gilman eds.), 5th edn., MacMillan, New York, pp. 1090-1112.

[3] Smith, H. (1977) *Antibiotics in clinical practice*, 3rd edn., Pitman Medical, Tunbridge Wells, Kent, pp. 20-22.

[4] Garrod, L. P. (1976) *Br. Med. J.*, **2**, 933-936.

[5] Binda, G., Domenichini, E., Gottardi, A., Orlandi, B., Ortelli, E., Pacini, B. and Fowst, G. (1971) *Arzneim. Forsch.* (*Drug. Res.*), **21**, 1907-1977.

[6] Porpaczy, P. (1970) in *Proceedings of the 6th International Congress of Chemotherapy*, University of Tokyo Press, Tokyo, vol. 1, pp. 895-896.

[7] Atlas, E. and Turck, M. (1968) *Am. J. Med. Sci.*, **256**, 247-254.

[8] Oosterlinck, W. and De Sy, W. (1972) *J. Urol. Nephrol.*, **3**, 233-240.
[9] Acar, J. F. B. and Brisset, J. M. (1975) in *Clinical use of combinations of antibiotics* (J. Klastersky and E. Jawetz eds.), Hodder and Stoughton, London, pp. 126-134.
[10] Kerry, D. W., Hamilton-Miller, J. M. T. and Brumfitt, W. (1975) *J. Antimicrobial Chemotherapy*, **1**, 417-427.
[11] Arioli, V., Berti, M., Carniti, G., Rossi, E. and Silvestri, L. G. (1977) *J. Antimicrobial Chemotherapy*, **3**, 87-94.
[12] Farrell, W., Wilks, M. and Drasar, F. A. (1977) *J. Antimicrobial Chemotherapy*, **3**, 459-462.
[13] Gruneberg, R. N. and Emmerson, A. M. (1977) *J. Antimicrobial Chemotherapy*, **3**, 453-457.
[14] Hamilton-Miller, J. M. T., Kerry, D. W. and Brumfitt, W. (1977) *J. Antimicrobial Chemotherapy*, **3**, 193-194.
[15] Palminteri, R. and Sassella, D (1978) *Chemotherapy*, **25**, 181-188.
[16] Shiffman, D. O. (1975) *JAMA*, **231**, 635-637.
[17] Acocella, G. and Scotti, R. (1976) *J. Antimicrobial Chemotherapy*, **2**, 271-277.
[18] Arioli, V. (1977) *The correlation between agar dilution and agar diffusion tests with rifampicin, trimethoprim and sulphamethoxazole*, Lepetit Files.
[19] Anderson, J. D., Lacey, R. W., Lewis, E.L. and Sellin, M. A. (1974) *J. Clin. Pathol.*, **27**, 619-622.

27

Bactericidal action of the combination rifampicin plus trimethoprim

B. P. Goldstein, V. Arioli

Chemotherapy Department, Gruppo Lepetit SpA, Via G. Durando, 38 - 20158 Milano, Italy

Summary

The combination of rifampicin (Ramp) plus trimethoprim (Tmp) was very effective in sterilizing bacterial cultures in human urine *in vitro*. Our data demonstrate that: (1) Ramp is bactericidal in various culture conditions; (2) Tmp is bactericidal in urine when the concentration of bacteria is not too high; (3) upon reduction of the number of bacteria by Ramp, cultures become susceptible to the bactericidal action of Tmp; (4) the bactericidal action of Tmp suppresses the emergence of Ramp-resistant bacteria; (5) the bactericidal activity of Ramp plus Tmp is greater than that of sulfamethoxazole plus Tmp.

Introduction

Rifampicin (Ramp) and trimethoprim (Tmp) are both wide spectrum antibiotics which, upon oral administration in man, give high urine levels and moderate blood levels [1]. They are therefore suitable drugs for the therapy of urinary tract infections.

Combinations of the two drugs have shown synergy or additivity in inhibiting the growth of many bacterial strains *in vitro* [6], [4]. In addition Ramp + Tmp was more effective than single drug therapy on experimental animal infections [4]. Unlike resistance to the individual drugs, the frequency of doubly resistant bacteria, in sensitive cultures, is immeasurably low [4]. Recently Ramp + Tmp has proved particularly effective in the treatment of urinary tract infection in man [8], [9].

Our aim, in this study, was to determine the bactericidal activity of Ramp + Tmp *in vitro* and the effect of this combination on the emergence of Ramp-resistant bacteria. As we wished to simulate the conditions occurring in urinary tract infection, we used high inocula of bacteria and varied the concentrations of Ramp and Tmp to reflect, approximately, their pharmacokinetics in the urine of volunteers treated once daily with 600 mg Ramp + 160 mg Tmp. In addition we used human urine as the culture medium; this choice was made because of the well-known variability of Tmp activity with the medium [2], [3].

Materials and Methods

Bacterial strains. The *Escherichia coli* strains used were: L1069 (a rifampicin-sensitive K12 strain) and L1068 (rifampicin-resistant derivative of L1069); L47 (SKF 12140) a strain pathogenic for mice; two clinical isolates (L766 and L778) from urology patients. Other species used were: *Proteus mirabilis* L987 (clinical isolate) and *Klebsiella pneumoniae* L142 (pathogenic for mice).

Determination of bactericidal effect. Experiments were performed in pooled, filter-sterilized urine of healthy laboratory personnel. Samples which precipitated or showed evidence of contamination were discarded. Urine was inoculated from fresh overnight stationary cultures in Nutrient or Penassay broth and incubated for 2 hours with agitation at 37° prior to the start of the experiment. During each 24-hour period, bacteria were centrifuged and resuspended in fresh medium at 16 and 24 hours. The volume of each culture was 20 ml.

Viable bacteria were generally determined by plating aliquots of appropriate dilutions by inclusion in soft Nutrient agar on Nutrient agar plates. For *P. mirabilis* the viable count was determined by the most probable number method after dilution in Nutrient broth (Meynell and Meynell, 1965). Ramp-resistant bacteria were counted by plating on Nutrient agar supplemented with 200 μg/ml of Ramp.

Antibiotic concentrations. Ramp, Tmp and sulfamethoxazole (SMX) concentrations were varied to reflect approximately their concentrations in human urine after a single dose of 600 mg Ramp + 160 mg Tmp [1] or of co-trimoxazole (800 mg SMX + 160 mg Tmp, [5]). For SMX, the concentration was held at 300 μg/ml and for Tmp at 100 μg/ml; Ramp was present (at 200 μg/ml) only during the first 16 hours of each day, unless otherwise indicated.

Results and Discussion

Fig. 27.1 shows a short preliminary experiment which illustrates the bactericidal effects of Ramp (200 μg/ml) and Tmp (100 μg/ml) alone and in combination at low (fig. 27.1 a) and high (fig. 27.1 b) inocula of *E. coli* L47. In this experiment there was no medium change and the initial Ramp concentration was maintained throughout.

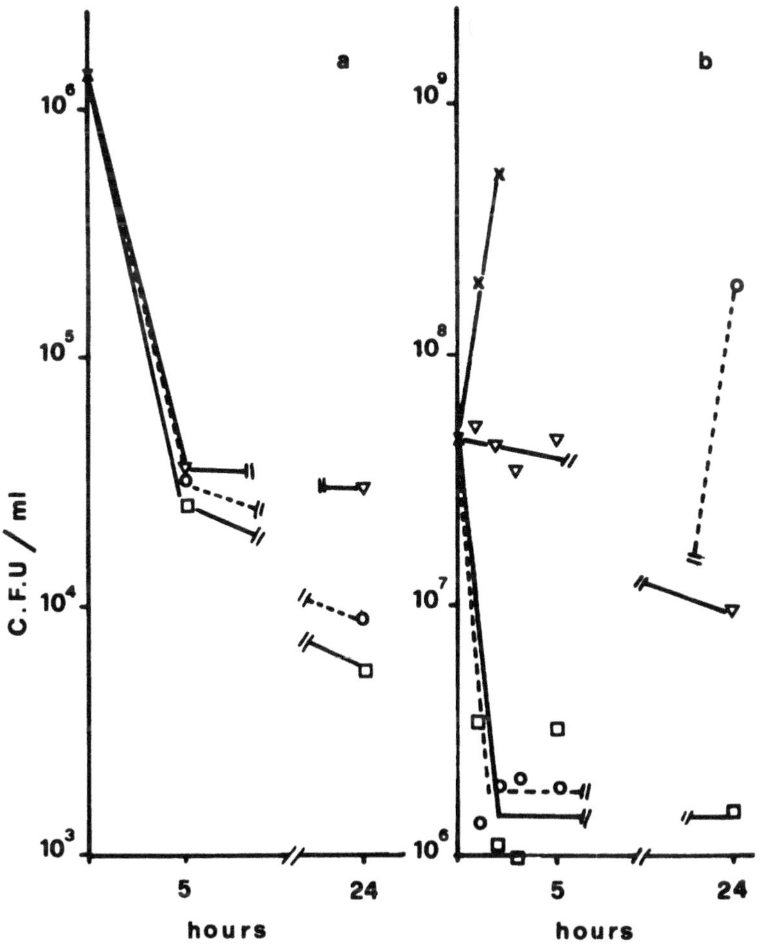

Fig. 27.1 - Bactericidal activity of Ramp, Tmp and Ramp + Tmp at 2 different bacterial inocula. Strain L47 was inoculated into urine so as to have, at time 0: a) 1.5×10^6 cells/ml (3×10^7 cells per 20 ml culture); b) 5×10^7 cells/ml (10^9 per culture). Additions: ○, 200 μg Ramp/ml; ▽, 100 μg Tmp/ml; □, 200 μg Ramp + 100 μg Tmp/ml; ×, 0-time, control. C.F.U. = colony-forming units.

Ramp alone was equally bactericidal at both bacterial concentrations, but at the high inoculum regrowth was observed at 24 hours. This was due to the emergence of a resistant population from the resistant bacteria present in the initial inoculum of 10^9 bacteria per culture. Tmp was less bactericidal at high than at low inoculum. In combination with Ramp, however, Tmp suppressed the emergence of a Ramp-resistant population (fig. 27.1 b).

Fig. 27.2 compares the bactericidal effects of Ramp + Tmp, Tmp alone and Tmp + SMX during 3 days of treatment of *E. coli* L778. The combination Ramp + Tmp reduced the viable count from 10^8/ml to 10^2/ml by the 16th hour of the second day. Tmp, alone or in combination with SMX, achieved this result only on the third day. Qualitatively similar results were obtained with *E. coli* L766 (data not shown).

Fig. 27.3 illustrates a similar experiment performed with *P. mirabilis* L987. Ramp + Tmp essentially sterilized the culture (< 10 bacteria/ml) by the 16th hour of the second day. At the end of 2 days of treatment 10^3 to 10^4 bacteria/ml remained in the culture treated with Tmp or with Tmp + SMX. Similar results were obtained with *K. pneumoniae* L142.

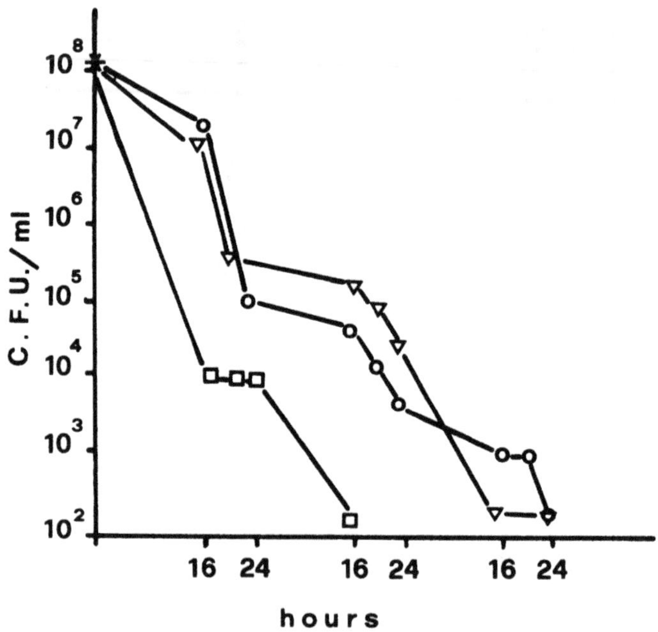

Fig. 27.2 - Bactericidal kinetics of *E. coli* L778 under various treatment schemes: ∗, zero-time control; ▽, 100 μg Tmp/ml; ○, 100 μg Tmp + 300 μg SMX/ml; □, 100 μg Tmp + 200 μg Ramp/ml alternated with Tmp alone (100 μg/ml) as described in Materials and Methods.

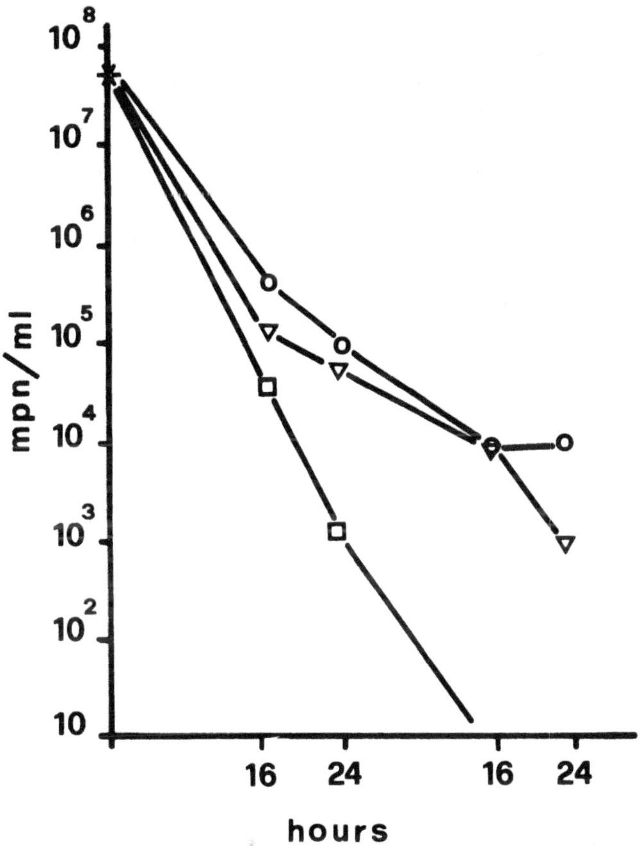

Fig. 27.3 - Bactericidal kinetics of *P. mirabilis* L987 under various treatment schemes. Symbols as in fig. 27.2 mpn = most probable number of viable cells.

Thus, concentrations of Ramp + Tmp easily attainable in urine after once daily therapy were very effective in sterilizing cultures of urinary pathogens. Tmp alone and Tmp + SMX were less effective; SMX did not add significantly to the bactericidal activity of Tmp alone.

Actually, some Ramp remains in the urine of volunteers given the single dose of Ramp + Tmp even during the latter part of the day [1]. Thus the complete absence of Ramp during the 16-24 hour periods of our experiments represents a conservative simplification. Preliminary experiments showed that if 50 μg/ml Ramp were present during the 16-24 hour periods, the rate of sterilization of cultures by Ramp + Tmp was only slightly increased.

Successful eradication of bacteria from dense cultures treated with Ramp + Tmp appears to be due to the bactericidal action of

Ramp and the ability of Tmp to suppress the emergence of Ramp-resistant (rif-r) bacteria.

Fig. 27.4 shows an experiment designed to study the effect of Tmp on rif^r bacteria. Sensitive and resistant bacteria were combined in the ratio 1000:1. In the culture treated with Ramp alone, killing of sensitive bacteria and growth of resistant cells led to rapid emergence of a resistant population. The action of Tmp blocked this in cultures treated with Tmp, alone or in combination with Ramp.

The results of these studies lead to the following picture of the

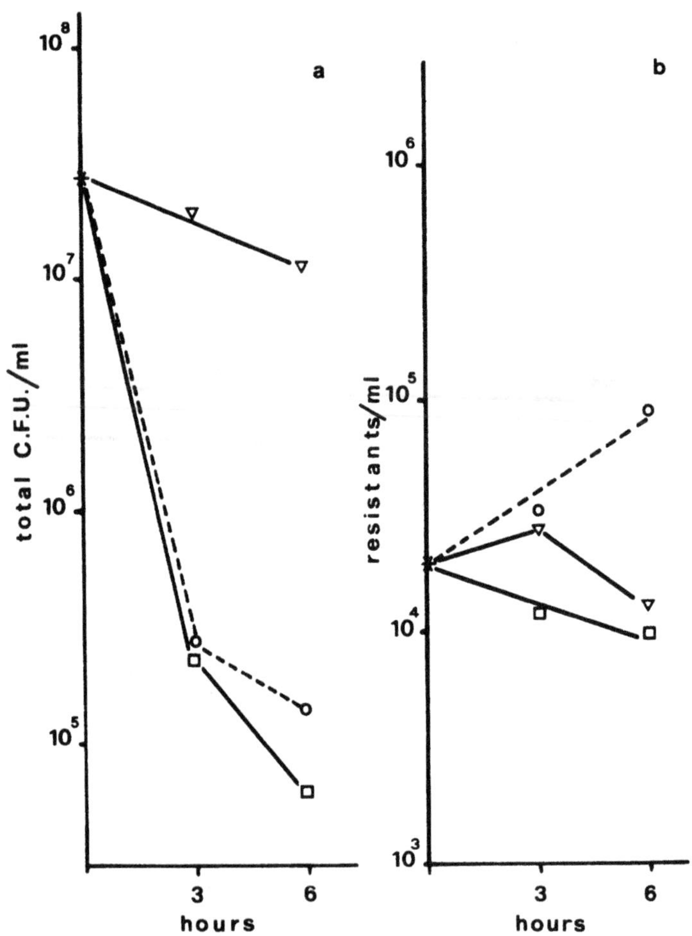

Fig. 27.4 - Effect of Ramp, Tmp and Ramp + Tmp on the selection of Ramp-resistant bacteria. Urine was inoculated with *E. coli* strains L1068 (Ramp-resistant) and L1069 (Ramp-sensitive) in the ratio 1:1000 and treated with: ○, 200 μg Ramp/ml; ▽, 100 μg Tmp/ml; □, 200 μg Ramp + 100 μg Tmp/ml.

interaction between Ramp and Tmp in sterilizing bacterial cultures. Ramp is bactericidal even at high inocula. Reduction of the inoculum by Ramp allows Tmp to express its bactericidal activity, particularly against Ramp-resistant cells. Thus both components contribute to the sterilization of the culture.

References

[1] ACOCELLA, G. and SCOTTI, R. (1976), "Kinetic studies on the combination rifampicin-trimethoprim in man. I. Absorption and urinary excretion after administration to healthy volunteers of single doses of the two compounds alone and in combination, and of the combination over a period of 1 week", *J. Antimicr. Chemo.*, **2**, 271-277.

[2] AMYES, S. G. B. and SMITH, J. T. (1974), "Trimethoprim-sensitivity testing and thymineless mutants", *J. Med. Microbiol.*, **7**, 143-153.

[3] AMYES, S. G. B. and SMITH, J. T. (1974), "Trimethoprim action and its analogy with thymine starvation", *Antimicr. Ag. Chemother.*, **5**, 169-178.

[4] ARIOLI, V., BERTI, M., CARNITI, G., ROSSI, E. and SILVESTRI, L. G. (1977), "Interaction between rifampicin and trimethoprim *in vitro* and in experimental infections", *J. Antimicr. Chemo.*, **3**, 87-94.

[5] BACH, M. C., GOLD, O. and FINLAND, M. (1974), "Absorption and urinary excretion of trimethoprim, sulfamethoxazole, and trimethoprim-sulfamethoxazole: Results with single doses in normal young adults and preliminary observations during therapy with trimethoprim-sulfamethoxazole", in *Trimethoprim-Sulfamethoxazole - Microbiological, Pharmacological and Clinical Considerations*, (M. Finland and E. H. Kass eds), University of Chicago Press, Chicago, pp. 152-166.

[6] KERRY, D. W., HAMILTON-MILLER, J. M. T. and BRUMFITT, W. (1975), "Trimethoprim and rifampicin: *in vitro* activities separately and in combination", *J. Antimicr. Chemo.*, **1**, 417-427.

[7] MEYNELL, J. J. and MEYNELL, E. (1965), *Theory and Practice in Experimental Bacteriology*, Cambridge University Press, London.

[8] SAVAZZI, G., SASSELLA, D. and TONI, M. (1979). "Comparison of rifaprim, a combination of rifampicin and trimethoprim, and cotrimoxazole in urinary tract infections", chapter 26 of this book.

[9] VISCONTI, A., MARINI, U., AZZOLLINI, F. and ROBOTTI, E. (1979). "Double blind clinical study of a combination of rifampicin-trimethoprim and rifampicin alone in urinary tract infections", chapter 24 of this book.

28

First choice indications for cotrimoxazole therapy

G. Visco, P. Sette

Ospedale "L. Spallanzani" - Via Portuense 292 - 00199 Roma - Italy

Summary

Cotrimoxazole proved more effective than pyrimethamine in the treatment of 11 patients with acquired toxoplasmosis and more effective than gentamicin in the treatment of 16 patients with "minor" Salmonella food infection gastroenteritis and septicaemia. Cotrimoxazole also proved just as effective as ampicillin or a combination of penicillin and chloramphenicol in the treatment of 40 patients with acute bacterial meningitis. In antibacterial sensitivity tests made before treatment, cotrimoxazole proved more active than other antibacterial drugs against isolates found in 1210 patients with acute respiratory or urinary tract infections. Cotrimoxazole proved particularly effective against upper respiratory tract infections due to *D. pneumoniae*, *Str. pyogenes* or mixed flora and against *E. coli* urinary tract infections. These results suggest that cotrimoxazole may be the drug of choice in the treatment of acquired toxoplasmosis, "minor" salmonella food infection gastroenteritis and septicoemia, acute bacterial meningitis affecting patients hypersensitive to β-lactam drugs, and acute respiratory and urinary tract infections.

Cotriomoxazole, a combination of trimethoprim (TMP) and sulphamethoxazole (SMZ), acts synergistically against a broad spectrum of gram-positive and -negative bacteria. Cotrimoxazole may be used to replace other antibacterial agents in certain disorders, and may be used, in our opinion, as the drug of choice in the following infections.

Acquired toxoplasmosis. Antibiotic sensitivity tests are not usually performed in cases of this infection. In searching for a drug that might replace pyrimethamine in the treatment of this infection, we compared the effects of cotrimoxazole with those of pyrimethamine in 11 patients affected by mild lymphatic acquired toxoplasmosis. Of these patients, 6 received TMP 80 mg and SMZ 400 mg/10 kg body weight/day p.o. for 20 days and 5 received pyrimethamine 1 mg/kg/day p.o. for 20 days. Both treatments proved effective. However, fever and ESR returned to normal quicker in the cotrimoxazole treatment group, and, in a follow-up examination 4 months after the end of treatment, 2 relapses were observed in the pyrimethamine treatment group and none in the cotrimoxazole group (Table 28.1).

"Minor" Salmonella food infection gastroenteritis with septicaemia. In cases of this infection, treatment generally has to begin before the results of antibiotic sensitivity tests are available [1]. In searching for drugs against this type of infection, we found that cotrimoxazole and gentamicin were probably the most effective [2]. We therefore decided to compare the effects of these drugs in 16 patients with Salmonella food infection gastroenteritis and symptoms of septicaemia. Of these patients, 7 received cotrimoxazole (TMP 40 mg and SMZ 200 mg/10 kg body wt./day i.v. for 5 days, followed by TMP 80 mg and SMZ 400 mg/10 kg body wt./day p.o. for 5 days), and 9 received gentamicin (5 mg/kg/day i.m.), for 10 days. Treatment began shortly after three time-spaced blood samples had been taken. Subsequent antibiotic sensitivity tests showed that isolates from at least 2 blood cultures of 6/7 patients (85.7%) given cotrimoxazole were sensitive to cotrimoxazole and of 7/9 patients (77.7%) given gentamicin were sensitive to gentamicin. Remission of all symptoms was observed in both treatment groups after 6 to 6½ days of treatment. Blood cultures made 20 days after the end of treatment proved negative in all patients of both treatment groups. In contrast, stool cultures made 20 days after the end of treatment proved negative in

Table 28.1 Comparative results of antiprotozoal treatment in 11 patients with acquired toxoplasmosis (mild lymphatic type), 6 of whom received cotrimoxazole (TMP 80 mg + SMZ 400 mg/10 kg body wt./day p.o.), *and 5 pyrimethamine* (1 mg/kg/day p.o.) *for 20 days.*

	Cotrimoxazole (6 patients)	*Pyrimethamine (5 patients)*
No. of days (means ± s.e.) before disappearance of fever	4.3 ± 0.7*	9.4 ± 1.2*
No. of days (means ± s.e.) before ESR returned to normal	9.5 ± 3.1	16.8 ± 5.7
No. of days (means ± s.e.) before remission of lymphadenopathy	19.17 ± 2.0	21.4 ± 2.5
No. of patients with relapses	0	2

* P< 0.05

Table 28.2 Comparative results of antibacterial treatment in 16 patients with salmonella food infection gastroenteritis, 7 of whom received cotrimoxazole (TMP 40 mg + SMZ 200 mg/10 kg body wt./day i.v. for 5 days, followed by TMP 80 mg + SMZ 400 mg/10 kg body wt./day p.o. for 5 days), *and 9 gentamicin* (5 mg/kg/day i.m.) *for 10 days.*

	Cotrimoxazole (7 patients)		*Gentamicin (9 patients)*	
	N.	%	*N.*	%
Patients infected by strains sensitive *in vitro* to the administered drug	6	85,7	7	77,7
Patients clinically cured	7	100	9	100
Patients with negative 20-day follow-up blood cultures	7	100	9	100
Patients with negative 20-day follow-up stool cultures	4	57.1	6	66.7
No. of days (means ± s.e.) before remission of symptoms	6.14 ± 2.1		6.67 ± 2.6	

4/7 patients (57.1%) given cotrimoxazole and in 6/7 patients (66.7%) given gentamicin (Table 28.2).

Acute bacterial meningitis. Cotrimoxazole may be considered the drug of choice in the treatment of patients with acute bacterial meningitis who are hypersensitive to β-lactam drugs. In a study we made some time ago [3] in 40 patients with acute bacterial meningitis, we compared the effects of cotrimoxazole in 20 patients with those of ampicillin in 12 patients and of a combination of penicillin and chloramphenicol in 8 patients. All three treatments proved highly effective. However CSF cell counts returned to normal quicker in the cotrimoxazole treatment group than in the other two treatment groups.

Acute respiratory and urinary tract infections. Antibacterial treatment of these infections should be related to the results of antibacterial sensitivity tests made before treatment. The results of these tests showed that cotrimoxazole was more active than other commonly used antibacterial drugs against isolates found in 1210 patients with acute respiratory or urinary infections [4], [5].

The results of cotrimoxazole treatment, over the last 8 years, in 488 patients with acute respiratory tract infections (Tables 28.3, 28.4, 28.5, 28.6, 28.7) showed that cotrimoxazole was:

Table 28.3 Etiology and diagnosis in 488 cases of acute respiratory tract infections.

Diagnosis	*No. of cases*	*% of cases*	*Most frequent isolates*
Pharyngitis and tonsillitis	31	6.3	*β-hemolytic group A Streptococci.* *St. aureus.* *D. pneumoniae.* Gram-neg. bacteria. Mixed flora.
Tracheitis and bronchitis	95	19.5	*β-hemolytic group A Streptococci.* *D. pneumoniae.* Mixed flora. Gram-neg. bacteria. *St. aureus.*
Bronchiectasis	44	9.0	Mixed flora.
Exacerbations of chronic bronchitis	208	42.7	*K. pneumoniae.* *Ps. aeruginosa.*
Pneumonia	110	22.5	Gram-neg. bacteria. *D. pneumoniae.*
Total	488	100	

Table 28.4 Disappearance of pathogenic bacteria from exudates in 488 cases of acute respiratory tract infection treated with cotrimoxazole.

No. of cases	*Diagnosis*	*Disappearance: % of cases*
31	Pharyngitis-tonsillitis	61
95	Tracheitis-bronchitis	52
44	Bronchiectasis	50
208	Exacerbations of chronic bronchitis	51
110	Pneumonia	64

a) very effective against upper respiratory tract infections due to *D. pneumoniae*, *Str. pyogenes* or mixed flora;
b) effective against pneumonia and exacerbations of chronic bronchitis;
c) partially effective against *Staph. aureus* and gram-negative infections, curing about half the number of cases involving these bacteria [4].

The results of cotrimoxazole treatment in 722 patients with

Table 28.5 Disappearance of pathogenic bacteria from exudates in 488 cases of acute respiratory tract infection treated with cotrimoxazole.

No. of cases	*Isolated bacteria*	*Disappearance: % of cases*
100	*D. pneumoniae* *Strep. haemol. group A*	88
33	*Staph. aureus*	51
132	*Klebsiella, Ps. aeruginosa*	51
36	Other gram-negative bacteria	47
187	Mixed flora	56

Table 28.6 Clinical results in 488 cases of acute respiratory tract infection treated with cotrimoxazole.

		Results (%)			
No. of cases	*Diagnosis*	*Very good*	*Good*	*Moderate*	*Negative*
31	Acute tonsillitis	74	10	10	6
95	Tracheo-bronchitis	54	5	10	31
208	Exacerbations of chronic bronchitis	44	9	5	42
44	Bronchiectasis	43	7	14	36
110	Pneumonia	52	5	11	32

Table 28.7 Clinical results in 488 cases of acute respiratory tract infection treated with cotrimoxazole.

		Results (%)			
No. of cases	*Isolates*	*Very good*	*Good*	*Moderate*	*Negative*
100	*D. pneumoniae* *Group A - haemol. str.*	84	7	6	3
33	*Staph. aureus*	43	7	10	40
132	*Klebsiella* *Ps. aeruginosa*	23	6	7	64
36	Other gram-negative bacteria	42	14	8	36
187	Mixed flora	52	6	11	31

Table 28.8 Sterilization of urines in 722 cases of urinary tract infection treated with cotrimoxazole.

No. of cases	*Isolates*	*Sterilization (% of cases)*
281	*E. coli*	73
105	*Str. faecalis*	60
171	Mixed flora	57
107	Other gram-negative bacteria	45
48	*Proteus* spp.	20
10	*Ps. aeruginosa*	10

urinary tract infections (Table 28.8) showed that cotrimoxazole sterilized urine in 73% of the *E. coli* infections, in 60% of the *Str. faecalis* infections and in 57% of mixed flora infections.

References

[1] CHERUBIN, C. E., NEU, H. C., IMPERATO, P. J. (1974) "Septicaemia with non typhoid Salmonella", *Medicine* (Baltimore), **53**, 365.

[2] ABBATE, G. F., ALAGIA, I., ALTUCCI, P., GUARASCIO, P., LEONESSA, V., NARCISO, P., VISCO, G., ZECHINI, F. (1977) "Transfer of resistance in *Salmonella wien*: Epidemiological implications", *Proc. 10th Intern. Congr. of Chemotherapy.*, Zurich, Sept., 18-23.

[3] VISCO, G., GUARASCIO, P., TORNESE, E., ZECCHINI, F. (1978) "Terapia delle meningoencefaliti purulente con cotrimoxazolo", *Gior. Mal. Infett. Parass.*, **2**, 163.

[4] VISCO, G., GUARASCIO, P., CARVELLI, C. (1978) "Esperienza terapeutica con cotrimoxazolo nelle infezioni batteriche respiratorie", *Gior. Mal. Infett. Parass.*, **2**, 165.

[5] GUARASCIO, P., VISCO, G., SETTE, P. (1978) "Terapia delle infezioni batteriche delle vie urinarie con cotrimoxazolo", *Gior. Mal. Infett. Parass.*, **2**, 171.

29

Antibiotic prescription monitoring in the Lombardy region of Italy

M. Mandelli, G. Tognoni

Laboratorio di Farmacologia Clinica - Istituto Mario Negri - Milano

M. L. Farina

Unità di cura intensiva neurochirurgica - Ospedale Maggiore - Milano

A. Malacrida, R. Vaiani

Servizio di Farmacia, Reparto di Patologia, Ospedale di Rho

Summary

In accordance with the antibiotic prescription monitoring program of the Lombardy Region in Italy after the introduction of therapeutic formulary, the case records of two surgical wards of an 800-bed hospital and of 15 intensive care units were reviewed. The results of these reviews showed that: (a) with a few exceptions, such as cotrimoxazole, fixed-dose combinations of antibiotics were no longer in use; (b) combined antibiotic therapy (whether based on the simultaneous administration of 2 antibiotics or on the sequential administration of 2 or more antibiotics, each administered singly for prescribed periods) still required to be approached more rationally from the point of view of drug selection, drug dosage and duration of treatment. It was found that antibiotic prescription reviews contributed significantly towards the better selection and use of antibiotics. These reviews are now beginning to be conducted on a routine basis in the Lombardy region, and the subject of antibiotic prescription monitoring is now being included in the training programs of medical and nursing staffs.

Introduction

As a result of the widespread abuse or misuse of therapeutic drugs in all countries, increasing attention is being paid to the problem of improving their selection and of controlling their consumption and use [1], [2], [3], [4].

The policy in the Lombardy region governing the consumption

and use of antibiotics in hospitals is based on the establishment of: (a) a regional formulary containing a limited number of approved drugs, including antibiotics (Table 29.1); (b) drug information programs designed to provide all concerned with information on the proper selection and use of therapeutic drugs; (c) effective clinical microbiological standard operating procedures in all hospitals; and (d) prospective and retrospective drug utilization surveys.

Consistent with this policy, regional policy governing the use of drug combinations provides that: (a) fixed-dose combinations (e.g. cotrimoxazole) may be used provided they act synergistically; (b) single drug therapy is generally preferable to therapy based on the *ad hoc* combination of drugs; (c) the use of *ad hoc* combinations should be subject to standard protocols and should be limited as far as possible to maintenance or chronic therapy; (d) the use of antibiotic combinations should be limited as far as possible to the treatment of severe or life-threatening infections; (e) antibiotic combinations should act synergistically against identified causative organisms.

Pursuant to these policies, we reviewed the prescription and use of antibiotics in two surgical wards of an 800-bed hospital over a period of 12 months and 15 intensive care units (6 months).

Method

A number of case records were selected at random. After noting the name and particulars of the patient, discharge diagnosis, and all information necessary to evaluate the effects of treatment of each case, the following data were collected: (a) the number of antibiotic prescriptions; (b) the number of cases in which antibiotics were prescribed, broken down into the number of cases in which antibiotics were prescribed for prophylactic reasons and for therapeutic reasons; (c) the number of cases in which fixed-dose combinations

Table 29.1 Therapeutic formularies in 45 hospitals. (Acceptance of the idea of a restricted list of drugs is borne out by the mere 10% of drugs selected by local Commissions outside the indications given in the Regional Formulary [PTOR]).

	*PTOR**	*Teaching*	*General*	*Community*
N° of drugs	573	276-320**	207-335**	129-208**
Antibiotics	81	34.8***	41.2***	45.6***

* PTOR = Prontuario Terapeutico Ospedaliero Regionale (Regional Formulary).
** Range.
*** Mean.

were prescribed; (d) the number of cases in which two or more antibiotics were prescribed for simultaneous administration; (e) the number of cases in which two or more antibiotics were prescribed for sequential administration (i.e. for the successive administration of different antibiotics, each administered singly for prescribed periods, the first period generally being short); (f) the number of cases in which antibiotics were prescribed, broken down into pathological categories; (g) the number of cases in which 1-2-3-4 or more antibiotics were prescribed, broken down into pathological categories.

Results

Surgical wards. Of 3331 surgical ward case records covering a period of 12 months, 333 were randomly selected and the data shown in Table 29.2 and 29.3 were collected.

Fixed-dose combinations were prescribed in only 5 cases. Even the classical mixture of bacitracin and neomycin was very rarely prescribed.

Antibiotic combinations (ampicillin + nitrofurantoin) were prescribed for simultaneous administration in only 2 cases.

Antibiotic combinations were prescribed for sequential administration in 19 cases. In 17/19 only 2 antibiotics were prescribed: in 6 cases ampicillin replaced rifampicin, in 1 case a cephalosporin replaced rifampicin, in 1 case tetracycline replaced rifampicin, in 9 cases of variety of antibiotics replaced a cephalosporin; in 11 cases the temperature profile did not seem to justify the sequential administration of 2 antibiotics; in 4 cases peak temperature was 39°. In 2/19 cases, involving patients with complicated infections, four antibiotics were prescribed for sequential administration.

Antibiotic sensitivity tests were never requested or reported.

In most cases (102/139), antibiotics were prescribed for non-specified prophylactic (antibiotic cover) reasons.

Intensive care units. Of 2512 intensive care unit case records covering a period of 6 months, 628 were randomly selected and the data shown in Table 29.4, 29.5 and 29.6 collected.

In the vast majority of cases (222/324), antibiotics were prescribed for prophylactic reasons.

In 198/324 cases only 1 antibiotic was prescribed.

In 90/324 cases 2 antibiotics were prescribed for simultaneous administration. This was expected since intensive care unit patients frequently require aggressive combined antibiotic therapy pending clarification of the clinical picture [5].

In 36/324 cases, 3 or more antibiotics (in 2 cases up to 9) were

Table 29.2 Surgical wards. Clinical records (3331).

Random sample 333 (10%)

With antibiotics → Cases 139 (41.7%) → Prophylaxis 102 (73,4); Therapy 37 (26,6)

With antibiotics → Prescriptions 161 (1.15/Patient)

Pathology (%)

Gallbladder	15.30	Minor surgery	17.75
Appendectomy	13.60	Trauma	11.60
Gastroduodenal ulcers	14.20	Vascular	1.50
Large bowel	3.70	Chest	0.90
Pancreas-Spleen	2.45	Others	15.45
Hepatic colics	3.60		

Table 29.3 Surgical wards: antibiotic combinations.

Antibiotics	*N. clinical records*	
Fixed-dose combination	5 (3.6%)	Cotrimoxazole (2); Neomycin + Bacitracin (3);
Multi-prescription		
– simultaneous	2 (1.4%)	Ampicillin + Nitrofurantoin;
– sequential	19 (13.7%)	in 17 cases two drugs; in 2 cases four drugs.

Table 29.4 Intensive care units. Clinical Records (2512).

Random sample 628 (25%)

With antibiotics → Cases 324 (51.6%) → Prophylaxis 222 (68.52%); Therapy 99 (30.55%); Not reported 3 (0.92%)

With antibiotics → Prescriptions 541 (1.67/Patient)

prescribed. No correlation with any particular pathologic category was noted.

Apart from 8 cases in which cotrimoxazole was prescribed, no fixed-dose combinations were prescribed.

Table 29.5 Intensive care units: number of antibiotics prescribed.

	N. of antibiotics prescribed				
	1	*2*	*3*	*4*	*> 4*
Clinical records (%):	61.1 *	27.4	4.9	2.1	4.2
Pathology (%):					
Trauma	42.86	39.28	8.93	3.57	5.35
Surgery	61.00	25.00	8.00	1.00	5.00
Esophageal intubation	76.92	19.23	–	–	3.84
Neuromuscular disorders	71.42	28.57	–	–	–
Cardiopulmonary disorders	63.10	29.76	–	4.76	2.38
Others	68.62	19.60	5.88	–	5.88

* Including fixed-dose combinations.

Table 29.6 Intensive care units: antibiotic combinations.

Antibiotics	*N. Clinical Records*	*%*
Fixed-dose combinations	8	2.4
Multi-prescription		
– simultaneous	68	20.9
– sequential	24	7.4
– simultaneous and sequential	31	9.5
– not specified	3	0.9

Comment

The results of our survey suggest that regional policy on the selection and use of antibiotics and of antibiotic combinations in hospital based on regional formulary is beginning to take effect. Apart from cotrimoxazole, fixed-dose combinations of antimicrobial agents are now hardly ever used in our hospitals, even though they are still important on the Italian market and some are included in the proposed formulary for out-patient practice in the Italian national health service [6].

Much has still to be done in implementing regional policy on the prescription and use of antibiotics in hospitals [3], [7], [8], [9], [10], [11]. The results of our survey suggest that it is urgently necessary to make all concerned more aware of the hazards of prophylactic therapy, of the need for antibiotic sensitivity testing whenever possible, and of the need for a more rational approach to the prescription of 2 antibiotics for simultaneous administration or of 2 or more antibiotics for sequential administration [12], [13], [14].

We feel that drug utilization reviews, such as those we conducted, help to make hospital staffs, at all levels, more aware of the real need for selecting and using antibiotics more judiciously.

References

[1] Kunin, C. M. (1977) *J. Am. Med. Assoc.*, 237, March 7-May 2.
[2] Skegg, D. C. G. and Doll, R. (1977) *Br. Med. J.*, **2**, 1523-1524.
[3] Achong, M. R., Wood, J., Theal, H. K., Goldberg, R. and Thompson, D. A. (1977) *Lancet*, **2**, 1118-1122.
[4] Jones, S. R., Barks, J., Bratton, T., McRee, E., Pannell, J., Yanchick, V. A., Browne, R. and Smith, J. W. (1977) *Am. J. Med. Sci.*, **273**, 79-85.
[5] La Force, F. M. and Eickhoff, T. C. (1977) *Anesthesiology*, **47**, 195-202.
[6] Klastersky, J. and Jawetz, E. (eds.) (1976) *Clinical use of combinations of antibiotics*, Hodder and Stoughton, London.
[7] Mc Gowan, J. E. Jr. and Finland, M. (1974) *J. Infect. Dis.*, **130**, 165-168.
[8] Mouton, R. P., Glerum, J. H. and van Loenen, A. C. (1976) *J. Antimicr. Chemo.*, **2**, 9-19.
[9] Brodie, D. C., Smith, W. E. Jr. and Hlynka, J. N. (1977) *Am. J. Hosp. Pharm.*, **34**, 251-254.
[10] Leach, F. N. (1978) *Br. Med. J.*, **1**, 766-768.
[11] *Morbidity and Mortality Weekly Report* (1978) **27**, 139-145.
[12] Roy, A. D. (1976) *J. Antimicr. Chemo.*, **2**, 233-238.
[13] Grossman, J. H. III, Adams, R. L. and Hierholzer, W. J. Jr. (1977) *Am. J. Obstet. Gynecol.*, **128**, 690-692.
[14] Wilkinson, P. J., Ball, A. J., Doran, J., Gillespie, W. A. and Orton, V. S. (1977) *J. Clin. Pathol.*, **30**, 417-420.

30

Antibacterial activity and *in vitro* interaction of antibiotics in conventional media and organic fluids

A. Rodríguez, M. C. R. Quecedo, T. Olay, M. V. Vicente

Instituto de Farmacología Española - C/Méndez Alvaro, 57 (Edificio CEPA) Madrid - 7, Spain

Summary

The MIC of ampicillin, cephalothin, cefoxitin, fosfomycin, gentamicin and amikacin against strains of gram-negative or gram-positive bacteria were generally lower in nutrient (NA) than in Mueller-Hinton (MH), human urine (HU) and rabbit peritoneal fluid (PF) agar. In vitro studies showed that combinations of fosfomycin and cefoxitin or gentamicin were more synergistic in MH than in NA, HU or PF against strains of *Proteus*, *Enterobacter* or *S. marcescens*. Combinations of cefoxitin and amikacin or gentamicin were synergistic in MH and PF but not in NA or HU against strains of *S. marcescens*.

Introduction

The antibacterial activity of antibiotics *in vitro* is usually determined in conventional culture media, such as Mueller-Hinton, nutrient, Brain-Heart Infusion or Tripticase Soy broths or agars.

It is well known that the medium composition influences the antibiotic activity. Changes in the pH and in the concentration of certain ions can increase or decrease the *in vitro* activity of some antibiotics [4], [7], [9], [10], [12].

It is also well known that *in vivo* conditions of bacterial infection are not the same as *in vitro* conditions, and that *in vivo* conditions differ considerably with regard to nutrients, ions and pH, depending on the site of infection. For this reason, a number of investigators [1], [3], [5], [8], [11] have used human serum and urine in their *in vitro* studies on the antibacterial activity of certain antibiotics.

In the present study we compared the *in vitro* antibacterial activity of a number of antibiotics, alone and in various combinations, in two conventional media (Mueller-Hinton and nutrient) and in two organic fluids (human urine and rabbit peritoneal fluid).

Materials and methods

Agar media. Mueller-Hinton (MH), pH 7.4 (± 0.1); nutrient (NA), pH 7.0 (± 0.1); human urine (HU), pH 5.7 (± 0.1); and rabbit peritoneal fluid (PF), pH 7.8 (± 0.1).

Minimum inhibitory concentrations (MIC). The MIC of ampicillin, cephalothin, cefoxitin, fosfomycin, gentamicin and amikacin against *E. coli*, *Klebsiella* sp., *Enterobacter* sp., *Serratia marcescens*, indole-positive *Proteus*, *Pr. mirabilis*, *Ps. aeruginosa* and *Str. faecalis*, were determined by the plate dilution method, according to the World Health Organization, in MH, NA, HU and PF agar.

Synergy studies. Two methods were used to test the *in vitro* interaction of antibiotic combinations: (a) The plate dilution checkerboard method in MH, NA, HU and PF agar, using a combination of fosfomycin and cefoxitin against *S. marcescens* and indole-positive *Proteus*. The effects of the combination were considered: synergistic if the fractional inhibitory concentration index (FIC) was $\leqslant 0.5$; partially synergistic if the FIC was > 0.5 and $\leqslant 0.75$; additive if the FIC was 1; indifferent if the FIC was 2; and antagonistic if the FIC was > 2. (b) The killing curve method in MH, NB, HU and PF using: fosfomycin, cefoxitin, and a combination of both drugs against *Proteus* No. 3; fosfomycin, gentamicin and a combination of both drugs against *Enterobacter cloacae* No. 9; cefoxitin, amikacin and a combination of both drugs against *Serratia marcescens* No. 16; and cefoxitin, gentamicin, and a combination of both drugs against *Serratia marcescens* No. 16. The effects of the combination were considered: synergistic if it produced an over 2-fold decrease in the colony count as compared with the colony count of the most active single drug; indifferent if it produced no decrease; and antagonistic if it produced an increase [2], [6].

Results

Minimum inhibitory concentrations. Median MIC values of fosfomycin against 5 strains of each bacterial species tested were very similar in MH and HU, lower in NA, and higher in PF, especially against *S. marcescens* (Fig. 30.1a).

Median MIC values of ampicillin against 5 strains of each bacterial species tested were very similar in the four media, except for a higher value in PF against *Str. faecalis* (Fig. 30.1a).

Median MIC values of cephalothin and cefoxitin showed no

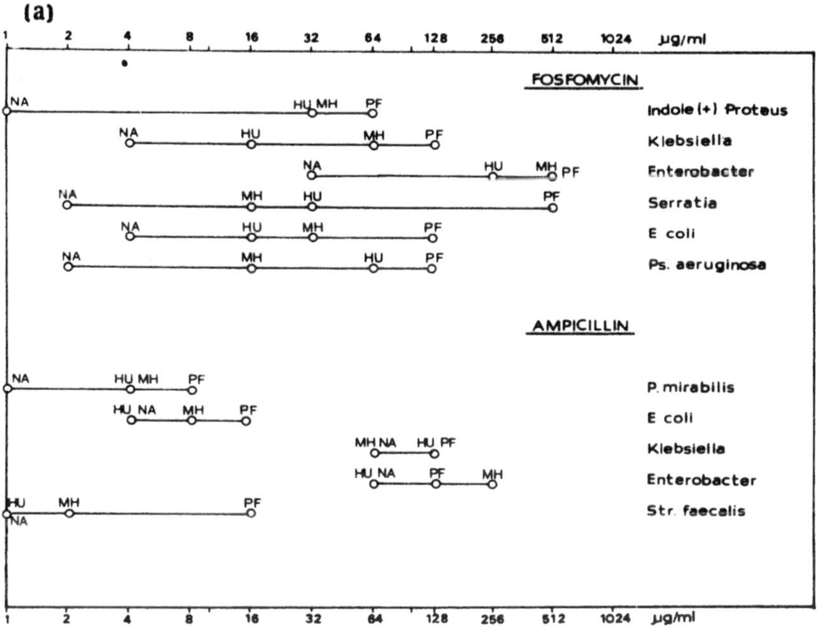

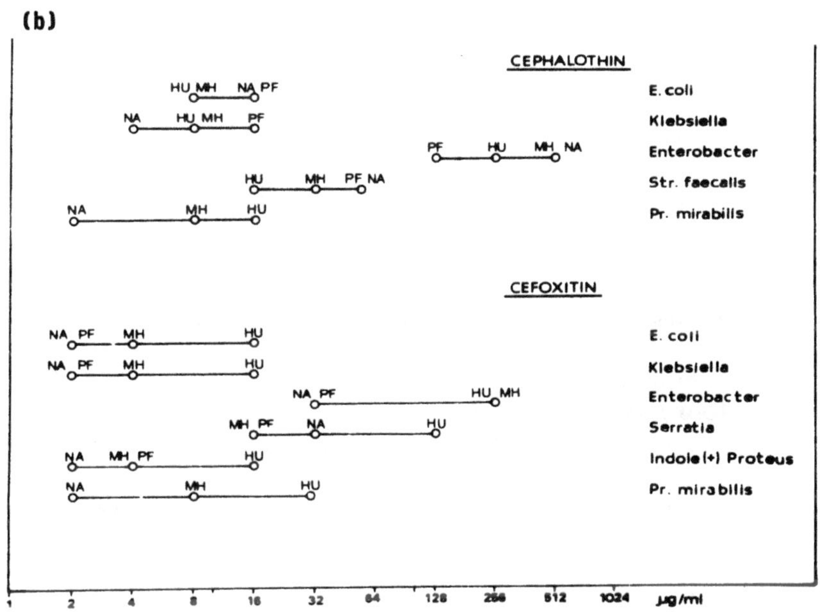

Fig. 30.1 - Medians of MICs in Mueller-Hinton (MH), human urine (HU), peritoneal fluid of rabbit (PF) and nutrient agars (NA) with fosfomycin and ampicillin (a), cephalothin and cefoxitin (b), gentamicin and amikacin (c, page 304) against several bacterial strains.

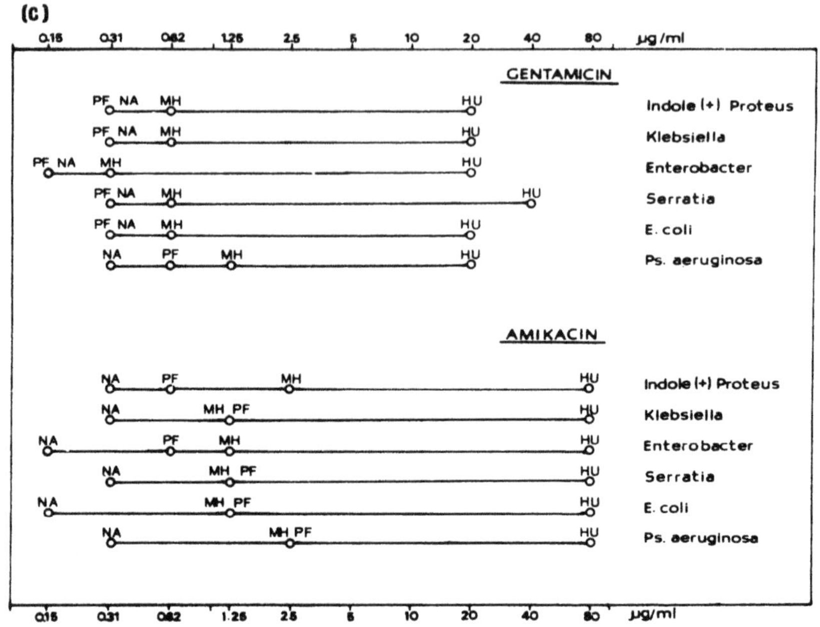

significant differences between the values obtained in the four media (Fig. 30.1b).

Median MIC values of gentamicin against 5 strains of each bacterial species tested were significantly higher in HU than in the other three media, values in HU ranging from 20 to 40 mcg/ml (Fig. 30.1c).

Median MIC values of amikacin against 5 strains of each bacterial species tested were also significantly higher in HU than in the other three media, values in HU reaching 80 mcg/ml (Fig. 30.1c).

Synergy studies. Studies performed with the checkerboard method showed that:

a) The combination of fosfomycin and cefoxitin acted synergistically in the four media against 5 strains of *S. marcescens*, with synergy more pronounced in MH than in the other three media (Fig. 30.2a).

b) The combination of fosfomycin and cefoxitin also acted synergistically in the four media against 5 strains of indole-positive *Proteus*, with synergy also more pronounced in MH than in the other three media (Fig. 30.2b).

In the studies performed with the killing curve method, although the antibiotic concentrations differed in each case because they

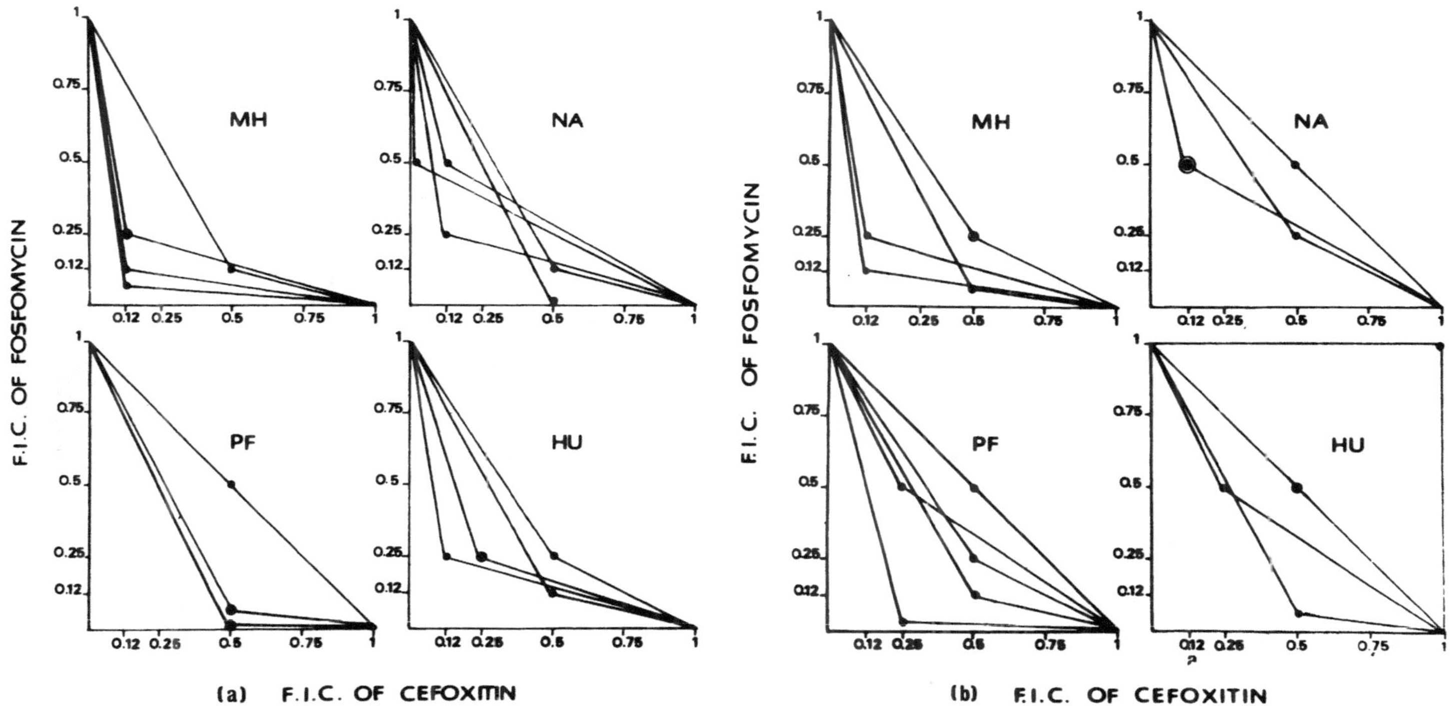

Fig. 30.2 - Isobolograms of FIC indexes of fosfomycin and cefoxitin combinations against 5 strains of *S. marcescens* (a) and indole-positive *Proteus* (b) by checker-board method in different solid media.

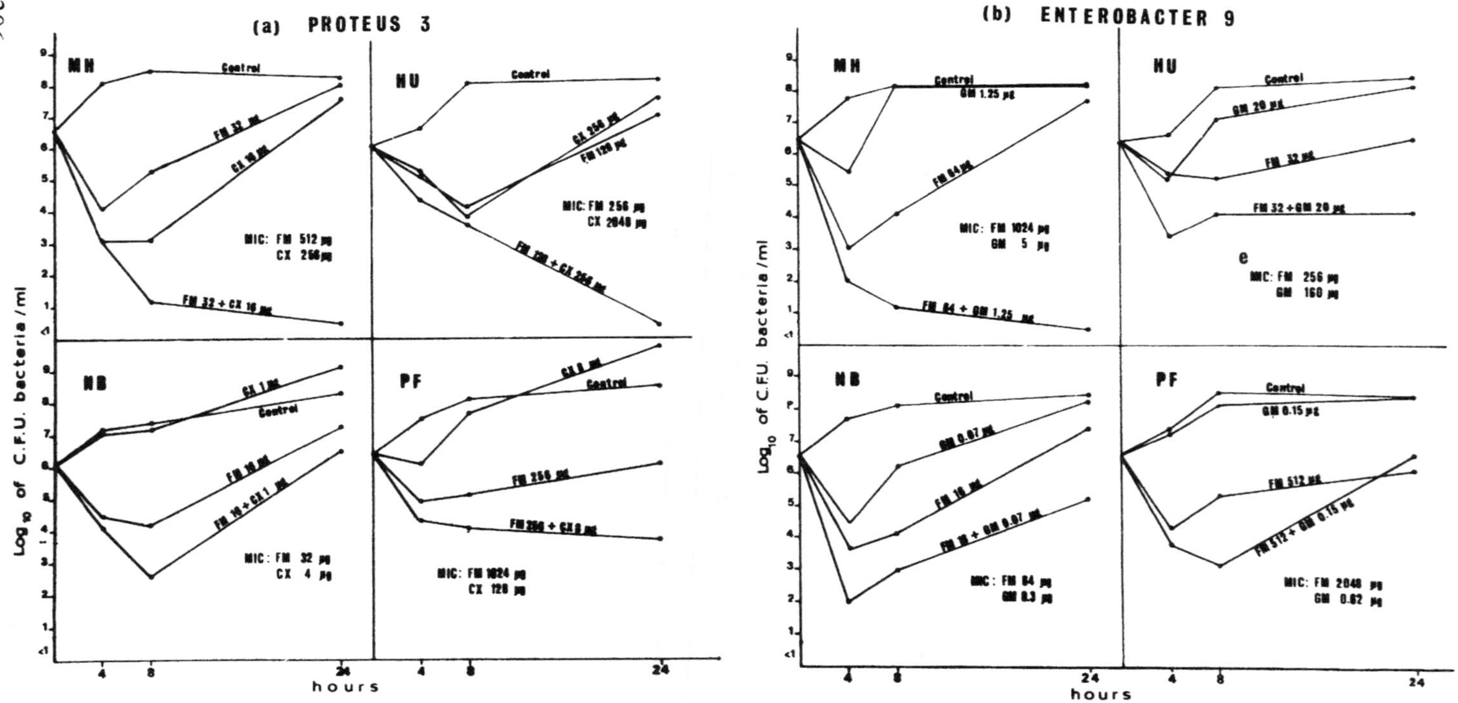

Fig. 30.3 - Killing-curves obtained with fosfomycin, cefoxitin and their combinations against indole-positive *Proteus* No. 3 (a), and with fosfomycin, gentamicin and their combinations against *Enterobacter cloacae* No. 9 (b) tested in different broth media.

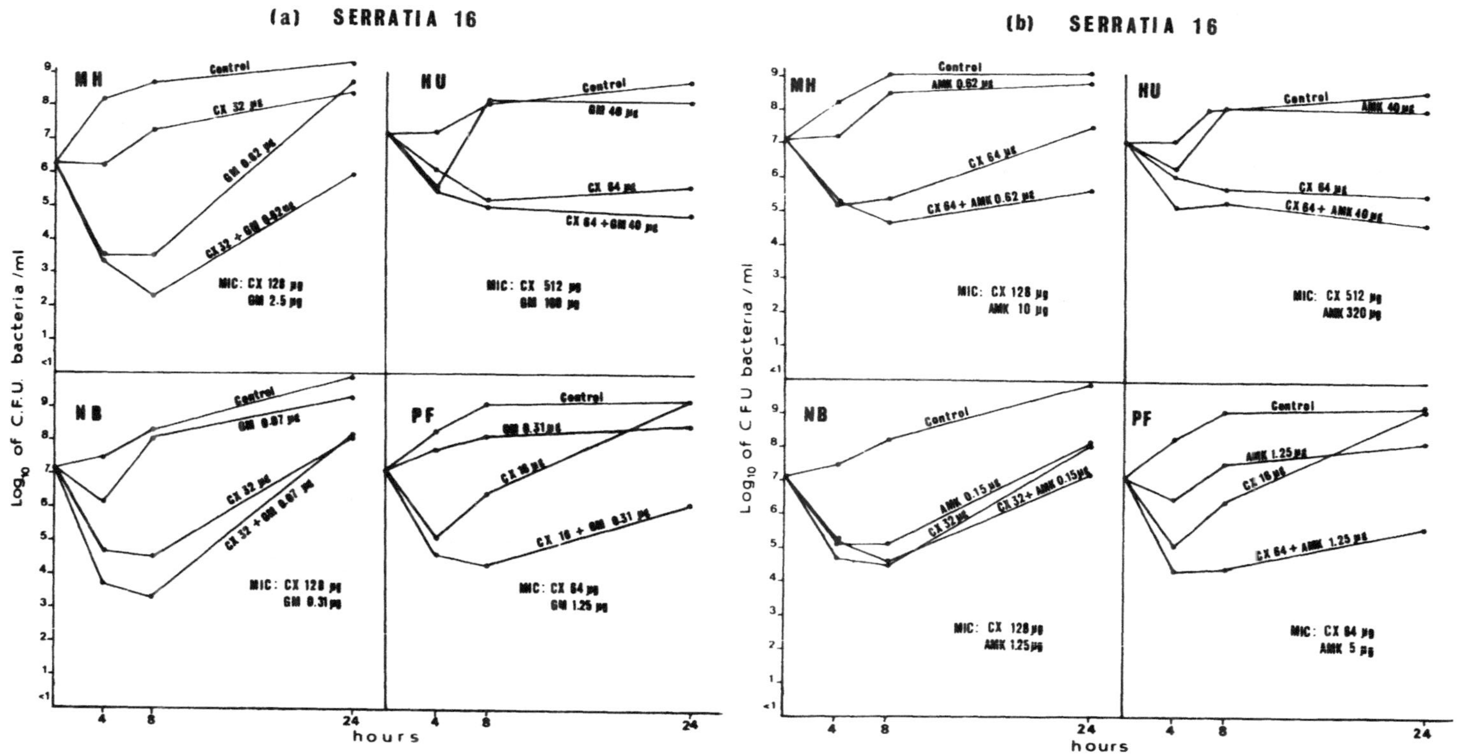

Fig. 30.4 - Killing-curves obtained with cefoxitin, gentamicin and their combinations (a), and cefoxitin, amikacin and their combinations (b), against *S. marcescens* No. 16 in different media.

depended on the MIC found in each medium, they were always less than their MIC. The results of these studies showed that:

a) The combination of fosfomycin and cefoxitin acted synergistically in the four media against indole-positive *Proteus* No. 3. Synergy was more pronounced in MH and HU than in PF and NB. Synergy in MH and HU showing a 6- to 7-fold decrease in the colony count after 24 h (Fig. 30.3a).

b) The combination of fosfomycin and gentamicin acted synergistically in the four media against *Enterobacter cloacae* No. 16. Synergy was more pronounced in MH than in the other three media. Synergy in MH showing a 7-fold decrease in the colony count after 24 h (Fig. 30.3b).

c) The combination of cefoxitin and gentamicin acted synergistically in MH and PF against *S. marcescens* No. 16, but not in NB and HU. Synergy in MH and PF showed a 2-fold decrease in the colony count after 24 h (Fig. 30.4a).

d) The combination of cefoxitin and amikacin acted synergistically against *S. marcescens* No. 16 in MH and PF but not in NB and HU. Synergy in MH and PF showed a 2-fold decrease in the colony count after 24 h (Fig. 30.4b).

References

[1] Klastersky, J., Daneau, D., Swings, G. and Weerts, D. (1974) *J. Infect. Dis.*, **129**, 187-193.

[2] Libke, R. D., Regamey, C., Clarke, J. T. and Kirby, W. M. M. (1973) *Antimicrob. Ag. Chemother.*, **4**, 564-568.

[3] Minuth, J. N., Musher, D. M. and Thorsteinsson, S. B. (1976) *J. Infect. Dis.*, **133**, 14-21.

[4] Molavi, A., Barza, M., Cole, W., Berman, H. and Weinstein, L. (1973) *Chemotherapy*, **18**, 7-16.

[5] Reller, L. B. and Stratton, Ch. W. (1977) *J. Infect. Dis.*, **136**, 196-204.

[6] Rosenblatt, J. E. and Stewart, P. R. (1974) *Antimicrob. Ag. Chemother.*, **6**, 84-92.

[7] Rubenis, M., Kozis, V. M. and Jackson, G. G. (1963) *Antimicrob. Ag. Chemother.*, pp. 153.

[8] Shah, P. M., Schirmer, K. P., Helm, E. B. and Stille, W. (1976) *Arzneim.-Forsch. (Drug Res.)*, **26**, 1636-1638.

[9] Starkey, D. and Iannetta, A. (1972) *Antimicrob. Ag. Chemother.*, **1**, 466.

[10] Starkey, D. and Iannetta, A. (1972) *Appl. Microbiol.*, **23**, 775.

[11] Stratton, Ch. W. and Reller, L. B. (1977) *J. Infect. Dis.*, **136**, 187-195.

[12] Waterworth, P. M. (1972) *J. Clin. Pathol.*, **25**, 979-983.

31

Comparison of a rifampicin-trimethoprim combination with rifampicin alone and with cotrimoxazole in urinary tract infections

E. Bartoli, R. Faedda

Istituto di Patologia Speciale Medica - Università di Sassari

Viale San Pietro 12, 07100 Sassari, Italy

S. Zanetti, P. Cappuccinelli

Istituto di Microbiologia - Università di Sassari

Summary

Forty-three patients with urinary tract infections (UTI) were studied in a controlled, prospective, randomized manner with respect to their response to Rifaprim® (an association of rifampicin with trimethoprim) as compared to either rifampicin alone or to cotrimoxazole. While Rifaprim® cured 50% of patients, urines became sterile in only 21% of cases with cotrimoxazole and 27% with rifampicin: this difference was statistically significant. In 83% of the Rifaprim® cases the bacterial species isolated initially disappeared from the urines as opposed to 45% with rifampicin alone and 43% with cotrimoxazole. Reinfection rates were not different in the three groups. However, the reinfecting agents were almost invariably sensitive to rifampicin and trimethoprim and resistant to the other drugs. The present data demonstrate that the association of rifampicin with trimethoprim does successfully decrease the rate with which resistant strain of bacteria emerge from the urines of patients treated for their UTI, whilst representing an effective association against infection per se as compared to single agents.

Introduction

Treatment failures during antibiotic therapy are mainly due to the emergence of resistant bacterial strains. Antibiotic associations can to a large extent prevent this undesirable phenomenon [1].

There are three main considerations which should guide the rationale for associating antibacterial agents: the mechanism of action should be different for each drug, they must be compatible and, finally, inhibition must take place at different steps along the same biosynthetic pathway. The association of trimethoprim (TMP) with rifampicin (RMP) satisfies these requirements since the drugs exhibit different mechanisms of action, the former being an inhibition of tetrahydrofolate reductase [2], the latter of RNA polymerase [3], although their action occurs on the same metabolic pathway (RNA synthesis). This hypothesis has been confirmed by *in vitro* experiments [4]: the effect of these two agents is enhanced when they are combined; moreover selection of resistant strains is reduced.

This study was performed in order to ascertain, by a randomized clinical trial, the effectiveness of the association of rifampicin with trimethoprim (Rifaprim®) as compared to rifampicin alone and cotrimoxazole in the treatment of urinary tract infections (UTI), where occurrence of bacterial resistance during treatment, as well as high rates of relapse, are common [5].

Material and methods

The studies were performed on 43 patients, 10 males and 33 females. Thirty-three were hospitalized and 10 were treated as outpatients. As soon as a UTI was suspected, urine cultures were taken by a sterile technique not requiring catherization, by meatal cleansing and midstream collection into sterile tubes. The urines were added to the appropriate culture media within minutes of their collection. Cultures were repeated within the third day and again within the fourth week following the end of therapy.

All patients had all routine laboratory examinations performed, and were carefully examined prior to, during and after treatment, when measurements of physiological variables were repeated. Standard techniques for culture, isolation, identification of bacteria and *in vitro* disc sensitivity testing were adopted.

The experimental design initially involved a randomized double blind comparison between Rifaprim® (RMP 450 mg and TMP 120 mg twice daily for 10 days) and RMP alone (450 mg, twice daily for 10 days), followed by a subsequent randomized comparison between Rifaprim® (RMP 600 mg and TMP 160 mg, once daily at bedtime for 7 days) and cotrimoxazole (sulphamethoxazole 800 mg and TMP 160 mg, twice daily for 7 days). Since there was no difference between results obtained during treatment with Rifaprim® in the two trials, results were pooled and the pooled data relative to Rifaprim® are shown in this study. In the paper the data are presented, for purposes of clarity, as if the experiment had been carried out from the beginning with the group subdivision indicated in the tables, although this was in fact done retrospectively.

The following criteria were used:

1. UTI was diagnosed (and the patient included in the trial) when at least one bacterial species was isolated from the urines of the proband in counts higher than $10^5 \times ml^{-1}$, or when the same germ was isolated more than once on repeated cultures, although in counts lower than $10^5 \times ml^{-1}$.
2. A patient was considered cured when his urines were entirely sterile at the last control after the end of treatment.
3. Reinfection was diagnosed when a species different from that isolated prior to treatment was present at the end of therapy in counts higher than $10^4 \times ml^{-1}$.
4. Success against bacteria was rated even when not all bacteria initially isolated had disappeared from the patient's urine: in this case the treatment was considered successful against those species that were no longer present, whilst a failure was registrered for the germs persistent.
5. Failure was registered when the same species initially isolated were still present, whatever their number.

Results

Overall 43 patients were available for the final analysis of antibacterial efficacy. Before treatment, 12 bacterial species were isolated (Table 31.1). Rifampicin alone was given to 11 patients, from whom 8 species were isolated prior to treatment: five of these species persisted at the end of therapy, with an overall cure rate of 3 out of 11 patients. Cotrimoxazole was given to 14 patients: the cure rate was 3 out of 14, the species isolated being essentially the same as for rifampicin. Thus, both the above mentioned treatments exhibited a poor response.

Rifaprim® was given to 18 patients, in whom 10 species were isolated before treatment. The cure rate (9 out of 18) was higher with respect to the other two groups combined ($\chi^2 = 3.114$, $P < 0.10$).

Table 31.2 also reports the failure rates for each treatment, as well as the cure rates and the reinfection rates. Rifampicin failed (according to the criteria presented in the previous section) in 6 out of 11 patients. The pathogens involved in these failures were the most common ones, both gram-positive and gram-negative, and associated disease was considered a likely cause of failure in two patients only, who were affected by renal stones and diabetes respectively.

Treatment failures were 8 out of 14 with cotrimoxazole, again due to common pathogens; two patients, treated with antiblastic agents owing to Hodgkin disease and carcinoma of the breast, were thought to have failed for reasons other than antibiotic treatment.

Table 31.1 The column on the left shows all bacterial species isolated in our patients. The numbers under the drugs used refer to the number of each species isolated. In the column "after treatment" reinfecting species are not considered.

Species	*Rifaprim group* *before*	*after*	*Rifampicin group* *before*	*after*	*Cotrimoxazole group* *before*	*after*
Klebsiella	3	0	2	2		
Escherichia	4	1	4	1	8	3
Proteus	5	0	4	2	5	2
Citrobacter	1	0	1	1	1	1
Pseudomonas	1	1				
Streptococcus	3	0	2	1	1	0
Staphylococcus	3	1	3	0	3	2
Mima	1	0				
Providencia	1	0				
Serratia	1	0				
Enterobacter			2	0		
Listeria			1	0		

Table 31.2 The column on the left refers to the number of patients treated and to the overall results of treatment. The results of individual treatments are given under the name of each drug used. The percentages shown refer to the numbers given in the left hand column.

No. patients		*Treatments* Cotrimoxazole		Rifampicin		Rifaprim	
43	Treated	14	32%	11	25%	18	41%
15	Cured	3	21%	3	27%	9	50%
17	Failures	8	57%	6	54%	3	16%
11	Reinfections	3	21%	2	18%	6	33%

Rifaprim® failed in only 3 out of 18 patients, with infections due to *P. aeruginosa*, *E. coli* and *Staphylococcus aureus* respectively. One of these patients had renal tubular acidosis, renal failure and multiple bilateral renal stones and calcifications.

Discussion

UTI are often difficult to eradicate owing to the occurrence of reinfections, and bacterial resistance as well as for other reasons, such as the presence of pyelonephritis, diabetes, renal stones and anatomical abnormalities.

In this controlled randomized study we used two treatments — rifampicin and cotrimoxazole — thought to be effective against UTI, and compared their results with those of the combination rifampicin and trimethoprim (Rifaprim®).

With the combination, the success rate was 50% in terms of patients cured, and 84% in terms of the total disappearance of bacterial species isolated before treatment, whilst reinfections were 33%. Treatment failure occurred in 3 cases, where persistance of bacteriuria may have been partially due to the presence of renal stones and diabetes.

Treatment with either cotrimoxazole or rifampicin was much less effective: the cure rate was 21% with the former and 27% with the latter, while 42% of the initial infecting agents had disappeared by the end of treatment with cotrimoxazole, and 45% with rifampicin. Rates of reinfection were 21% and 18% respectively, treatment failures 57% and 54%.

Thus, the association appears to be significantly more effective than either agent alone. This seems to be due to a decreased rate of bacterial resistance, which represents the theoretical background for experimenting the association.

The difference between the number of patients on the one hand and the number of cures plus failures on the other is due to reinfections, which are shown in Table 31.2. Two patients had reinfections whilst under treatment with rifampicin; both reinfecting agents (*Staph. aureus* and *Klebsiella*) were resistant to rifampicin. Two patients treated with cotrimoxazole were considered to have bacterial contamination and three additional patients to be reinfected by *Staph. aureus* and *E. coli*, which were resistant to cotrimoxazole. Six patients became reinfected after treatment with Rifaprim®: those due to *P. aeruginosa* and *E. coli* were resistant to the therapeutic agent. Thus, it seems that the association does not abolish reinfection with resistant agents.

Table 31.3 gives data on the sensitivity to each component of Rifaprim for those strains which were isolated from reinfection cases. With rifampicin 2 out of 7 species were sensitive to both components of Rifaprim®, while all of them were sensitive to both in reinfections during treatment with cotrimoxazole. Thus, it would appear from retrospective analysis that treatment would have been much better if it had been carried out with the association, rather than with a single agent.

The effectiveness of Rifaprim® is in all likelihood attributable to the synergism between the two components besides the reduction in the emergence of mutant resistant strains. In fact *in vitro* studies have

Table 31.3 The table shows the number of reinfections with the species indicated for each treatment group. All reinfecting agents were sensitive to rifampicin and trimethoprim except those marked with an asterisk, which were isolated from the urines of patients given rifampicin and were rifampicin resistant.

Species	*Rifaprim*	*Rifampicin*	*Cotrimoxazole*
Proteus	1	1	1
Escherichia	2	2*	1
Klebsiella		3*	
Shigella	1		
Neisseria			1
Streptococcus	1	1	
Staphylococcus	1		

demonstrated that the occurrence of mutant bacteria resistant to the drugs is remarkably reduced as compared to each agent alone [4]. In addition, the pharmacokinetic properties of each agent in the Rifaprim® combination are such as to maintain suitable antibacterial concentrations within the kidney and urines [6].

The main reason for suggesting the association in the treatment of UTI is born out by our study, which demonstrates that a high proportion of reinfecting strains were sensitive to Rifaprim®, when tested *in vitro*; thus, they would have responded and presumably healed, had they been treated with the association. Actually, all failures in both the rifampicin and the cotrimoxazole group were accompanied by the emergence of resistant mutants, whilst failures with Rifaprim® were, in at least two cases, accounted for by reasons other than the ineffectiveness of the antibiotics.

Thus, it would appear that the only possible objection to using Rifaprim® in UTI could arise from the concern that resistance to rifampicin may occur in an increasing number of *Mycobacteria* strains, rendering the agent progressively less active against tuberculosis. However, recent results obtained in countries where rifampicin has been used extensively for infections other than TB indicate that the statistical likelihood of the emergence of resistant mycobacteria is the same as that measured in countries where rifampicin has been restricted to the treatment of TB [7]. Thus, although this concern is reasonable, it does not appear to be supported by experimental and statistical data, at least up to the present time. For these reasons we believe that Rifaprim® can be considered an extremely suitable association for the treatment of UTI, capable of significantly reducing bacterial resistance without the risk of encouraging the emergence of strains of *Mycobacterium tubercolosis* insensitive to rifampicin.

References

[1] PRATT, W. B. (1977) *Chemotherapy of infection*, Oxford University Press, New York, pp. 1-462.

[2] BURCHALL, J. J. and HITCHINGS, G. H. (1965) *Mol. Pharmacology*, **1**, 126-137.

[3] LANCINI, G., PALLANZA, R. and SILVESTRI, L. G. (1969) *J. Bacteriology*, **97**, 761-768.

[4] KERRY, D. W., HAMILTON MILLER, J. M. T. and BRUMFITT, W. (1975) *J. Antimicrob. Chemother.*, **1**, 417-427.

[5] KUNIN, C. M. (1972) *Detection, prevention and management of urinary tract infections*, Lea & Febiger, Philadelphia, pp. 1-230.

[6] ACOCELLA, G. and SCOTTI, R. (1976) *J. Antimicrob. Chemother.*, **2**, 271-277.

[7] ACOCELLA, G., HAMILTON MILLER, J. M. T. and BRUMFITT, W. (1978) *Lancet*, **1**, 740-742.

Closing remarks

D. S. Reeves

I have been asked by Professor Nicolis to produce a brief summary of the meeting. This meeting has set out to discuss combined chemotherapy, and that we've certainly done. No overall conclusions have been reached but, there again, no one would expect this on such a contentious and broad subject as combined chemotherapy. However, I think the true value of the meeting lies in the fact that we have defined and posed to ourselves a number of questions which are going to need answering in the future; I have long been convinced that meetings are as valuable for asking questions as answering them. Perhaps we ought first to ask ourselves, before we consider the use of combined chemotherapy, whether the patient needs any antibiotic therapy at all? Then, if we are to use combined chemotherapy, can we possibly treat the patient with a single drug to the same effect? I understand from the feeling of the meeting that we all accept the use of unfixed combinations for serious unknown sepsis and for mixed infections. One of the main points to come out of this meeting is that people are hesitant to introduce the idea of further fixed combinations, and obviously feel that for a combination to be introduced for clinical practice, it should be shown to be better than any single drug available at the present time. Furthermore, the combination should preferably be less toxic and certainly no more toxic than the comparable single agent. To determine this clinically requires of course a considerable amount of clinical and laboratory investigation, but I'm sure that with the support of the proper authorities this can be carried out.

Another important aspect is the pharmacokinetics of combination of antibiotics. There is a need to investigate more often the kinetics of drugs in combination in actual patients. One of the most interesting papers for me as someone interested in pharmaco kinetics in patients, was perhaps the experiment on Professor Ghione's mice. We are woefully ignorant at the present time about the effects of pathophysiology on drug kinetics in patients, and in particular, this is true of antibiotics. I have no comments to make really about the position of combined therapy for tuberculosis as it is well accepted.

We have heard strong arguments on both sides as to the introduction of the new combination of trimethoprim and rifampicin. Although we may have emotional feelings both for or against introducing new combinations, what is really required is a cool, calm look at the new combinations, or other proposed fixed combinations, and to take an objective and scientific viewpoint.

Before finishing I would like to thank very much, on behalf of the participants of this meeting and my co-Chairman Professor Grassi, the Society for Applied Pharmacological Sciences for so kindly arranging this meeting in this beautiful and stimulating venue of Rome and their President Professor Nicolis for being such a charming host. I would also like to thank on your behalf the supporters of the meeting, the pharmaceutical industry. It is important to thank Dr. Curcio for all the hard work she has done in organizing the meeting, making sure that things ran as smoothly as they did. Mrs. Vocaturo and her staff have also made a great contribution towards that end.

Finally, on behalf of us all I would like to congratulate the speakers and participants on their splendid contributions, without which the meeting would have not been so successful.

Index

Finito di stampare nel mese di settembre 1979 dall'OTV STOCCHIERO - Vicenza

GPSR Compliance
The European Union's (EU) General Product Safety Regulation (GPSR) is a set of rules that requires consumer products to be safe and our obligations to ensure this.

If you have any concerns about our products, you can contact us on

ProductSafety@springernature.com

In case Publisher is established outside the EU, the EU authorized representative is:

Springer Nature Customer Service Center GmbH
Europaplatz 3
69115 Heidelberg, Germany

www.ingramcontent.com/pod-product-compliance
Lightning Source LLC
Chambersburg PA
CBHW071239230926
51849CB00001B/161
* 9 7 8 9 0 2 4 7 2 2 8 0 8 *